Intellectual Property in the Life Sciences

A Global Guide to Rights and Their Applications

Consulting Editor **Paul England**

Consulting editor
Paul England

Publisher
Sian O'Neill

Commissioning editor
Katherine Cowdrey

Editor
Carolyn Boyle

Marketing manager
Alan Mowat

Production
Russell Anderson

Publishing directors
Guy Davis, Tony Harriss, Mark Lamb

Intellectual Property in the Life Sciences: A Global Guide to Rights and Their Applications
is published by
Globe Law and Business
Globe Business Publishing Ltd
New Hibernia House
Winchester Walk
London SE1 9AG
United Kingdom
Tel +44 20 7234 0606
Fax +44 20 7234 0808
Web www.globelawandbusiness.com

Printed and bound by CPI Group (UK) Ltd., Croydon, CR0 4YY

ISBN 9781905783571

Intellectual Property in the Life Sciences: A Global Guide to Rights and Their Applications
© 2011 Globe Business Publishing Ltd

DISCLAIMER
This publication is intended as a general guide only. The information and opinions which it contains are not intended to be a comprehensive study, nor to provide legal advice, and should not be treated as a substitute for legal advice concerning particular situations. Legal advice should always be sought before taking any action based on the information provided. The publishers bear no responsibility for any errors or omissions contained herein.

MIX
Paper from
responsible sources
FSC® C013604
www.fsc.org

Table of contents

Foreword

David Rosenberg
GlaxoSmithKline

The contribution made to public welfare by biopharmaceutical innovation since the latter part of the last century is undeniable. Indeed, one analyst has calculated that approximately 40% of the two-year increase in life expectancy measured from 1986 to 2000 can be attributed to the introduction and use of new medicines. HIV/AIDS is no longer an untreatable fatal disease. Childhood vaccination has all but eradicated some previously fatal diseases and significantly reduced infant mortality.

Biopharmaceutical innovation takes many forms. It includes the development of first-in-class medicines and vaccines, competitive innovative products which have subtle but medically significant differences from the first-in class, new combinations displaying synergistic effects and new formulations which may, for example, increase patient compliance.

These advances have taken place in a context where, on average, it takes in excess of 10 years and costs in excess of $1.3 billion to bring a new product to market and where less than half of new products launched are profitable. It is trite for those who will be reading this book to say that it is hard to imagine that these welfare benefits would not have occurred were it not for the existence of the incentives to invest and take risks provided by robust yet flexible intellectual property systems that provide a period of exclusivity to the innovator for a defined period.

Patents are at the heart of these incentives. It can be argued that patents in the biopharmaceutical industry are of more importance to innovation and of greater value than in any other sector. The relatively recent introduction in many countries of patent term restoration is an implicit acknowledgement of the importance of patents as an incentive to innovate in this sector.

The significance of data exclusivity, which prevents 'piggy-backing' on the extensive data generated by innovators to satisfy regulators that a product is safe and effective, should not be underestimated. It can provide a period of protection from competition from copies (but not other innovative products) in the event that, for whatever reason, the investment undertaken by an innovator does not benefit from effective patent protection.

As is the case in other sectors, trade marks provide the advantages associated with branding. In addition, and importantly, they are a vital tool in the fight against counterfeiting which, in this sector, is a public-health rather than a commercial imperative.

The world has seen remarkable change over the last decade or so, driven to a significant degree by globalisation and advances in many fields of technology.

Pharmaceutical innovation models and the biopharmaceutical industry have both led and adapted to those changes.

Regulatory hurdles in the developed world have increased, leading to increased R&D costs and decreasing numbers of new molecules being approved. Pressure on prices (exacerbated by the economic crisis) and 'patent cliffs' have constrained the amount of money available in many companies for R&D.

Emerging markets are now seen as key growth drivers for many in the industry.

Personalised medicines, often based on genetic diagnostics, are becoming more and more significant therapeutically and, in respect of some therapy areas, commercially. Advances in the field of biotechnology have been such as to lead to predictions that in the next few years biotech drugs will dominate the league of bestsellers.

An increasing recognition of the importance of science emanating from academia and smaller companies is leading to research models changing to embrace the concept of open innovation. No longer will it be the norm (if it ever was) that a company will 'simply' discover a molecule in its own labs and take it through development to launch and marketing using only its own skills and knowledge. Instead, there is a rapid growth in collaboration between academia and small and large industry in which the talents of key stakeholders are brought together to maximise the efficiency and quality of innovation. As a consequence, new and sophisticated models of sharing intellectual property are developing.

The industry has recognised the moral imperative of improving the access of poorer sectors of society to vital medicines and the need to develop medicines for neglected diseases and has taken significant steps in this direction. Far from being evidence, as some claim, of the international intellectual property system in this field being an obstacle to innovation and access to medicines, this is evidence of the flexibility of intellectual property as a tool for driving innovation and access. Rather than innovation and intellectual property being seen as somehow opposing forces, the innovation driven by intellectual property is a vital part of the continuum that leads to better public welfare.

This book addresses the international intellectual property system, the law in many emerging markets, second generation inventions, specific issues in the areas of biotechnology and collaborative models, in addition to more traditional subjects such as the European framework. As such, it reflects the changes that are taking place in law in this area. It is a welcome contribution to an understanding of the law which helps shape a field of vital public importance.

David Rosenberg is Vice President of Corporate IP Policy at GlaxoSmithKline (GSK).

After several years in the IP group of Clifford Chance, David joined GSK to undertake an IP policy function. He has worked on many aspects of IP policy both within and outside the European Union, including issues relating to access to medicines.

David was a member of the group comprising representatives of industry and academic stakeholders that, in conjunction with the European Commission, drew up the IP policy of the Innovative Medicines Initiative. Furthermore, he represents GSK on the IP committees of several pharmaceutical and cross-industry organisations. He also chairs the IP committee of the European Federation on Pharmaceutical Industries and Associations (EFPIA) and was heavily involved in the industry-led inquiry into DG competition.

David is currently deeply involved in the discussions to create a unitary European patent and unified patent court.

David is a regular speaker (and very occasional author) on all of these issues.

Introduction

Paul England
Simmons & Simmons LLP

A changing industry, and changing rights

There are arguably three major forces that are shaping the development of intellectual property law in the life sciences at the moment: the diversification of pharmaceutical businesses into new areas of innovation; the advance of technology; and the increasingly global reach of those businesses. Diversification has come in the context of challenging economic circumstances, greater pressure on drug pipelines due to the stringencies of regulation and the costs of R&D, increasing generic competition and patent expiry. Indeed, in respect of the latter, estimates abound on the loss of sales that will result from the expiry of patents protecting blockbuster drugs over the next few years, with a recent estimate placing the figure in excess of $260 billion between 2011 and 2016.[1] Furthermore, the march of technology is driving innovation in areas little imagined just a dozen years ago, before the sequencing of the human genome, such as the impact of advanced DNA sequencing and bioinformatics methods on personalised medicine. Hence, these are times of great challenge for the life sciences sector and the traditional, small-molecule blockbuster business model may soon prove to be just one of a number of other sustainable possibilities.

However, there is one thing that is likely to remain a constant through this diversification. It applies to businesses from university start-ups to the established pharmaceutical giants – the need for market exclusivity. Without it, the costs of research into new drugs could never be financed:

> *It is the patent system, which has made the advances in medicines possible. Although economists sometimes debate whether the patent system is useful generally, no one has ever seriously challenged its place for medicines. And that is because it is so obvious that without a reliable patent monopoly there is simply no incentive to invest. The entire period of the command economy of the Soviet Union did not produce, so far as I know, any major pharmaceutical product. And in the West, even in those rare cases of an important invention being made at a University (eg the cephalosporin antibiotics) or by a small inventor or company (eg the anti-epilepsy drug sodium valproate), it has been the pharmaceutical industry which has undertaken the risk of the considerable costs of development.*[2]

---

1   EvaluatePharma®, *World Preview 2016*, dated May 2010.
2   The Right Honourable Sir Robin Jacob, "Patents and Pharmaceuticals" – a paper given on November 29 2008 at the presentation of the Directorate General of Competition's Preliminary Report of the Pharma-sector inquiry.

This is where the protection of intellectual property rights – whether they are traditional patents and trade marks, or relatively new forms of protection such as supplementary protection certificates – is important because, with every new direction that technology takes, there is a need for intellectual property laws to keep up. However, the pace at which technology develops is much faster than that of the law (and the ability to enforce it) and this is often found lagging behind. Combined with diversification, this is now a big challenge for the life sciences industry, the legislator and the intellectual property practitioner alike. As Figure 1 shows, there are many types of products now emerging from the life sciences industry (left-hand column), but do the rights available (right-hand column) adequately protect them?

Take for instance supplementary protection certificates (SPCs). As the authors of Chapter 3 point out, the SPC regime works well for a product consisting of a novel and single active ingredient that is covered by a patent and a first marketing authorisation. However, the healthcare sector has increased in complexity since the original SPC legislation was introduced in 1993[3] and the current regime does not cater for this. The problem is acute for combination drugs, such as vaccines, and a raft of references has been made to the Court of Justice of the European Union asking for the correct interpretation of the SPC Regulations on a number of questions that arise for such products. For example, is a marketing authorisation for product A+B also an authorisation for A alone?[4] And can an SPC have a zero or negative term in order to allow for an application for a paediatric extension?[5] These questions are pushing the SPC Regulation beyond the limits for which it was originally conceived.

Consider, also, the development of computer-assisted methods for analysing genetic data – bioinformatics – mentioned above. These 'in-silico' methods increasingly rival traditional 'wet-lab' techniques for the identification of genetic markers of disease. Such is the speed at which an individual's DNA can be sequenced and screened that this technology opens up the possibility of finding the root causes of disease in specific groups rather than addressing merely the symptoms of whole populations (so-called personalised medicine). But can the data underpinning these drugs be protected? First, there is the issue of whether patents should be allowed for isolated forms of naturally occurring DNA molecules. This somewhat fundamental question is still a live issue in the United States. The recent Court of Appeals of the Federal Circuit decision on the subject in *The Association for Molecular Pathology v US Patent and Trademark Office et al*[6] has held in essence that they are, but it also found that claims to screening DNA sequences are patent eligible only if they involve a transformative step beyond simply 'comparing' or 'analysing' two gene sequences (see the chapter on the United States). However, this is unlikely to be the end of the matter in that jurisdiction. In the United Kingdom, a related issue arose in *Eli Lilly & Company v Human Genome Sciences Inc.*[7] This concerns how much information a

---

3   Council Regulation (EEC) No 1768/92 which came into force on January 2 1993, now replaced by Regulation (EC) No 469/2009 of May 6 2009 (the 'SPC Regulation').
4   *Medeva BV's SPC Applications Case* C-322/10.
5   *Merck & Co Inc v Deutsches Patent und Markenamt*, Case 15 W (pat) 36/08, CJEU case C-125/10.
6   No 2010–1406, 2011 WL 3211513, *20-21 (Fed Cir Jul 29, 2011).
7   [2010] EWCA Civ 33; Supreme Court judgment of November 2 2011.

patent for a gene and the protein it expresses must disclose about its therapeutic application to satisfy the requirements of industrial applicability. This can be a problem if the link between a disease and a sequence is based on *in-silico* comparative screening methods of DNA sequences, where therapeutic application is inferred from DNA sequence homology with genes of known function, rather than traditional wet-lab techniques in which the therapeutic protein and its application are known, or demonstrated, to some extent from the beginning.

Figure 1: Can all innovations be protected by one or more intellectual property rights?

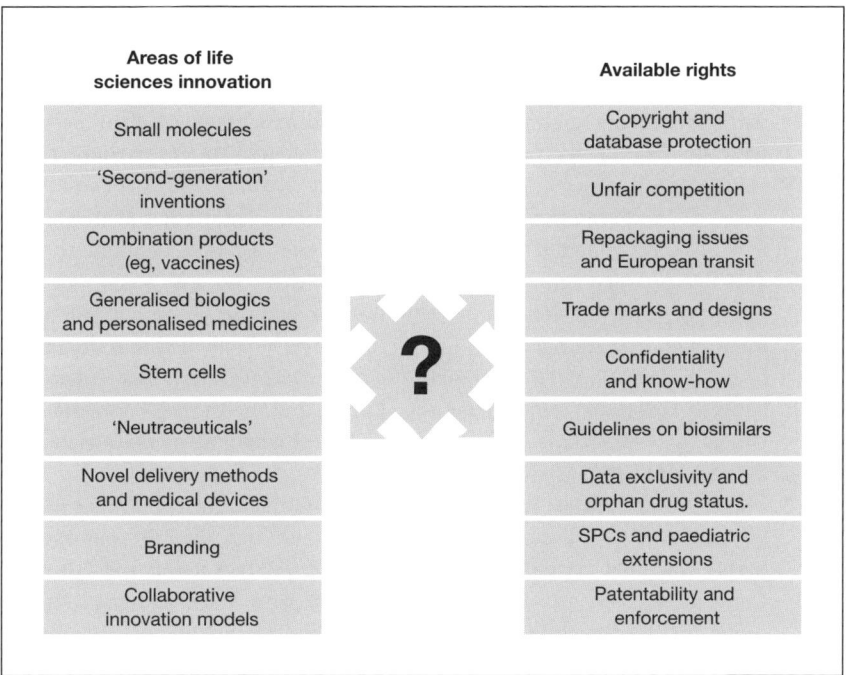

Areas of life sciences innovation		Available rights
Small molecules		Copyright and database protection
'Second-generation' inventions		Unfair competition
Combination products (eg, vaccines)		Repackaging issues and European transit
Generalised biologics and personalised medicines		Trade marks and designs
Stem cells	**?**	Confidentiality and know-how
'Neutraceuticals'		Guidelines on biosimilars
Novel delivery methods and medical devices		Data exclusivity and orphan drug status.
Branding		SPCs and paediatric extensions
Collaborative innovation models		Patentability and enforcement

What about protection for the DNA sequence and other data that is the basis of these discoveries? European database right has so far proved narrow in application although cases in the bioinformatics context have not yet come before the Court of Justice of the European Union. Similarly, the extent to which data copyright may afford protection remains speculative (see Chapter 2).

It would also be a mistake to assume that the important intellectual property rights applicable to life sciences are confined to patents and attendant means of protection such as SPCs. Other rights are also of tremendous importance, particularly trade marks. This is evident from the way these rights have been the focus of parallel importation disputes over the years (see Chapter 2). However, there is a growing realisation, in the light of the pressures being brought about by patent expiry, that branding and design of the drugs themselves, their packaging, and/or their delivery methods are important in their own right as means to distinguish the products of innovators from their generic rivals, build brand loyalty and even provide market exclusivity – take for instance, AstraZeneca's 'Purple Pill' and Pfizer's 'Little Blue Pill' for Viagra. Trade marks, trade dress, passing off, unfair competition and rights protecting designs are therefore essential rights in the portfolio of the life sciences company. Here too there are challenges, however: whilst trade mark protection can in principle extend to medical devices (see the Canada chapter, for example), there are limitations connected to functionality, an issue of particular difficulty in Europe (see, for example, the chapters on Denmark and France).

The globalisation of life sciences and intellectual property law
What is the relevance of pharma's global reach, in this context? The global marketplace in which business now operates is a familiar aspect of today's world, but it poses special challenges for intellectual property lawyers and their commercial partners working in life sciences. This is because intellectual property rights are typically national in scope. Their protection therefore ends at the borders where they are registered. But even if rights are registered in every country for which protection is needed, and even though some rights require no registration at all to subsist, there is a another difficulty – the often significant disparity between the protection available between one country and another, either because some rights simply do not have an equivalent in all countries or because those equivalent rights are interpreted by national courts in differing ways. How is the life sciences industry to operate internationally under all of these different regimes? This remains an issue as between European countries, between Europe and the United States and between the United States and Canada, never mind the great emerging markets of the future such as India and China.

But, again, the law is catching up – slowly – to recognise that trade is increasingly borderless. This is most apparent in Europe. Although the European bloc remains a highly fractured system, pan-European instruments, such as the European Patent Convention, and EU laws governing marketing authorisations, supplementary protection certificates, border enforcement and biotechnology inventions, seek to create uniformity. The Community Trade Mark is the exemplar of this, being a unitary right that covers all member states of the European Union. Chapter 2 provides an overview of European-wide rules governing intellectual property and its enforcement as well as issues that arise from the function of the European Union as a free-trade bloc with rules allowing parallel trade within the European Union and border enforcement as against the rest of the world. Chapter 3 deals with those regulatory rules that are less often thought of as intellectual property laws, as such, but allow the extension of certain monopoly protection or *de facto* market exclusivity. These rights are now taking on a particular significance, as so many patents protecting blockbuster drugs face expiry.

The European Union is not the only area where common rules apply to intellectual property. There is also the Eurasian bloc of nine states that are signatories to the Eurasian Patent Treaty, currently: Russian Federation, Belarus, Kazakhstan, Armenia, Azerbaijan, Moldova, Turkmenistan, Tajikistan and Kyrgyzstan. The

Eurasian Patent Office was set up as a common channel through which to grant equivalent patents protecting the countries within this vast geographical territory (see Chapter 1 and the chapter on Russia). There are similar groupings in the Gulf and Africa.

Then there are of course the treaties that lay down rules for the establishment and treatment of intellectual property on a cross-continental scope. As pointed out in Chapter 1, globalisation is nothing new in this respect, with international rules going back to the 1880s. However, it was not until the recent TRIPS agreement that the global harmonisation of rules for the obtaining and enforcement of intellectual property has been driven forward on a truly international basis.

The challenge of true harmonisation

Does the commercial globalisation of the life sciences business and the international reach of the legal framework mean that intellectual property is now largely harmonised from one country to another? The answer is no. The reasons for this are complex, but they include differences of implementation, or lack of implementation, of international instruments into national laws, different legal traditions with competing influences and, not least, different approaches in national courts. This is perhaps most obvious in the European context. Take, for example, an issue fundamental to patent protection – the approach taken to the scope of protection of patent claims. For the countries subject to the European Patent Convention (EPC), Article 69 of the EPC and its attendant Protocol on interpretation describe how the scope of protection of a claim is to be determined:

Article 1

General principles

Article 69 should not be interpreted as meaning that the extent of the protection conferred by a European patent is to be understood as that defined by the strict, literal meaning of the wording used in the claims, the description and drawings being employed only for the purpose of resolving an ambiguity found in the claims. Nor should it be taken to mean that the claims serve only as a guideline and that the actual protection conferred may extend to what, from a consideration of the description and drawings by a person skilled in the art, the patent proprietor has contemplated. On the contrary, it is to be interpreted as defining a position between these extremes which combines a fair protection for the patent proprietor with a reasonable degree of legal certainty for third parties.

Article 2

Equivalents

For the purpose of determining the extent of protection conferred by a European patent, due account shall be taken of any element which is equivalent to an element specified in the claims.

However, reconciling the Protocol with the existing approaches of the courts of European countries is difficult. The most obvious manifestation of this can be seen in their attitude towards the so-called 'doctrine of equivalence' – construing a claim to include embodiments of the invention that are not literally included in the claim but are nonetheless equivalent to literal embodiments by function, means, result or other technical characteristics. As Figure 2 shows, within Europe the UK jurisdictions stand out for not applying a doctrine of equivalence, despite Article 2 of the Protocol. Nonetheless, it has been held by the House of Lords (now replaced by the UK Supreme Court) that the purposive approach adopted by the UK courts is in keeping with this provision.[8] Whilst the German courts and Dutch courts will apply the doctrine of equivalence in appropriate cases, the ways they do so are quite different too, even though they also are at pains to follow the Protocol. There are other important differences – in Germany there is no doctrine of file wrapper estoppel, but in the Netherlands amendments made to a claim in prosecution may be referred to when construing the granted claim of a Dutch patent.

However, the European courts in which patents are most often litigated appear to be increasingly conscious that this situation needs to change. The recent statement of Jacob LJ in *Grimme Maschinenfabrik GmbH v Scott*[9] illustrates this:

> *Broadly, we think the principle in our courts – and indeed that in the courts of other member states – should be to try to follow the reasoning of an important decision in another country. Only if the court of one state is convinced that the reasoning of a court in another member state is erroneous should it depart from a point that has been authoritatively decided there.*

Indeed, shifts in national approaches are still occurring in this important area. In particular, at the time of writing, the Bundesgerichtshof has made an important decision in *Occlusion Device*[10] in which it overturns the very broad scope of claim protection applied by the lower courts in favour of applying a narrower, claim-centred approach, consciously bringing it closer to its European neighbours. The result of the case – non-infringement – is the same as the cases on equivalent European patents heard in the English and Hague Court of Appeals.

As Chapter 2 explains, moves are currently afoot in the European Commission to establish a Unitary Patent that will have a singular, protective effect over all the qualifying European countries designated. Under further plans, disputes concerning this Unitary Patent, as well as European patents, are intended to be litigated in a Unified Patents Court. As well as the much-vaunted costs savings that this system is intended to have, its proposed system of mixing judges from different jurisdictions together in the panels hearing cases may accelerate the harmonisation of approaches to patent law. But how successful this will be and, indeed, when the court will be established remain unknown.

What about further afield than Europe? Returning to the doctrine of equivalence, a slightly different picture emerges outside Europe, which says much about the complexities of international patent law and the barriers to true harmonisation. Figure 2 shows that the United Kingdom is not actually alone in disavowing the doctrine of equivalence as a means to patent claim construction; neither do Australia, Canada or India adopt it. This exposes one of the competing traditions that lies beneath much intellectual property law: Commonwealth countries have close

---

8   *Kirin-Amgen Inc and Others v Hoechst Marion Roussel Ltd and Others* [2004] UKHL 46.
9   [2010] EWCA Civ 1110.
10   May 10 2011 – X ZR 16/09 – *Occlusion Device*.

Figure 2: The approach to the doctrine of equivalence, reach-through claims and methods of treatment in sample EPC signatory countries and elsewhere

Jurisdiction	Doctrine of equivalence	Reach-through claims	Methods of treatment
EPC and EU countries			
Denmark	Yes	Not unpatentable *per se*, but unlikely to be valid.	Not patentable.
Germany	Yes	Not unpatentable *per se*, but not valid.	Not patentable.
The Netherlands	Yes	Not unpatentable *per se*, but unlikely to be valid.	Not patentable.
Spain	Yes	Not unpatentable *per se*, but of doubtful validity.	Not patentable.
The United Kingdom	No – purposive construction.	Not unpatentable *per se*, but unlikely to be valid.	Not patentable.
Countries that are non-EPC and non-EU			
Australia	No – purposive construction.	Not unpatentable *per se*, but of doubtful validity.	Patentable.
Canada	No – purposive construction.	Not unpatentable *per se*, but unlikely to be valid.	Not patentable.
China	Yes	Judicial interpretations suggest these may be valid.	Not patentable.
India	No – tends towards literal construction.	These have been patented, but untested in case law.	Not patentable.
Israel	Yes	Not unpatentable *per se*, but validity untested.	Not patentable.
Japan	Yes	Not unpatentable *per se*, but unlikely to be valid.	Not patentable.
United States	Yes	Not unpatentable *per se*, but validity currently subject to doubt.	Patentable.

and recent connections with English common law and many of the concepts and approaches once shared have remained. English cases in these countries may still be influential and vice versa. By contrast, the United States, also a common-law country but one that has had more than two centuries of travelling in its own direction, has its own, rival influence in the world. This influence is particularly noticeable in China. Both the advent of TRIPS (see Chapter 1, section 13(a)) and the drive to create innovation-based life sciences sectors in China – benefiting from huge domestic markets, as well export opportunities – is bringing about rapid development in the area of intellectual property law and its associated regulation there. It is a breathtaking transformation. As Judge Jiang Zhipei, former Justice of the IP Tribunal, Supreme People's Court of China, recently explained,[11] China has spent three decades on the road to completing the modernisation of its IP regulatory system, whilst many Western countries achieved the same over two or three hundred years. The judge added that the Chinese courts are not working in isolation. They are following and studying the decisions and procedures of other countries, particularly the United States, on matters such as claim construction, to see how they can improve the judicial interpretation of Chinese law.

But the complexities of international intellectual property law resist any easy analysis. Column 3 of Figure 2 shows that there are unexpected approaches to be found in the emerging economies. Of the countries in the table, it is only India and China where it currently seems most likely that reach-through claims could survive a validity challenge. Then again, look at the fourth and final column. Here, the EPC countries are unified in their exclusion of method-of-treatment claims, by virtue of Article 53 EPC:

European patents shall not be granted in respect of:

...

(c) methods for treatment of the human or animal body by surgery or therapy or diagnostic methods practised on the human or animal body; ...

Likewise, they are excluded from patentability in India, Japan, China, Israel and Canada. But in Australia and the United States there is no such exclusion.

The role of this book
It is because of these enduring differences and the reducing size of the world in commercial terms that the question of whether and how intellectual property law harmonisation can be achieved is in the air at the moment. At a recent University College London seminar, David Kappos, Director of the USPTO, raised the subject of *global* harmonisation of intellectual property law, no less.[12] Kappos was keen to point out that the United States was doing its bit to reach such a goal, with work afoot to deal with first-to-invent, enforcement of the 18-month publication requirement and other 'deficiencies' of the US system. This must be a worthwhile goal for all concerned: it will allow more simplicity for life sciences businesses operating in more than one country and greater certainty and consistency of protection in global markets. But it is just a goal at present.

Until such a time, there is a need for a book that presents a comprehensive overview of the rights protecting the products of the life sciences industry. This must reflect the new areas of innovation to which diversification strategies are leading the

---

11   Speaking at Simmons & Simmons LLP on April 4 2011.
12   Speaking at University College London on April 4 2011.

sector, as well as covering the more conventional areas of drug development. It should also be global in scope and include those countries where markets are most promising for the future, as well as the European and North American jurisdictions where the relationship between intellectual property and the life sciences is more established. It also needs to describe the international and European regimes within which they all operate. Perhaps above all, such a book should be arranged in a practical form so that it is useful to those working in the industry, experienced private practitioners and the curious alike. This means arranging the subject matter according to the product areas with which the life sciences are concerned, and the key issues they need to know about how intellectual property applies to them. Hopefully, this is that book.

Paul England's principle expertise is in life sciences patent litigation and associated regulatory issues. He has a particular interest in multijurisdictional patent litigation strategy, the preparation of case theory and the handling of expert evidence. Paul is the author of Expert Privilege in Civil Evidence *(Hart Publishing, Oxford) and publishes regularly in practitioner publications and peer-reviewed journals.*

Prior to joining Simmons & Simmons LLP in 2010, Paul spent over 10 years at another patent practice in the City of London. Before turning to law, he gained an honours degree in chemistry at Edinburgh University, followed by a doctorate in biochemistry and molecular biology from Linacre College, Oxford University. Paul also has a Diploma in Intellectual Property Law from Bristol University.

Paul has directly experienced the life sciences sector, having worked for the Human Genome Organisation and the former Wellcome Foundation (now merged with gsk).

Chapter 1: An introduction to international intellectual property instruments relevant to life sciences

Nick Bassil
Kilburn & Strode LLP

Any discussion of intellectual property in the field of life sciences needs to be seen in the proper international context. Globalisation might be seen as a relatively modern phenomenon, but in the field of intellectual property the roots of international harmonisation go back to the late 1880s.

Successful companies in the pharmaceutical industry have used the different aspects of the international intellectual property system to achieve their business goals. Patent protection for innovative new treatments is generally seen as fair compensation for the investment needed to take a medicine through the various national clinical trials regimes. Registered trade-mark protection provides users of medicines with assurance over the origin and quality of a medicine. Registered design protection protects the investment of a company in the design process. There are other intellectual property rights which may also be of relevance to the life sciences industry and some of these are covered in this chapter.

Despite attempts at international harmonisation over many years now, there remains a large array of international treaties all with different and often overlapping memberships. This is particularly true in Europe where a complex web of international agreements and treaties now exists.

This chapter attempts to provide an overview of the main intellectual property instruments relevant to the life sciences in the international field. By definition, some simplification has had to be adopted and the precise historical background is not covered in detail.

At the outset, it should also be pointed out that intellectual property rights are exclusive or exclusionary rights. The grant of a registered intellectual property right does not provide any permission to the owner to exercise or to use the right, merely the ability to prevent others from carrying out activities without permission that would otherwise infringe the registered right.

It should also be noted that the intellectual property system is not without its critics. However, some steps have been taken to use the system for charitable purposes. For example, GlaxoSmithKline plc (GSK) recently set up the 'The Pool for Open Innovation against Neglected Tropical Diseases' which aims to use IP generated in companies and universities in the fight against tropical diseases which might otherwise be overlooked.

1.   General instruments

1.1   Paris Convention

The Paris Convention (or more properly 'The Paris Convention for the Protection of Industrial Property') was originally signed on March 20 1883. It has been the subject of several revisions and is now administered by the World Intellectual Property Organisation (WIPO) which is a specialised agency of the United Nations.

The most recent revision of the Paris Convention was signed in Stockholm on July 14 1967 and was further amended on September 28 1979.

The Paris Convention was set up to enable applicants for intellectual property rights in one country to obtain a corresponding filing date in a country other than their home country simply through a mutual recognition of the filing of the initial application.

When the Convention was first established, it was designed to address a very practical consideration over how to put on file corresponding applications in different countries at the same time in order to establish a 'priority date' with respect to any competing third-party applications. With the advent of modern communications such as fax and e-mail this rationale has become less important, but the Convention still provides applicants with some very useful features.

Once an applicant for a patent application has established a priority date, the date becomes the point in time against which the patentability of all subsequent applications claiming priority are judged with respect to the 'state of the art'. The 'state of the art' is the term used to describe the sum total of publicly available materials that qualify as prior art (eg, published patent applications, scientific papers, textbooks etc). In some countries the list of prior art includes oral disclosures, poster displays as well as internet publications.

In the life sciences field, many patent applications are based on an initial scientific observation which led to the making of the invention claimed. However, in a fast-moving research field, it is also important to file an early patent application to secure a filing date. The danger in filing an application too early is that there will not be enough experimental data in the application to provide the necessary support and sufficient written description.

The Paris Convention priority system enables an applicant to file an initial priority application based on a limited disclosure of the invention and then to file a follow-up patent application claiming priority from the first filing which can include new experimental results and/or written description.

The system requires that the later patent application relates to the 'same invention'; otherwise the right to claim the benefit of the priority date will not exist. In Europe there is now a stricter interpretation about the 'prior art' effect of intervening disclosure published in the interval between the priority date of the first application and the filing date of the subsequent application if the content of the invention claim has changed between the two patent filings.

The ability to claim priority from one local application in a subsequent foreign application has enabled the growth of the intellectual property system in its modern form and is the basis for the various current international systems. The priority system established by the Paris Convention is now an essential part of the process of filing and obtaining intellectual property rights.

For a patent application, it enables an applicant a period of 12 months to carry out further experiments or design additional prototypes before making a full patent filing. Given the costs and complexities of the patent system, such a delay is advantageous as an applicant can also carry out further due diligence by way of searching for any potentially relevant prior art (if not already carried before making the priority filing).

The Paris Convention is now used to support a fairly standard patent application filing strategy in combination with the Patent Cooperation Treaty (PCT) which provides applicants with up to 30 months before significant costs start to arise with the filing of separate national patent filings (see below).

For trade marks and registered designs (sometimes referred to as industrial designs), the priority period is six months. For such applications, there is usually no need for further work on the application itself, but the delay provided by the priority period enables the application filing strategy to be decided upon and put into effect.

1.2	**World Intellectual Property Organization Convention**

The World Intellectual Property Organization (WIPO) is an organisation of the United Nations based in Geneva, Switzerland. WIPO itself is responsible for maintaining certain treaties relating to intellectual property rights. There are 24 treaties in all, including the Paris Convention, the Berne Convention, the Patent Cooperation Treaty, the Hague Agreement, and others. WIPO in its modern form is governed by the WIPO Convention which is open to members of the United Nations.

1.3	**Other agreements giving rise to recognition of priority claims**

The modern history of the Paris Convention and its administration by WIPO has shown some limitations to the Convention, concerning states or territories that are not recognised by the United Nations.

Some territories, such as Taiwan, have opted to conclude separate bilateral deals with important trading partners. An alternative system which provides for claims to priority to be recognised has grown up with the World Trade Organization (WTO).

(a)	***Agreement on Trade Related Aspects of Intellectual Property Rights (TRIPS)***

The Agreement on Trade Related Aspects of Intellectual Property Rights, also known by the acronym 'TRIPS', arose from the negotiations under the General Agreement on Tariffs and Trade (GATT) in 1994 (at the end of the so-called Uruguay Round). It is a treaty administered by the WTO.

The provisions of TRIPS apply to all members of the WTO and relate to all major types and classes of intellectual property rights, including confidential information, as well as to procedures for enforcement of rights.

Membership of the WTO is different from the Paris Convention (and also from other international intellectual instruments). Most industrialised countries are members, with the remainder (eg, Russia, Iran, Belarus, and Kazakhstan) and other developing nations (eg, Sudan, Ethiopia) as observers. A few countries are neither members nor observers and are unlikely to achieve either status in the foreseeable future (eg, North Korea and Somalia).

The importance of the TRIPS agreement is that it has helped drive international attempts to achieve harmonisation in the process of obtaining intellectual property rights, as well as in the enforcement of such rights.

For states that have not ratified the Paris Convention (for whatever reason), membership of the WTO provides an alternative procedure for claiming the benefit of a domestic filing date in a subsequent foreign or international patent application (filed via the Patent Cooperation Treaty (PCT)).

In addition to providing the basis for priority claims, the WTO also provides for minimum standards for IP protection and enforcement.

(b)   Bilateral agreements

Since the international systems for recognition of claims to priority are administered by WIPO (a United Nations organisation), or the World Trade Organization, there is a problem for states which exist *de facto*, but which have no *de jure* international recognition.

Taiwan occupies a slightly anomalous international position. The island is self-governing and refers to itself as the Republic of China. The mainland People's Republic of China sees the island as a breakaway province. Taiwan is unable to be a member of the various international conventions which recognise the People's Republic of China as the sole government for the territory of China.

However, Taiwan has negotiated bilateral agreements with many national governments in order for national patent applications to be accorded a right to claim priority in Taiwan. The Taiwanese Patent Office has also recently concluded a similar agreement with the State Intellectual Patent Office (SIPO) in China on a similar basis.

Similarly, in Kosovo (which is not recognised by all states and consequently is also not a member of the United Nations), it is possible to file an application and to have a right to priority recognised from an earlier filed application in a Paris Convention or WTO state under the relevant IP laws in Kosovo.

2.   Patent instruments

2.1   Patent Cooperation Treaty

The Patent Cooperation Treaty (PCT) is an international treaty administered by the World Intellectual Property Organisation (WIPO).

The PCT is a system for simplifying the procedures for filing corresponding patent applications in different countries. The PCT provides applicants with a mechanism for filing a single so-called international patent application at a local national patent office (or directly with the International Bureau of WIPO). The application is examined for formalities requirements and is then subject to a search and preliminary examination procedure which results in the issue of an international search report and a Written Opinion.

The international patent application never proceeds as far as being granted by

WIPO, but the results of the search and the comments in the Written Opinion mean that an applicant who uses the PCT can determine whether it is worth proceeding with the filing of separate local national patent applications subsequently.

Filing an international patent application under the PCT is therefore a very popular filing strategy for applicants in all technical fields. However, in the life sciences field where specifications can sometimes be quite lengthy (with corresponding translation costs to consider), the ability to receive a preliminary search and opinion from a central source can be very useful in developing a patent filing strategy.

An international patent application filed under the terms of the PCT will pend for a period of at least 30 months from the declared priority date (or filing date if no earlier priority date is claimed). The applicant may claim the benefit of a priority date of an earlier application by virtue of the Paris Convention or the provisions of the GATT/TRIPS agreement under the World Trade Organization.

There is a facility for filing amendments to the claims after receipt of the international search report. In addition, it is possible to file a response to the comments in the Written Opinion, provided that a request for international preliminary examination is filed.

Since the PCT is administered by WIPO, which is a UN organisation, membership is only open to states which are full members of the United Nations. As a result, Taiwan falls outside the scope of the treaty. In addition, there remain many states which have not signed up to the PCT. However, the growing importance of trade agreements under the WTO has meant that more states have joined the PCT in recent years.

A PCT patent application may be filed in the languages specified by the national patent office where the application is filed (the Receiving Office), or by the International Bureau of WIPO. Subsequently, the PCT application must be translated into one of the following languages for publication (if not already filed in one of the specified languages): Arabic, Chinese, English, French, German, Japanese, Korean, Portuguese, Russian or Spanish.

At the end of the so-called international phase of the PCT application, the applicant must decide where to file national or regional phase patent applications based on the PCT application. The content of such applications will be the same as the PCT application and in many states the outcome of the PCT search and examination procedure will be taken into account by the local patent office where the national or regional patent application is filed.

The PCT system enables an applicant to select a variety of national patent filings in combination with some regional patent filings. A regional patent application is made to a central patent office which administers a special treaty or arrangement for the central examination and grant of a patent application. At present, the PCT offers applicants the opportunity to designate regional patent filings via the European Patent Convention at the European Patent Office (EPO), the Eurasian Patent Convention at the Eurasian Patent Office (EAPO), the African regional Intellectual Property Office (ARIPO) and the African Intellectual Property Organization (OAPI).

However, there are some restrictions in the states available via a PCT patent application. For example, while it is possible separately to designate the United Kingdom via a national patent application or via a European patent application, other European states have closed the national route and patent protection is not available via a European patent application from a PCT application.

In the life sciences field it can sometimes be advantageous to consider filing national patent applications in certain European jurisdictions, so as to achieve the early grant of a patent which can be enforced or registered elsewhere (eg, filing a UK national application as well as an EPO application designating the United Kingdom).

As indicated above, not all countries are members of the PCT. Such countries cannot, therefore, be designated as part of a PCT application and consequently national patent applications cannot be filed at the expiry of the international phase.

It should also be noted that the Gulf Cooperation Council (GCC) regional patent system is not currently a party to the PCT and so cannot be selected from within a PCT patent application.

2.2   European treaties

(a)   European Patent Convention

After World War II, various agreements relating to trade were made in Europe, in order to assist the recovery of Europe's national economies and to bind the countries together so as to reduce the chances of another conflict emerging.

In relation to intellectual property, the main agreements of interest are the Treaty of Rome (and subsequent treaties of the European Economic Communities (EEC) which later become the European Union) and the European Patent Convention.

One of the agreements that was finalised relatively early was the Treaty of Rome, which laid the basis for the European Union (EU) in 1956. However, until the coming into force of the Single European Market Act in 1987, the European Union was not concerned with intellectual property rights. Instead, a parallel agreement was executed in 1973 between various countries in Western Europe which became known as the European Patent Convention (EPC).

The separate development of the European Union and the European Patent Convention has meant that it is possible for different states to belong to each organisation. All member states of the European Union belong to the European Patent Convention.

It is important to note that the EPC is not a treaty of the European Union. The contracting states to the European Patent Convention include countries which belong to both the European Union and EPC, but there are also states which are party to the EPC only which have yet to join the European Union (or which in some cases may never do so). However, for states now wishing to be members of the European Union, it is a requirement that the state also accedes to the European Patent Convention (EPC) as part of the application process.

The purpose of the EPC is to provide a single authority – the European Patent Office (EPO) – for examining and granting patents across Europe. However, although a single patent application is filed at the EPO, the terms of the EPC mean that the patent which is granted only takes effect in the national European states party to the

EPC which were initially designated in the application and where appropriate procedures have been taken on grant to validate the patent at the national level.

The EPO has offices in The Hague, Berlin and Munich. European patent applications may be filed at any of the offices of the EPO, or through a local national patent office (for an initial filing only). Divisional European patent applications must be filed directly with the EPO.

The filing of a European patent application at the EPO is therefore equivalent to taking an option on a collection of related national patents. Rather than filing separate individual national patent applications at national patent offices with all the issues of local language translations and the inevitable duplication of search and examination, the European patent application provides a convenient mechanism to obtain grant of a patent which can be validated on grant in the required territories.

As indicated above, one of the key advantages of using the EPO is to avoid the need for local language translation for a patent application. A European patent application may be prosecuted in one of English, French or German. Upon grant, it is necessary to file a translation of the granted claims into the other two languages not used for prosecution of the application.

The national validation procedure after grant before the local national patent offices requires the completion of various formalities according to local laws. The formalities may include a full translation of the patent specification as granted, the claims as granted, filing a power of attorney and/or recording a local address for service.

However, recent developments have meant that the requirement to file a full translation of the patent specification upon grant have been removed in many countries that are party to the EPC by virtue of the London Agreement which entered into force on May 1 2008. The provisions of the London Agreement mean that certain EPC contracting states can dispense with the need for a translation entirely, or that states can insist on a translation of the claims only into the local language.

As of July 1 2011, the EPC states which have implemented the London Agreement are: Croatia, Denmark, France, Germany, Hungary, Iceland, Latvia, Liechtenstein, Lithuania, Luxembourg, Monaco, the Netherlands, Slovenia, Sweden, Switzerland and the United Kingdom.

(i) **EPC contracting states:** As of July 1 2011, the contracting states to the EPC are: Albania, Austria, Belgium, Bulgaria, Croatia, Cyprus, Czech Republic, Denmark, Estonia, Finland, France, Germany, Greece, Hungary, Iceland, Ireland, Italy, Latvia, Liechtenstein, Lithuania, Luxembourg, Malta, Macedonia, Monaco, the Netherlands, Norway, Poland, Portugal, Romania, San Marino, Serbia, Slovak Republic, Slovenia, Spain, Sweden, Switzerland, Turkey and the United Kingdom.

Payment of a single designation fee will now include the designation of all contracting states to the EPC in a single European patent application.

A number of other states have concluded 'extension agreements' with the EPO as a preliminary step to becoming full EPC contracting states. Such states may be added to a European patent application by payment of an 'extension fee' per country required. As of July 1 2011, the states with current extension agreements with the EPO are Bosnia and Montenegro.

Other states have signed agreements to recognise the grant of a European patent, which are expected to come into force in due course. It is expected that some form or local registration may be required. At present, the EPO has signed such agreements with Morocco and Tunisia (although no date for coming into force has yet been notified).

The territorial extent of a European patent includes certain dependent territories outside Europe of the contracting states of the EPC. A European patent may also be used as the basis for a local patent registration in Hong Kong and/or Macau.

A European patent granted which was validated in the United Kingdom may be used to obtain local patent registration in certain territories dependent on the United Kingdom, as well as some former UK colonies and dependent territories.

(ii) Particular features of the EPC relevant to the life sciences: The EPC and its counterpart provisions in the law of the various member states provides the basis for certain exclusions to patentability which are important for obtaining rights for pharmaceutical inventions (eg, methods of medical treatment of the human or animal body by therapy or by surgery, or a method of diagnosis practised on the human or animal body).

Instead, the EPC will recognise the patentability of the use of a substance or a composition in a method of therapy or surgery. Methods of diagnosis may be patentable provided that there is no interaction with the body of a patient and a medical diagnosis is not immediately part of the claimed invention.

The exclusions arise from considerations of public policy. The EPC has been a model for similar exclusions in the patent law of other states around the world – the main exceptions being the United States and Australia, where method-of-treatment claims are not excluded.

Other statutory exclusions to patentability in the EPC have been introduced by the European Union Directive on the Legal Protection for Biotechnological Inventions (98/44/EC) discussed herein.

(iii) Amendments to the EPC: The EPC originally entered into force on October 5 1973. Minor amendments were made to the Articles and Rules of the Convention of 1973 (EPC 1973) by means of changes agreed within the Administrative Council of the European Patent Organisation, but substantive changes were only possible by means of an intergovernmental conference. Eventually, it became clear that further more wide-reaching amendments were required and the various contracting states decided to make a major revision to the Convention (known as 'EPC 2000'). The amended Convention entered into force on December 13 2007.

(iv) The Boards of Appeal at the EPO: The Boards of Appeal at the European Patent Office are legally separate from the EPO under the terms of the EPC. The Boards of Appeal are set up to hear appeals in relation to legal or technical matters arising from the actions of the EPO in performing its duties. The Enlarged Board of Appeal at the EPO is able to provide rulings to clarify any divergences between decisions of the Technical or Legal Boards of Appeal, or on a reference from the President of the EPO.

The Enlarged Board of Appeal can also hear petitions from parties to an appeal if a fundamental procedural violation has occurred during the appeal process.

The decisions of the Boards of Appeal at the EPO are not binding on national patent offices or courts, but in practice there is a strong persuasive authority from such decisions where similar facts are at issue.

Since the EPO is not an organisation of the European Union, the decisions of the Court of Justice of the European Union are not binding on the Boards of Appeal at the EPO.

(b)   ***Unitary European Patent***

The European Union has discussed a unitary European patent for many years. The project was initially referred to as the Community Patent Convention (CPC), but has more recently become known as the 'Unitary European Patent'.

The project would see a single unitary European patent application which would include all EU member states. Upon grant there would be no need for any national validation procedure as for the European patent application currently administered by the EPO. The new IP right would therefore resemble the Community Trade Mark and Community Design systems administered by the Office for Harmonization in the Internal Market (OHIM) in Alicante, Spain.

The current proposal for a Unitary European Patent would see the EPO administer the new unitary patent in an analogous manner to the search and examination of European patent applications as at present. The Unitary European Patent would therefore be filed and examined in one of English, French or German.

However, the project has encountered two main stumbling blocks. First, some member states of the European Union oppose the system, since it will not recognise their languages. A second problem has emerged with respect to the court system to be set up to hear complaints of patent infringement and/or validity.

At the moment such legal actions are heard before local national courts. This has led to some divergences in outcomes for parties and also to a certain amount of forum shopping. Despite the fact that the basic texts of substantive European patent law are now harmonised by virtue of the European Patent Convention along with attempts by patent judges to recognise and respect each other's decisions, a degree of variability remains in the current system.

The proposal for a common patent court in Europe to deal with matters arising from the Unitary European Patent has, however, thrown up its own problems and it remains to be seen whether it will be brought into effect at the same time or subsequent to the Unitary European Patent.

Many companies would prefer to be able to apply for a single 'unitary' European patent. However, it will be necessary to provide assurances over the translation issue (ie, minimal or no translations as under the London Agreement) and an appropriate court system (ie, a system which involves expert judges and again minimises language issues).

At the time of writing in July 2011, there remains a large degree of political will to see the Unitary European Patent put in place, but those EU states in favour of the proposal have sought to use a new feature of EU procedures to enable them to proceed without those states which have raised questions. Again, whether this approach will ultimately be successful remains to be seen.

As of July 1 2011, the European Union member states are: Austria, Belgium, Bulgaria, Cyprus, Czech Republic, Germany, Denmark, Estonia, Spain, Finland, France, Greece, Hungary, Ireland, Italy, Latvia, Lithuania, Luxembourg, Malta, Netherlands, Poland, Portugal, Romania, Slovenia, Slovak Republic, Sweden and the United Kingdom.

(c)   ***European Union legislative initiatives***
The European Union has responsibility for the single European market in goods and services and as a result has initiated some legal developments which created new intellectual property rights.

(i) Supplementary Protection Certificates (SPCs): The European Union requires that certain products are subjected to stringent safety checks prior to being launched on the market. The procedures affect medicinal products and plant safety products. The clinical trials needed to satisfy the regulatory authorities therefore can reduce the amount of effective patent term available to a patent proprietor. As a result, the European Union like other countries has devised a system to 'compensate' for the regulatory delays.

For medicinal products, the relevant legislation is Regulation 469/2009, which codifies and replaces the initial Regulation 1768/1992. For plant protection products, the relevant legislation is Regulation 1610/1996.

The European Union has decided that for an authorised product it is possible to obtain a Supplementary Protection Certificate which provides for additional patent term protection (subject to a cap of five years) with respect to the authorised product, provided that the product is protected by a basic patent in force and that the product has not already been the subject of an SPC.

The EU system therefore creates a new IP right that, while linked to a basic patent, must be applied for separately at the national patent office in the EU member state concerned. The EU system therefore differs in a fundamental way from other systems adopted by counties such as the United States to compensate for regulatory delays by way of 'patent term extension'.

The SPC system also occurs in states of the European Economic Area (EEA). The European Economic Area was formed on the basis of a special trade agreement between the European Union and Iceland, Norway and Liechtenstein (which are members of the European Free Trade Association – EFTA). As a result the relevant EU law governing SPCs is also in force in the states of the European Economic Area.

Switzerland is not a member of the European Economic Area, but has independently instigated its own SPC system which resembles the EU/EEA system. Other states with SPC laws include Croatia, Albania and Macedonia.

The term of an SPC is calculated as follows. The period of protection is defined as being equal to the period which has elapsed between the date on which the application for a basic patent was lodged and the date of the first authorisation to place the product on the market in the Community, reduced by a period of five years (where the duration of the certificate may not exceed five years from the date on which it takes effect, with one exception described below).

The 'Community' in this instance means a state of the European Economic Area. One curious side effect is that Liechtenstein does not grant marketing authorisations itself and instead relies on authorisations made by the authorities in Switzerland. Consequently, an authorisation granted by Switzerland can control the term of an SPC granted by a state of the European Economic Area! However, once this anomaly was realised, the government in Liechtenstein introduced a 12-month delay such that Swiss authorisations do not have this immediate effect any more.

An SPC must be applied for within six months of the date of the grant of the marketing authorisation, or within six months of the date of grant of the patent (if the marketing authorisation is granted first).

The effect of the SPC legislation is therefore quite different from the patent term extension legislation enacted in other countries, where the grant of a product licence for a medicine is used to extend the life of the patent itself. There are also other differences which are beyond the scope of this work.

The precise interpretation of the EU SPC legislation has also been the subject of challenges in the national patent offices and before the courts of the national states, leading to a series of referrals to the Court of Justice of the European Union. Given the relatively obscure language of the original EU Regulation, it seems likely that such challenges will remain a feature of practice in this field for some time to come.

(ii) **The paediatric extension to SPCs:** The European Union subsequently modified the SPC legislation by means of Regulation EC 1901/2006 to allow for a six-month extension of term for an SPC if the authorised medicine was for a paediatric indication. The change was part of a drive to spur innovation in relation to the development of medicines for children.

The drafting of the initial SPC legislation and the paediatric extension legislation has many gaps, which have led to some interesting outcomes. One such outcome is the possibility of obtaining the grant of a 'zero-term' or 'negative term' SPC which enables an applicant to still obtain a six-month paediatric extension. Since every day of patent protection can be potentially extremely lucrative, applicants have found that it is worth studying the legislation critically to look for new or favourable interpretations.

(iii) **Regulatory data exclusivity and 'orphan drug' exclusivity:** The European Union has also introduced legislation to provide the manufacturers of innovative medicines with periods of market exclusivity to control the market entry of generic manufacturers. The ability to protect the clinical trial data required to put a novel product on the market in the European Union can be very valuable in its own right. In some respects, clinical trial data submitted in a dossier for a marketing authorisation can be considered to be an intellectual property right.

(iv) **Directive 98/44/EC on the Legal Protection for Biotechnological Inventions:** The European Union has also attempted to harmonise national patent law between its member states. The stated objective was to avoid differences at the national level that might lead to a distortion of the single European market. However, the process of introducing the legislation also provided opponents of biotechnology with an opportunity to insert limitations into the categories of patentable subject matter and the scope of patent protection under a patent.

As discussed herein, the European Patent Office is not an organisation of the European Union and is not bound by any EU legislation or decision of the Court of Justice of the European Union. However, the EPO came to the view that the content of EU Directive 98/44/EC was essentially in conformity with the case law of the Technical Boards of Appeal at the EPO and decided to incorporate the text of the Directive into the Rules of the EPC.

The introduction of Directive 98/44/EC (the 'Biotech Directive') into the rules of the EPC has had some serious consequences for the patentability of inventions relating to human embryonic stem (hES) cells which are of interest for therapeutic purposes as well as for use in screening new chemical compounds. The Biotech Directive prohibits the patenting of inventions which concern uses of human embryos for industrial or commercial purposes. This exclusion (rather than the pre-existing general morality clause in the EPC) has been used to deny the grant of early patent applications in this field. A recent referral of a case from the German courts to the Court of Justice of the European Union may clarify the wording of the Directive on this point for the national patent offices and courts in the member states of the European Union. However, it is far from clear whether the EPO would follow such a decision automatically, even though the text of the Directive now appears in the Rules of the EPC.

Directive 98/44/EC has also been the subject of comment by the Court of Justice of the European Union with respect to the extent of protection for patents containing claims to nucleic acid sequences. Although, the particular case concerned transgenic soya meal and the decision confined to the specific facts, there is a lesson to patent applicants in drafting patent claims appropriately. Transgenic plants have value as raw materials in processes for the production of pharmaceuticals as well as functional foodstuffs (although transgenic foodstuffs are controversial in many countries). Transgenic rats and mice have been widely used as experimental models in pharmaceutical drug discovery. Production of pharmaceutical products in agricultural animal species has also been a goal for many companies.

(v) Community Design: In addition to national law concerning registered designs, the European Union has also enacted Regulation 6/2002 to allow for the application of a Registered Community Design (RCD). An application for a Community Design is made at OHIM. The right, once granted, is a unitary right effective across the entire European Union.

Designs protection is relevant in the context of the life sciences to products such as diagnostic test kits and laboratory or research materials. Many such products are the result of a lengthy design process, and the overall design can be worth protecting.

In the context of a registered design application, a 'design' is taken to mean the appearance of the whole or a part of a product resulting from the features of, in particular, the lines, contours, colours, shape, texture and/or materials of the product itself and/or its ornamentation. A 'product' is taken to mean any industrial or handicraft item, including packaging, get-up, graphic symbols or typographic typefaces, but excluding computer programs.

Applications may be filed within one year of being published at a national patent office, the Community Designs Office (at OHIM) or at the Benelux Patent Office.

The Regulation also provides for a system of unregistered Community design right, with protection of three years from the date of publication of the design.

(vi) Community Trade Mark: In addition to national law concerning registered trade marks, the European Union has also enacted laws to allow for the application of a Community Trade Mark (CTM). The basic legislation was brought forward in Regulation 40/94. This has now been replaced by Regulation 207/2009, which codifies and updates the basic law. An application for a Community Trade Mark is made at OHIM. The right, once granted, is a unitary right effective across the entire European Union.

In the context of a trade mark application, a 'trade mark' is a mark or sign capable of distinguishing the goods or services of an applicant for a defined class of goods or services, without being descriptive or deceptive, and also being capable of being represented graphically. Trade marks therefore also include shapes, sounds, colours and smells, as well as more usual words, names, logos and signs. A three-dimensional or shape trade mark therefore has some overlap with a registered design.

Trade marks are naturally extremely valuable IP rights in the life sciences area. Pharmaceutical products are only protected by patents and other exclusionary rights (under data exclusivity) for a limited number of years – whereas, in theory, a trade mark if used properly can last for ever.

Many pharmaceutical companies have therefore become expert in selecting a brand name to register as a trade mark which is easy to pronounce and remember. As products start to come 'off-patent' and lose protection, the company needs to make sure that users of those drugs (prescribing doctors as well as patients for over-the-counter remedies) are familiar with the brand name products. The reputational value of a well-recognised trade mark can be very important in securing revenue streams after patent expiry.

(vii) Community Plant Variety Right (PVR): By means of Regulation EC 2100/94, the European Union has also enabled applicants to seek protection for plant variety rights at the level of the European Union in a unitary PVR application. The PVR will protect a plant variety provided it is distinct, uniform and stable. New plant varieties have relevance to the life sciences industry in terms of raw materials in processes for the production of pharmaceuticals, as well as for functional foods.

(d)   *Other regional patent agreements*

Other regional patent agreements exist. Some, such as the Eurasian Patent Office (EAPO), ARIPO and OAPI agreements, can be accessed via a PCT patent application. Others, such as that involving the Gulf Cooperation Council, require an application to be filed at the same time as filing a PCT application (with the option of making a claim to priority under the Paris Convention).

(i) Eurasian Patent Convention: The Eurasian Patent Convention is analogous to the European Patent Convention and covers some former states of the USSR. It provides for a single patent application to be filed which, once granted, extends to those states selected by the proprietor which are members of the Eurasian Patent Convention. The present member states are: the Russian Federation, Armenia, Azerbaijan, Belarus, Kazakhstan, Kyrgyzstan, Moldova, Tajikistan and Turkmenistan.

The Eurasian Patent Convention is a special regional patent agreement within the meaning of the Paris Convention and the PCT, so patent applications may be filed at the Eurasian Patent Office (EAPO) from a PCT application.

(ii) Gulf Cooperation Council: The Gulf Cooperation Council (GCC) is a regional patent system enabling a single patent application to be filed at the GCC Patent Office in Saudi Arabia. A GCC patent, once granted, is effective in all designated states. As discussed herein, the GCC is not a party to the PCT but does recognise claims to priority under the Paris Convention.

As of July 1 2011, the member states of the GCC are: Bahrain, Kuwait, Oman, Qatar, Saudi Arabia and the United Arab Emirates.

(iii) OAPI: The African Intellectual Property Organization (OAPI) is a regional patent system which covers mostly francophone countries in Africa. It is a party to the PCT and can be designated in a PCT application.

As of July 1 2011, the member states of OAPI are: Benin, Burkina Faso, Cameroon, Central African Republic, Chad, Congo, Côte d'Ivoire, Gabon, Guinea, Equatorial Guinea, Mali, Mauretania, Niger, Guinea Bissau, Senegal and Togo.

(iv) ARIPO: The African Regional Intellectual Property Office (ARIPO) is an alternative regional patent system in Africa and includes principally anglophone states. It is complementary to the OAPI system. It is also a party to the PCT and can be designated in a PCT application.

As of July 1 2011, the member states of ARIPO are: Botswana, Gambia, Ghana, Kenya, Lesotho, Malawi, Mozambique, Sierra Leone, Somalia, Sudan, Swaziland, Uganda, United Republic of Tanzania, Zambia and Zimbabwe.

3.   Other intellectual property instruments

The Madrid System for trade mark registrations and the Hague Agreement for Registered Designs provide international systems for obtaining trade mark or design registration in an analogous manner to patents under the PCT (but with some key differences).

However, many applicants still opt to use local national filings for strategic reasons, rather than use the Madrid System or the Hague Agreement.

3.1   The Madrid System

The Madrid System for international trade mark applications has two main elements. First, there is the Madrid Agreement Concerning the International Registration of Marks of 1891, and secondly there is the Protocol Relating to the Madrid Agreement of 1989 (for states unable to accede to the Madrid Agreement).

The Madrid Agreement and the Madrid Protocol provide a system for international trade mark registrations to be applied for and granted.

3.2 The Hague Agreement Concerning the International Registration of Industrial Designs

The Hague Agreement dates from 1925 and has been subject to several amendments. The current versions in force are the 1999 Act, the 1960 Act and the 1934 Act. In essence, the Agreement provides a system for international design registration, much like the provisions of the PCT system for patents.

3.3 The Berne Convention for the Protection of Literary and Artistic Works

The Berne Convention is concerned with the protection of copyright works. The treaty is administered by WIPO. Copyright is relevant to the field of life sciences in respect of written works, such as scientific papers, books and manuals of procedure, as well as computer programs amongst other things. Copyright in the life sciences field can therefore actually be quite important and should not be neglected by life sciences companies.

The Berne Convention was first established in 1886 and for a time was only adopted by certain states, with developing nations adhering to the Universal Copyright Convention (UCC) separately administered by UNESCO. However, more recently, since most states are members of the WTO (or aspire to become members of the WTO), the provisions of the Berne Convention have assumed a greater importance and the UCC has become less relevant.

The point of treaties such as the Berne Convention and the UCC is to provide for mutual international recognition of copyright protection for original published copyright works in the signatory states of the convention for the nationals of any signatory state. The recognition of such rights is therefore included in the national law relating to copyright of the signatory states.

Many states have amended their laws to vary the protection afforded to copyright works. Typically, the protection is not applied for but subsists through the act of creating an original qualifying work. However, the existence of the right is only ever tested in court in actions for infringement.

4. Other international treaties

Where matters of validity of intellectual property rights or infringement are considered before a court, questions often arise with regard to the interpretation of the treaty or Act from which the rights are derived.

4.1 Vienna Convention on the Law of Treaties of 1969

The Vienna Convention is a treaty established and administered by the United Nations which governs the applicability and of international treaties and agreements.

4.2 The 'Brussels Regime' on jurisdiction in civil and commercial matters

Within the member states of the European Union and the states of the European Free

Trade Association, there have been a recognition and a desire to regulate the conduct of litigation in civil matters. This has led to three separate legal instruments being brought into effect.

These instruments are the Brussels Convention (the Brussels Convention of September 27 1968 on Jurisdiction and the Enforcement of Judgments in Civil and Commercial Matters), the Lugano Convention (Convention of September 16 1988 on Jurisdiction and the Enforcement of Judgments in Civil and Commercial Matters) and the Brussels Regulation (Council Regulation (EC) No 44/2001 of December 22 2000 on Jurisdiction and the Recognition and Enforcement of Judgments in Civil and Commercial Matters).

4.3   Harmonisation of enforcement procedures in Europe

The European Union has also introduced legislation designed to bring an approximation of laws into effect with respect to the enforcement procedures for intellectual property rights across the various member states. The 'Enforcement Directive' (Directive 2004/48/EC of the European Parliament and of the Council of April 29 2004 on the enforcement of intellectual property rights) was an instruction to the member states to amend national laws as necessary to achieve the effects set down in the Directive and has now been brought into force.

Chapter 2: An introduction to European intellectual property rights

Scott Parker
Adrian Smith
Simmons & Simmons LLP

1. Patents

1.1 Patentable inventions

The requirements for patentable inventions are set out in Article 52 of the European Patent Convention (EPC). Patentability requires that the invention is new, involves an inventive step and is susceptible of industrial application. An 'invention' is not defined. However, it is clear from EPO case law that inventions must have a concrete and technical character and this is consistent with the non-exhaustive list of 'non-inventions' in Article 52(2) EPC including: discoveries, scientific theories, aesthetic creations, business methods, and programs for computers. The exclusion of business methods and programs for computers in particular has given rise to an important body of case law (both at the EPO and in the national courts) concerning the scope of these exclusions. Of greater relevance to pharmaceutical patenting are the exceptions to patentability set out in Article 53 EPC, namely that European patents shall not be granted: if their exploitation would be contrary to *ordre public* or morality, for plant or animal varieties or essentially biological processes for the production of plants or animals and methods of treatment of the human or animal body by surgery or therapy, and diagnostic methods practised on the human or animal body (in contrast to other jurisdictions such as the United States, where patents for methods of treatment are allowed).

The exclusion of inventions the exploitation of which would be contrary to *ordre public* or morality (defined by the Technical Board of Appeal in *Harvard/Onco-mouse* T356/93 as "not in conformity with the conventionally accepted standards of conduct pertaining to this culture") and for plant and animal varieties and essentially biological processes has been considered further in the context of the Biotechnology Directive 98/44/EC (see below) and it is hoped that the decision of the Court of Justice of the European Union (CJEU) in *Brüstle v Greenpeace* (case C34/10) will clarify the position with regard to the patentability of inventions using embryonic stem cells. This decision is likely to have important consequences for the European biotechnology industry.

The methods for treatment by surgery or therapy and diagnostic methods set out in Article 53(c) EPC are excluded from patentability as a matter of policy (ie, to protect clinicians and veterinarians from falling foul of patent laws). Previously, this exclusion was expressed on the basis that such methods are not susceptible to industrial application. However, this fiction has been corrected by EPC 2000. The scope of this exclusion, which is to be interpreted narrowly, has been considered in a number of EPO cases (summarised in the EPO Guidelines for Examination).

Importantly, the exclusion of Article 53(c) EPC does not apply to products for use in such methods and thus pharmaceutical products may be patented for multiple uses (ie, a patentable invention may reside in the product itself or the use of a known product for a new medical use). This is set out in the terms of Article 54(4) EPC, which states that the fact that a product may be part of the state of the art "shall not exclude the patentability of any substance or composition … for use in a method referred to in Article 53(c) EPC, provided that its use for any such method is not comprised in the state of the art". Whereas such second (or further) medical use claims were previously only permitted when drafted in the 'Swiss' style (eg, the use of X in the manufacture of a medicament for the treatment of Y), after the implementation of EPC 2000 this has no longer been necessary and indeed such claims are no longer accepted by the EPO (although old Swiss-style claims remain valid and enforceable).

The patentability of known products for medical use is not restricted to new therapeutic indications. The EPO has held that novelty may also reside in a new dosage regime or means of administration. However, the new use must satisfy the inventive step requirement and must be more than a mere discovery about an already known use. This distinction has been considered by both the EPO and the national courts.

1.2 Industrial application

As indicated above, patents shall only be granted for inventions that are susceptible to industrial application, defined in Article 57 EPC to mean inventions that "can be made or used in any kind of industry, including agriculture". Relative to the other requirements of patentability set out in Article 52 EPC, there is a paucity of decisions on industrial application. This is unsurprising as the term 'industry' is construed broadly and in most areas of technology the mere fact that a patent is worth applying for is of itself an indication that the invention has industrial value. However, for biotechnology inventions in particular the threshold may be harder to satisfy (ie, if the practical use to which the new technology will be put has not been identified at the date of the application).

The European, US and UK law of industrial application was reviewed by the UK courts in *Eli Lilly v Human Genome Sciences* (including an analysis of the following EPO case law: *Max-Planck* T 870/04, *Johns Hopkins* T1329/04, *Genentech* T604/04, *ZymoGenetics* T898/05, *Bayer* T1452/06 and *Schering* T1165/06). A number of common principles were identified, including the propositions that 'industry' must be construed broadly, that the industrial application must be derivable from the patent application (read with common general knowledge), the need for a sound and concrete basis for recognising that the contribution could lead to practical application in industry and that "the patentee must make a full disclosure of his invention, including a practical use to which it can be put. It [is] not a hunting licence to find such a use." In this case, in which the invention was the identification using bioinformatics techniques of a novel protein (neutrokine-α) through its homology to the TNF superfamily, the court decided on the facts that an industrial application for the gene sequence had not been made plausible by the specification. Although the UK court (both at first instance and on appeal) approved the EPO's approach to industrial applicability, the EPO Technical Board of Appeal upheld the same patent – an example of different findings of fact leading to different results.

1.3 Novelty

Novelty is dealt with in Article 54 EPC, which provides that "an invention shall be considered to be new if it does not form part of the state of the art". The state of the art comprises everything made available to the public (anywhere in the world) whether by written or oral description, by use or in any other way before the filing or priority date of the application. After the entry into force of EPC 2000, for the assessment of novelty (but not obviousness) the state of the art includes the content of European patent applications having an earlier priority date but published after the application in question (ie, co-pending patent applications). A co-pending PCT application may also form part of the state of the art so long as it has been published in one of the official languages of the EPO or its translation into one of these languages has been filed with the EPO and published and the national fee has been paid.[1] Article 55 EPC provides a limited six-month 'grace period' for disclosures made in consequence of "an evident abuse in relation to the applicant or his legal predecessor" (eg, where the disclosure is made in breach of a duty of confidence owed to the inventor) or for disclosure of the invention at officially recognised international exhibitions.

The onus is on the party seeking to revoke the patent to prove that the disclosure or prior use was made available to the public before the priority date and further that a skilled person would have been able to put the prior art into practice in such a way as to carry out the invention. Interpretation of the disclosure is by reference to the knowledge of the skilled person in the field at the relevant date. Importantly, the purpose of the prior art disclosure is irrelevant for assessment of novelty and thus a disclosure in an unrelated technical field, which may be directed at a completely different technical problem, may still constitute an 'accidental' anticipation (even if the same disclosure would be irrelevant for assessment of inventive step).

The test for novelty is a stringent one. For a disclosure or prior use to anticipate a claim it must disclose all of the features of the claim (ie, only if the invention disclosed by the prior art would infringe the claim in question, if performed post-grant, will it deprive that claim of novelty). The test is not simply that the prior product or process was available to the public but that the information conveyed by that product or process made the invention available. For example in *G2/88 MOBIL/Friction reducing additives*, the use of the additives in question, which were already known for one use, would necessarily have achieved the new use as well. Although the new use would have been inherent in the old use, this would not have been evident and so the novelty attack was rejected. Further, for a claim to be

---

1   Rule 165, Implementing Regulations to the Convention on the Grant of European Patents.

anticipated, it must be inevitable that following the disclosure of the prior art something within the scope of that claim will result. The test of 'inevitability' is strictly applied (*Union Carbide* T396/89).

One aspect of novelty that is of particular relevance to pharmaceutical inventions is where the novelty resides not in the product but in its use. As indicated above, 'second (or further) medical use' claims are permissible in Europe and provided that the other requirements of patentability are satisfied novelty may reside in a new indication, dosage regime or means of administering a known product. A second aspect of novelty that often arises in a pharmaceutical context is the patentability of a sub-group from within a previously disclosed class. The EPO approach is that the patented thing only lacks novelty if it is individually disclosed in the prior art, and this is not usually the case where a selection is made from more than one list (or, in the case of a 'Markush formula', from one list of substituents at one position and another list of substituents at another position, etc).

The old German law used to be that disclosure of a class entailed disclosure of all members of the class, so all members of that class lacked novelty. English law took a similar view but tempered by the concept of selection patents: a selected sub-class was patentable if it enjoyed an advantage (taught in the later patent) not enjoyed by other members of the main class. In recent years the German and UK courts have both rejected the old approach and adopted the 'individualised description' approach of the EPO (ie, distinguishing what is within the scope of the prior art from what has actually been taught). In doing so, the English court also rejected the rules previously applied to selection patents. In *Dr Reddy's v Eli Lilly* the UK Court of Appeal followed the EPO approach, summarised by Lord Justice Jacob as follows: "It regards what can fairly be regarded a mere arbitrary selection as obvious. If there is no more than an arbitrary selection then there is simply no technical contribution provided by the patentee."

The settled EPO jurisprudence that disclosure of a racemate is not (by itself) a novelty-destroying disclosure of the component enantiomers is now followed by national courts (although in the absence of particular difficulties in resolving the racemate, establishing 'inventive step' (see below) may be difficult).

1.4 Obviousness

Lack of inventive step (also referred to as 'obviousness') is the most common means of attacking the validity of a patent. The approach taken by the EPO to assessing inventive step (followed to a greater or lesser extent by most EPC countries) is the 'problem-solution approach' (ie, identify the 'closest prior art', establish the 'objective technical problem' and then ask whether the claimed invention would have been obvious to a skilled person starting from the closest prior art and the problem to be solved). The approach most commonly followed by the UK courts is the *Windsurfing* test (reformulated in *Pozzoli*), although in recent years the English courts have also shown a willingness to follow the EPO approach. Both are means of applying a structured approach in order to avoid an *ex post facto* analysis and in most cases the 'mechanics' of the approach followed is unlikely to result in a different outcome (save perhaps where identifying the problem is itself part of the invention);

whatever the approach followed, the key question remains: 'Is the invention obvious to a person skilled in the art?'

It is established case law of the EPO that the question is whether the skilled person *would* have arrived at the invention in the expectation of the improvement or advantage actually achieved (not whether he *could* have done so). However, the weight to be attached to motivation and the expectation of success may vary from case to case.

Until the mid-2000s the English courts were widely perceived as being out of step with the rest of Europe in terms of the approach to obviousness, particularly as applied to pharmaceutical inventions. In recent years the English courts have been adopting a more patentee-friendly approach to 'secondary' patents, and patents to enantiomers, combinations, dosage regimes and new crystal forms have all been upheld, where in the past (until the mid-2000s) one would have expected that these patents would have been revoked for lack of inventive step (on the basis that they were 'obvious to try'). The English court is now more aligned with the rest of Europe in focusing its attention on what the skilled person *would* have done, taking into account considerations such as the motive to find a solution to a problem, alternative avenues of research, the effort involved and the expectation of success, rather than looking at what (with hindsight) he could have done.

A developing area of the law of obviousness (and one that overlaps with insufficiency issues – considered below) is that a patent which claims classes of things in respect of which no technical effect is disclosed, or where the technical effect is purely speculative, lacks an inventive step (so-called 'plausibility' obviousness). In *Conor v Angiotech*, the English House of Lords approved the European case law of *AGREVO* T939/92 and *Johns Hopkins* T1329/04.

1.5   Insufficiency

Article 83 EPC sets out the requirement that "the application must disclose the invention in a manner sufficiently clear and complete for it to be carried out by a person skilled in the art" and this corresponds to the ground for revocation at Article 100(b) EPC. This 'sufficiency' requirement is an important means of maintaining the balance between, on the one hand, encouraging investment in innovation and rewarding invention and, on the other hand, ensuring that others can work the invention after the patent has expired. In combination with Article 84 EPC, the requirement for clarity and that the claims are "supported by the description", Article 83 also addresses the policy requirement that the patentee is only entitled to claim the contribution he has made to the art and taught in the patent. As stated in *Genentech/Polypeptide* expression I T292/85 there is a "general legal principle that the extent of the patent monopoly, as defined by the claims, should correspond to the technical contribution to the art in order for it to be supported or justified".

It is settled law that the disclosure of the application must be sufficient to enable the skilled person to perform the invention across the scope of the claim (eg, *EXXON/Fuel Oils* T409/91). However, where the patentee has taught a principle of general application, teaching one way of performing the invention may be enough, even if the claim is of broad scope. Put at its simplest, the claim scope needs to be commensurate with the technical contribution made.

As at the date of the application the skilled person, having read the application as a whole and in light of the common general knowledge, must be able to put the invention into practice without 'undue effort' (*Genentech/Human* t-PA T929/92). This will be assessed on the facts of the case but a degree of trial and error will be allowed so long as this does not require an inventive step.

It is difficult to point to material differences in the law of insufficiency as applied across Europe. Nevertheless, it remains the case that insufficiency attacks hit the mark with greater frequency in some countries (most notably the United Kingdom) than in others.

1.6 Inequitable conduct

The list of grounds for opposition set out in Article 100 EPC is exhaustive. Unlike in the United States and certain other jurisdictions, inequitable conduct is not a ground to revoke or render unenforceable a patent in Europe. The European Patent Convention does not deal with misrepresentation or fraud.

1.7 Patent infringement

Infringement requires first that a prohibited act is carried out whilst the patent is in force and in the territory of the patent. Furthermore, the product or process that is subject of the act must fall within the scope of the claims.

The infringing acts, which differ depending on whether the claim is for a product or a process, are set out in Article 25 of the Community Patent Convention.[2] Article 25 defines the prohibited acts as:

(a) *making, offering, putting on the market or using a product which is the subject-matter of the patent, or importing or stocking the product for these purposes;*

(b) *using a process which is the subject-matter of the patent or, when the third party knows, or it is obvious in the circumstances, that the use of the process is prohibited without the consent of the proprietor of the patent, from offering the process for use within the territories of the Contracting States;*

(c) *offering, putting on the market, using, or importing or stocking for these purposes the product obtained directly by a process which is the subject-matter of the patent.*

In respect of infringement by offering (etc) the product obtained directly by a patented process, it is important to note that the requirement that the product is a *direct* result of the process has been strictly applied, severely curtailing the ability of patentees to take enforcement action where the process is performed abroad and then a product of that process is imported (ie, if the product has been further processed or altered in some way after performance of the claimed method, then its importation and sale in a patent protected country will not be an infringement). For example in the UK case of *Monsanto v Cargill*, the court held that the progeny of a genetically transformed plant were not the direct products of the claimed method for

---

2   Although all signatory states did not ratify the Community Patent Convention and it never came into force, equivalent provisions are enacted in the laws of a number of European states and the Convention itself remains widely influential in judicial decision making. Closely similar provisions are currently to be found in Article 6 of the draft Regulation 9224/11 which proposes provisions for the creation of unitary patent.

producing a genetically transformed plant.

As well as the primary acts of infringement identified above, Article 26 of the CPC provides for indirect infringement where a party supplies or offers to supply means relating to an essential element of the invention. However, unlike the primary acts of infringement for which knowledge or intention is irrelevant (save for offering a process for use), indirect infringement also requires the patentee to establish that the alleged infringer had the requisite knowledge (ie, that those means are suitable for putting and intended to put the invention into effect). For this requirement the objective knowledge of the reasonable person will suffice. The territorial requirements of indirect infringement also require close attention.

Of course, for there to be an infringement the relevant product or process must fall within the scope of the claims. The extent of protection conferred by the claims is determined by reference to Article 69 EPC and the Protocol on its interpretation. In essence this provides that the description and drawings shall be used to interpret the claims and that a balance must be struck between a very narrow 'literal' interpretation of the claims (that would not give fair protection to patentees) and a broad interpretation in which the claims serve only as a guideline (that would not give third parties a reasonable degree of certainty as to the extent of the patent).

How this guidance has been interpreted by the national courts differs. Whereas in most European countries non-literal infringement is considered by applying a 'doctrine of equivalents', in the UK there is no doctrine of equivalents and Article 69 EPC is satisfied by a 'purposive' construction of the claims.

2.   The European patent litigation system

As has been explained in the previous chapter, the EPC establishes the EPO as the granting body for European patents. Once granted, a European patent is to be treated in each of the designated countries as having "the effect of and be subject to the same conditions as a national patent granted by that state" (EPC Art 2). Thus a European patent is often equated to a bundle of national patents, each one identical in form and granting the patentee a monopoly right in respect of the invention in question in that particular designated EPC signatory state.[3] That said, it should be noted that post-grant amendments at a national level may result in different designations of the same European patent having different claims.

Patents (whether national or European) only have effect within the borders of that country (although in some cases the decision of one national court may be persuasive in others). Accordingly, if a pharmaceutical company wants to 'clear the way' to launch a new product in Europe it may have to bring proceedings in multiple jurisdictions to revoke a blocking patent. Similarly, if a potentially infringing product is launched across Europe, then infringement suits will need to be brought in multiple jurisdictions. This provides litigants with the opportunity to forum shop (ie, to choose to litigate in the court(s) that best suit that party's commercial objectives).

In practice, for reasons of cost and efficiency, patents are not litigated in every

---

3   Note that the EPC signatory states are not limited to the European Union but also include, for example, Switzerland, Norway and Turkey. As at July 2011 there were 38 EPC signatory states.

country in Europe. However, it is not uncommon for patents for pharmaceutical products to be litigated in (say) three or four jurisdictions – on the basis that once the arguments have been tested in these countries, a settlement will be reached that will cover the others. The legal, procedural and strategy issues that guide the selection of the countries in which to bring proceedings are of course fundamental to successful litigation. Relevant factors may include: the specialist nature of the court, the precedent value of a decision, the rigorousness of the procedure (including the ability to obtain disclosure and cross-examine witnesses), the value of being able to separate infringement and validity in the bifurcated systems of Germany and Austria, availability of interim remedies (including injunctions), application of the law and the perception of how 'patent friendly' a particular court may be, speed and cost (including the ability to recover legal costs from the other side if successful). It is most common for patent litigation to be brought in one or more of the following countries, in which the court systems are seen as being the most sophisticated when it comes to patent matters: Germany, the United Kingdom and the Netherlands (followed by France and Italy).

Against this backdrop there is always the possibility of conflicting decisions in different jurisdictions. Notwithstanding that the underlying law is derived from the EPC, the way in which this has been applied by judges and procedural differences (in particular the greater emphasis placed on disclosure of documents and cross-examination of witnesses in some jurisdictions) means that under the current system there will always be conflicting decisions in different member states. An example of this is the recent pan-European *Novartis v Johnson & Johnson* contact lens litigation. Novartis' patent was revoked for insufficiency by the English court and for anticipation in Germany, but has been upheld in the Netherlands and France.

In recent years there has been a judge-led move towards harmonisation (either as a stepping stone towards a Community patent or a common European litigation system, or as a reaction to the lack of progress in this direction). For example, a number of recent cases exemplify the rapprochement between the English courts and the EPO, and in the recent UK Court of Appeal judgment in *Grimme Maschinenfabrik v Scott* guidance was given on the relevance of decisions of the courts of other European countries. The principle across Europe should be "to try to follow the reasoning of an important decision in another country" and "only if the court of one state is convinced that the reasoning of a court in another member state is erroneous should it depart from a point that has been authoritatively decided there". There is a perception that judges across Europe are paying greater respect both to EPO case law and to the decisions of courts in the other major European patent litigation countries. As a result it is hoped that in future conflicting decisions in respect of the same European patent will become less common.

There are three caveats or points to note in respect of the general applicability of the statement above that, within Europe, patents (whether 'European' or national) need to be enforced or challenged at a national level. These are set out next.

2.1 EPO opposition
The EPC provides that, for a period of nine months following grant of a European

patent, the validity of that patent may be challenged in opposition proceedings at the EPO. The result of a successful challenge is that the patent is invalidated in all designated states (in which case the decision of the EPO, once the right to appeal to the EPO Technical Boards of Appeal has been exhausted, is final[4] – ie, the patent cannot be 'resurrected' in proceedings before the national courts). The alternative outcomes are that the patent may be upheld in amended form (typically narrowed in scope), or it may be maintained as granted.[5] In either case the patent may then be challenged at a national level. In contrast, for infringement matters the national courts have exclusive jurisdiction. The EPO will not hear infringement proceedings.

An opposition may be filed by 'any person' (save for the patentee). There is no test for standing. Indeed, an opposition may be brought by a firm of patent attorneys (allowing their client to remain anonymous). The notice of opposition must contain a statement of the extent to which the European patent is opposed (ie, which claims are being challenged) and of the grounds on which the opposition is based, as well as an indication of the facts and evidence presented in support of these grounds (Rule 76(2)(c) EPC). As for the grounds on which the opposition may be based, these are set out in Article 100 EPC:

- that the subject matter of the patent is not patentable within the terms of Articles 52 to 57 EPC (ie, novelty, inventive step, exceptions to patentability, lack of industrial application and 'non inventions');
- that the invention is not disclosed clearly and completely enough for a person skilled in the art to carry it out; and
- that the subject matter extends beyond the content of the application as filed.

An opposition will first be examined for its admissibility on procedural grounds. Admissible oppositions will then be the subject of substantive examination. The opposition procedure is *inter partes* and both patentee and opponent(s) have the opportunity to submit their written arguments and to argue their case at oral proceedings. Although expert declarations are allowed, there is no scope for cross-examination of the 'other party's' expert. That said, provided that notification is given, an accompanying person (typically a scientist or inventor) may play an important role at an oral hearing. Another procedural point worth highlighting is that a patent may be amended during opposition in order to meet a ground for opposition. Indeed, it is fairly common for the patentee to use a cascading series of auxiliary requests as a means to advance alternative positions as to the scope of the claims sought (in a way that does not feature to the same extent in most national court systems).

It is inherent in the system that national proceedings will run in parallel with EPO oppositions. Given that the typical EPO opposition may last around three years

---

4   Save for the possibility of judicial review by the Enlarged Board of Appeal under EPC Art 112a where it is alleged that the outcome of the opposition was influenced by a fundamental procedural defect.

5   According to the EPO's annual report for 2009, in approximately 44% of oppositions the patent was revoked, in 30% of oppositions the patent was maintained in amended form and in 26% of cases the opposition was rejected.

(five or six years if the decision is appealed), this is not uncommon. Different countries take different approaches to the question of whether national proceedings should be stayed pending resolution at the EPO. Although in certain countries and in certain circumstances this may arise (ie, the national court will exercise its discretion to stay the proceedings), the general trend is towards continuing with the national and EPO proceedings in parallel. The principal exception is in Germany where a nullity action cannot be commenced until the opposition period has lapsed or EPO opposition proceedings have been concluded. There are encouraging signs that the EPO is trying to respond to the criticism it has received over the duration of opposition proceedings and also that the EPO and national courts are prepared to engage with each other and adopt a more flexible approach to timing in order to avoid potentially unnecessary duplication. For example, in the recent case of *Eli Lilly v Human Genome Sciences*, the EPO was prepared to accelerate its procedure in order to hear an appeal within only five months of receiving a request from the UK Court of Appeal (allowing the UK to fix a later hearing date and avoid the possibility of the UK court upholding the patent only for it to be revoked centrally shortly thereafter). The flip-side of the pressure on parties to request acceleration (where appropriate) and the reluctance of national courts to stay proceedings is that it is becoming increasingly difficult to achieve a tactical delay.

The primary advantages of the opposition procedure are certainty – if successful the patent will be knocked out across Europe – and cost, which is low compared with litigation in any one of the major patent litigation jurisdictions.[6] On the other hand, one needs to act quickly in order to file an opposition during the nine-month period following grant and one needs to be prepared to wait for the opposition procedure to run its course or initiate national proceedings in parallel and accept the risk of inconsistent decisions and unnecessary costs (eg, if a patent is upheld by a national court only to be revoked or narrowed by the EPO).

That said, only a small number of European patents are opposed (less than 5% in 2009). The perception is that the opposition procedure is more commonly used by originator pharmaceutical companies than by generic companies, who because of the lag between patent grant and commercial success of the product will often only identify a target market and blocking patent after expiry of the opposition period.

2.2   Cross-border injunctions and 'torpedo' actions

Some national courts (most notably the Dutch and Italian courts) have been prepared to interpret EC Regulation 44/2001 (Brussels I) in such a way as to permit them to decide patent issues outside their own borders (eg, to grant cross-border injunctions against a single defendant or group of related defendants). This is so notwithstanding that the European rules on jurisdiction provide that patent validity (which is almost always raised as a defence to infringement) can only be determined

---

6   Only rarely does the cost of an opposition exceed €100,000. By comparison, the cost of litigating a pharmaceutical patent in the UK High Court will cost at least £500,000 (with the cost of cases in other national courts to be added).

by the national court in the country to which that part of the patent relates (Art 22(4) of EC Regulation 44/2001).

However, following the ECJ rulings in *GAT/LuK* and *Roche/Primus* in 2006 the practice of awarding cross-border injunctions has become increasingly rare. In *GAT/LuK* the ECJ held that the validity of a patent cannot be determined by the court in another jurisdiction and that this rule is mandatory and could not be derogated from. Indeed, where a court is seized of a claim that is primarily concerned with a matter over which the courts of another state have exclusive jurisdiction, the seized court is required to decline jurisdiction and as a consequence cross-border injunctions are not possible where validity is raised as a defence.

In *Roche/Primus* the ECJ held that where it is alleged that a European patent is infringed in multiple countries, infringement in all countries cannot be considered by a single court on the basis that there are multiple defendants within the same group (eg, sister companies) and the company domiciled in that country was responsible for determining a coordinated policy of carrying out the same allegedly infringing acts by the others (the 'spider in the web' doctrine). Nevertheless in certain circumstances the Dutch courts have indicated a continued willingness to grant cross-border injunctions, although the extent of this practice is now considerably diminished.[7]

In Italy the historical willingness of the courts to seize jurisdiction for patent infringements across Europe combined with the slow pace of Italian litigation gave rise to the practice known as the 'Italian torpedo', in which a party fearing an infringement suit would file a claim for a declaration of non-infringement in Italy (or another slow jurisdiction such as Belgium), the intended effect being that if the patentee brings the same infringement dispute before the courts of another member state the court of that country would decline jurisdiction and stay the infringement proceedings under EC Regulation 44/2001.

However, in recent years the force of the torpedo strategy has diminished considerably as a result of several factors, including: courts in Germany and France ruling that in some circumstances a torpedo may be an abuse of law; the increasingly sceptical attitude of the Italian courts to the practice; the decision of several European courts (including in Italy) to decline jurisdiction on the grounds that no wrongful act had been committed in the member state of the court at issue; and the *GAT/LuK* and *Roche/Primus* decisions mentioned above. At the same time as the torpedo tactic has appeared to fall from favour in some countries, in the United Kingdom, where the prevailing view of practitioners had been that the filing of such proceedings is considered an abuse by the courts, it has received some surprising support in the Court of Appeal: in *Research in Motion UK v Visto Corporation* [2008] EWCA Civ 153 Jacob LJ has said: "much ingenuity is expended on all this game playing. Despite the temptation to do otherwise, it is not easy to criticise the parties

---

7   At the time of writing (July 2011), further questions have been referred to the CJEU by the District Court of the Hague that relate to the jurisdiction of national courts to grant cross-border injunctions. These concern the interpretation of 'irreconcilable decisions' in Article 6(1) and the interpretation of Article 22(4) of EC Regulation (EC) 44/2001: see *Solvay SA v Honeywell Fluorine Products Europe BV et al*, District Court The Hague, the Netherlands, December 22 2010, Case No 09-2275.

or their lawyers for this. They have to take the current system as it is and are entitled (and can only be expected) to jockey for what they conceive to be the best position from their or their client's point of view." However, these comments were made *obiter* in a decision that concerned the application of Article 28 of EC Regulation 44/2001. The extent to which the court would follow this view in other cases is therefore uncertain.

2.3   Community patent and UPLS

For almost 40 years there has been movement towards a Community patent system in which a single unitary Community patent would have equal effect across Europe. The aims behind this drive are to make patent litigation in Europe more predictable, quicker, cheaper and ultimately to increase patent quality and boost European innovation and competitiveness. However, notwithstanding these important goals, the common threads that run throughout this story are disputes between member states surrounding language, translation costs and the judicial system. Unfortunately, whether a Community patent and a common court system is any closer to fruition now than 40 years ago remains a matter for debate.

The first step towards a unified patent system for Europe was the Community Patent Convention (CPC), which had at its heart the creation of a Community patent. It was signed in 1975 but not ratified by all member states and so did not come into force. Several other initiatives followed in the wake of the CPC, most notably the Council Regulation on the Community patent and the Council's Common Political Approach. Ultimately these initiatives became mired in the same difficulties and were undone.

In parallel, momentum was gathering behind an integrated European patent litigation system and a draft protocol for common rules of procedure for patent litigation and the establishment of a common European appeal court was proposed (the 'European Patent Litigation Agreement' or 'EPLA'). Throughout the last decade, as repeated attempts to progress a Community patent failed, a general acceptance took hold that in the short term the focus should be on driving forward the EPLA, with the Community patent to follow. However, despite success in attracting participation from the European Commission, divisions resurfaced between those member states wanting a more integrated, EU-based system and those wanting to retain greater autonomy. As a result the EPLA failed and in 2007 the Commission proposed the Unified Patent Litigation System (UPLS) as a compromise.

The UPLS would introduce a single unitary patent covering all states. It also proposes a single system of European patent courts that would have jurisdiction for the infringement and validity of both existing European patents and for future Community patents. The favoured approach is based on a largely decentralised first instance (regional courts) applying common procedural rules, and on a single appeal court. In essence the proposal provided that the European Union accedes to the EPC, allowing the EPO to grant Community patents and setting up an EU patent court system.

Notwithstanding that the UPLS has the support of the majority of stakeholders, its implementation gives rise to several fundamental issues of constitutional law concerning the relationship between the European Union and the EPO, and the compatibility of the UPLS with the EC Treaty. In March 2011 the CJEU ruled that the UPLS is not compatible with EU law for the reason that member states cannot divest to an international body their powers over matters that require the implementation of EU law, nor can they deprive the CJEU of its powers to give preliminary rulings on matters of EU law. This decision dealt a major blow to the achievement of the goal to the UPLS proposals, not least because the fundamental issue is one of judicial control and a scenario in which the proposed patent system would be presided over by the Court of Justice is considered undesirable by the majority. However, it appeared that alternatives to an EU patent-centred system presided over by the CJEU were running out.

Most recently, in May 2011 the Commission proposed a way around the constitutional deadlock by giving the Commission the right to bring proceedings against member states should the new patents court fail to refer EU law issues to the CJEU and not requiring the European Union to accede to the EPC. Importantly, however, this proposal would keep the new courts within the EU judicial system but should not increase the involvement of the CJEU in patent matters. The substantive proposals of how the new patent system would work remain largely unchanged. Observers see this as a pragmatic proposal and it is expected that with this 'design around' it will not need to go back before the CJEU. Once again optimism is on the rise and momentum is gathering to overcome the language obstacles to agreeing an EU-wide 'unitary' patent and the system under which such patents (and existing European patents) would be litigated. Most notably the 'enhanced cooperation' route is being used in order to sidestep the concerns of Spain and Italy to the proposed three-language regime of the unitary patent (subject to ongoing challenge by Spain of the Council of the European Union's decision to authorise the enhanced cooperation procedure).

The Commission's goal is for the first unitary patents to be issued in 2013. This would have a profound effect on patent litigation in Europe, but whether it can be achieved remains to be seen.

3.   The Biotechnology Directive

Directive 98/44/EC on the protection of biotechnological inventions (the 'Biotech Directive') deals with the scope of patent protection for biotech inventions. Its goals were to harmonise the relevant law so as to strike a balance between the promotion of the European biotech industry and respect for ethical rules.

After 10 years at the drafting stage the Biotech Directive was finally adopted in 1998. It was unsuccessfully challenged before the ECJ by the Dutch government (with the support of Italy and Norway) and implementation into national law across all member states was not completed until 2007.

The Biotech Directive provides clarity as to what is and is not patentable. Article 3 provides that (i) the ordinary test of patentability shall apply to products that consist of or contain biological material or a process by means of which biological material is produced, processed or used; and (ii) biological material which is isolated from its natural environment or produced by a technical process may constitute a patentable invention. Plant and animal varieties *per se* are not patentable; neither is the human body nor the "simple discovery of one of its elements, including the sequence or partial sequence of a gene". Importantly, however, Article 5 goes on to provide that, once isolated from the human body or otherwise produced by a technical process, an element from the human body including a full or partial gene sequence may constitute a patentable invention, even if the structure of that element is identical to that of a natural element but provided that the industrial application for the gene sequence is disclosed in the application. Articles 13 and 14 set out the procedures for depositing biological material with a recognised depositary institution prior to the filing date.

Article 6 includes ethical exclusions for (amongst other things) human cloning and the use of human embryos for industrial or commercial purposes. Different patent offices have interpreted this exclusion differently. The EPO's position, as set out by the Enlarged Board of Appeal in *WARF* (G02/06), is that the intention of the legislation was to exclude from patentability (on moral grounds) inventions which necessitate the destruction of a human embryo. In the *WARF* case the claims themselves were not for a process involving the destruction of human embryos but as at the filing date a human embryo would have been the only possible starting material. Accordingly, the Enlarged Board ruled that the Biotech Directive prevents the patenting of inventions (such as human stem cell cultures) that could only be obtained by a process involving the destruction of human embryos (whether or not that forms part of the claim and regardless of whether after the filing date an alternative starting material could have been used). However, the position of the EPO post-*WARF* seems to be that inventions are not deemed immoral if the invention involves the use of human material that can be obtained without requiring the destruction of a human embryo (eg, if human embryonic stem cell lines were available at the filing date). The validity of these patents remains in doubt pending the decision of the CJEU in *Brüstle v Greenpeace* (expected later in 2011), particularly in light of Advocate General Bot's preliminary opinion, which proposes a wider definition of excluded inventions than is currently being applied by the EPO.

There have been relatively few court decisions in which the Biotech Directive has been applied. So far as validity cases are concerned the most significant is *Eli Lilly v HGS* on the subject of industrial applicability and the Article 5(3) requirement that the industrial application of a sequence or partial sequence of a gene must be disclosed in the application (considered above).

There have been very few infringement cases concerning biotechnology patents in Europe (although this is likely to change with the growth of biosimilars). It is therefore difficult at the present time to discern any particular trends regarding the scope given to biotechnology patents when enforced in the national courts.

In its first decision on the infringement provisions of the Biotechnology Directive, the CJEU gave its judgment in the case of *Monsanto Technology v Cefetra*. The CJEU held that Article 9 of Directive 98/44/EC does not confer protection on a patented DNA sequence when it is contained in soy meal ('dead' material), where it does not perform the function for which it was patented. This is regardless of the fact that it did perform that function previously in the soy plant, of which the meal is a processed product, and that the DNA could again perform its function after extraction from the soy meal and insertion into the cell of a living organism. Thus, the CJEU applied a strict approach to the requirement of Article 9 that the protection available for genes only extends to material in which the genetic information 'performs its function'.

4.   Trade marks

The legal regime in the European Union for the protection of registered trade marks is dominated by two EU instruments – (1) First Council Directive 89/104/EEC of December 29 1988, to approximate the laws of the member states relating to trade marks (variously known as either the Trade Marks Directive or the Harmonisation Directive); and (2) Regulation 207/2009 – the Community Trade Marks Regulation (CTMR).

Prior to the implementation of the Trade Marks Directive, the protection of registered trade marks in the European Union (in common with the general position elsewhere) was strictly territorial (ie, through trade mark registrations in each of the national trade marks registries of the EU member states). Before the Directive there were also many significant differences between the laws and procedures applying to trade marks in each of the individual member states. The purpose of the Harmonisation Directive (as its name indicates) was to harmonise the national laws of the EU member states in relation to the protection of trade marks. The result is that, whilst the national legislation in place in each of the EU member states regarding trade marks is still far from identical, there are core substantive aspects of each of those national laws which (on paper at least) conform with the Harmonisation Directive.

Although registrations in the national trade marks registries of the EU member states remain of importance, the dominant feature of the trade marks system in Europe (and this is true in particular for pharmaceutical companies and other life sciences businesses) is the Community Trade Mark. The Community Trade Mark (CTM) is a unitary right governing the entire European Union (and which has been extended to include additional countries which have joined the European Union by accession in the past). The CTMR was originally Regulation 40/94 of December 1993. However, this had been subject to a number of amendments and was reissued as Regulation 207/2009 of February 2009. The CTM system is operated by OHIM (also known as OAMI) based in Alicante, Spain. The key feature of the system is that by a single application to OHIM, a CTM registration can be obtained which protects the mark across the entire European Union. The pricing of CTM registrations is very cost-effective as compared with obtaining registrations in a number of national member states and the system has been highly successful in the sense that it has attracted large numbers of applications.

Upon application for a CTM, OHIM examines the application for registration of the trade mark in question for registrability on 'absolute grounds' (see below), and marks are sometimes refused by the office at this stage. Article 2 of the Trade Marks Directive provides that for a sign to be a trade mark, it must be "capable of distinguishing the goods or services of one undertaking from those of other undertakings". This requirement, often referred to as the 'origin function' has been held to be 'the essential function' of a trade mark. The most common bases on which marks are refused registration on absolute grounds are those which (in summary) prevent registration of signs which are not capable of fulfilling the essential function because they are insufficiently distinctive and/or are descriptive – Trade Marks Directive, Article 3(1)(b), (c) and (d). In particular, Article 3(1)(b) precludes registration of signs which are 'devoid of distinctive character'.

There is now an extensive (and continually growing) body of European case law on this issue (which is beyond the scope of this chapter), dealing with the full range of different kinds of marks. By way of example, in relation to marks having more than one element (eg, word marks consisting of more than one word, or a combined word and device mark), it is now clear from CJEU case law that assessment of distinctive character must be based on the overall perception of the mark by the relevant public rather than on an evaluation of each of the elements of the mark, separately. It follows that a composite mark consisting of elements, each of which, considered individually, lacks distinctive character, may still, taken as a whole, be considered to be sufficiently distinctive to be registered (*Eurohypo* C-304/06P). See below for a brief comment on the state of the law on assessment of distinctiveness and registrability of another type of mark of particular relevance in the life sciences field, namely product and packaging shape.

Provided that no objection to a CTM application is raised by OHIM on absolute grounds (or, if raised, is overcome), then the application will proceed to be advertised for opposition purposes. OHIM does not itself block applications on 'relative grounds' (ie, on the basis of the existence of earlier registrations for the same or similar marks). Rather, it will notify the proprietors of earlier similar or identical CTMs and the onus is on those proprietors to oppose the later application if they wish. A CTM registration can be opposed on the basis of an earlier CTM for an identical or similar mark or on the basis of an earlier national right, whether an earlier registration in one of the member states or an unregistered right relating to the trade mark in question in a member state – for example, on the basis that use of the mark applied for in the United Kingdom would amount to passing-off. As with the basis on which OHIM may refuse marks on absolute grounds, there is now a large body of European case law addressing the question of when marks will be held to conflict with one another. There is considerable crossover, in that many of the decisions are relevant in the context of action by owners of earlier rights to oppose registration of later marks or to challenge existing registrations, as well as being relevant to determining infringement cases.

In summary, registration of a sign may be refused on opposition by the owner of a relevant earlier right, or its use prevented through infringement action by the owner of such a right where:

- the sign is identical and is being applied for/used (as appropriate) in relation to goods or services identical to those covered by the earlier mark; or
- one or both of the sign or the goods/services are only similar (as opposed to identical) to the earlier mark and its goods/services, and there exists a likelihood of confusion which includes the likelihood of association with the

earlier mark; or

- the sign is either identical or similar to the earlier mark and is applied for/used in relation to goods/services of any kind, provided that the earlier mark has a reputation in the Community and the use of the later sign, "without due cause takes unfair advantage of, or is detrimental to, the distinctive character or repute" of the earlier mark.

There has been quite a considerable amount of recent, notable development in the law relating to these tests and which addresses issues including the protectable functions of trade marks other than the origin function, what constitutes 'use' as a trade mark and what amounts to taking 'unfair advantage'. Significant decisions which incorporate reviews of much of the earlier case law on the relevant issues include the CJEU decisions in *L'Oréal/Bellure* (C-487/07), *Google France/Louis Vuitton* (C-236/08); and *L'Oréal/eBay* (C-324-09). As may be apparent from the identity of some of the parties to these cases, online activity has been testing the boundaries of European trade mark law and seems set to continue to do so. At the time of writing, the CJEU's decision is expected imminently in the *Interflora/Marks & Spencer* case (C-323/09), referred from the English High Court, which is likely to give important guidance on the extent to which use by a third party of trade marks as internet keywords is actionable by the trade mark proprietor.

A particular feature of the CTM application system is that the basic application fee covers registration of a mark in up to three classes of goods and services under the Nice Classification System. There is much debate over whether this feature has led to 'cluttering' of the CTM register (ie, registrations for marks in classes that are not strictly required for protection of the trade mark proprietor's interests but which have been included in the application on the basis that they are effectively 'free'). Whether and to what extent that analysis is correct, it is a fact that, whilst obtaining a CTM registration provides the proprietor with a powerful right – giving protection to the mark across the entire European Union – the size of the EU market, combined with the fact that a CTM can be opposed on the basis of either a pre-existing CTM or an earlier national right in one of the member states means that the task of identifying and clearing a new trade mark to be registered as a CTM for use across the European Union is a difficult one. That task is very challenging in certain business sections, in particular for pharmaceuticals. In that context there is, almost inevitably, a requirement for a single mark to cover the European Union and the dual requirements of finding a mark which is acceptable on a regulatory basis (as the invented name for the product) and which is also registrable and available for registration as a CTM present particular difficulties.

Should a CTM applicant, or a party to proceedings before OHIM (eg, opposition, invalidation or revocation proceedings) wish to appeal a decision of the Office, then in the first instance that appeal is made to the OHIM Boards of Appeal which are constituted specifically for this purpose. The quality and consistency of the decisions of the OHIM Boards of Appeal is one of the features of the CTM system which has attracted a fair amount of criticism in the past. Subsequent appeals are made to the European Court of Justice General Court (formerly called The Court of First Instance) and to the full Court of Justice of the European Union, formerly known as the ECJ. As already mentioned, since the Harmonisation Directive and the CTMR there has been a large number of cases referred to the General Court and the CJEU and a substantial body of case law exists which has interpreted how the harmonised trade mark laws and the CTM system are to be applied and operated in practice.

The European case law referred to above (which has on occasions been quite difficult to reconcile with earlier decisions) has often led to aspects of the relevant laws and procedures being interpreted in ways that may not have been anticipated upon a straightforward reading of the provisions of the Trade Marks Directive or the CTMR. One example of this is in connection with the registrability of non-traditional trade marks. When the Trade Marks Directive was introduced it was anticipated that it would open the door to easy registration for many different kinds of trade mark (beyond the traditional kinds – eg, words and logos) such as product shapes, colour combinations and single colours, sounds, smells etc. Although examples of all these kinds of marks have been registered, there has been a distinct retrenchment through the case law over the years, such that registration of many such non-traditional trade marks has become far more difficult than was originally the case when the Trade Marks Directive was introduced and, in some cases, is now a virtual impossibility. The Trade Marks Directive Article 3(1) states that signs shall not be registrable (or if registered shall be liable to be declared invalid) if they consist exclusively of (i) shapes which result from the nature of the goods themselves, (ii) shapes necessary to obtain a technical result and (iii) shapes which give substantial value to the goods. The way in which this restriction has been interpreted – in particular as regards points (i) and (ii) – has, in practice, made it extremely difficult to obtain (or retain) registrations of marks for the shapes of products or product packaging. A series of ECJ decisions, in particular: *Philips/Remington* (C-299/99), *Linde* (C-53/01, C-55/01) and, most recently, *Lego/Mega Brands* (C-48/09P) have resulted in the position being that a shape will be considered 'necessary' to achieve a technical result even if there are other shapes which would achieve that result. Furthermore, when deciding whether a sign consists 'exclusively' of the shape of goods which is necessary to obtain a technical result, the fact that it may also include elements with no technical function will be disregarded if those are considered 'non-essential'. The prospects for product/packaging shape registrations is made still more difficult because even the fact that a shape-mark has acquired distinctive character through use will not render it registrable if it is refused on these grounds.

The difficulties of registering the appearance of products (eg, pills, capsules or packaging shapes) as trade marks may have affected significantly the interests of life sciences companies such as producers of pharmaceuticals and medical devices. Arguably, the difficulties in protecting this kind of matter may, however, have been alleviated somewhat by the introduction of the Community Registered Design system (see below).

At the time of writing (July 2011), the trade mark system in Europe is in something of a state of flux due to a major study of the European trade mark system conducted by the Max Planck Institute at the behest of the European Commission and an associated consultation exercise with users and user groups. The study and consultation exercise has been a very substantial piece of work and has led to much discussion and a number of suggestions. However, in truth it is impossible to predict at all reliably at this stage what the changes to the system will be which will flow from this exercise, and all life sciences businesses should continue to monitor proposals and make use of opportunities to make their views known – whether directly or through industry bodies or relevant associations.

5. Community designs – registered and unregistered

Although EU member states retain national laws for the protection of registered designs, and in some cases (such as the United Kingdom) unregistered designs also, the protection of designs in the European Union is dominated by Unregistered Community Designs (UCDs) and Registered Community Designs (RCDs). Both these rights were created by Council Regulation (EC) No 6/2002 and implemented by Commission Regulation (EC) No 2245/2002. Unregistered Community Design protection has existed since March 2002, whilst RCD protection commenced on April 1 2003. Like Community Trade Marks, Community Designs are unitary rights covering the whole European Union.

For the purposes of both UCD and RCD protection, 'design' means "the appearance of the whole or a part of a product resulting from the features of, in particular, the lines, contours, colours, shape, texture and/or materials of the product itself and/or its ornamentation". If this definition sounds very wide, that is because it is. Examples of protectable objects/materials would include: the form of blister packs for pharmaceuticals; the overall shape, lines and materials etc of medical devices or (in the case of a complex devices) their visible parts; the corporate or product logos of life sciences businesses; or even the appearance of their web pages.

In order to be protected, designs must be 'new' and have 'individual character'. 'New' in this context means that no identical design has been made available to the public before the date on which, in the case of UCD, the design for which protection is claimed was first made available to the public, or, in the case of RCD, the date of filing of the application for registration.

A design will be taken to have 'individual character' if it produces a different 'overall impression' on 'the informed user' as compared with the overall impression produced by other designs available. In deciding whether a design has individual character, the degree of freedom of the designer is to be taken into account. The latter proviso is a potentially important one for many products in the life sciences arena, given the significant regulatory and practical constraints on the design of such products.

Unregistered Community Design protection exists (as its name indicates) without the need for any registration. To this extent the protection is similar to copyright, in that it comes into existence automatically. Unlike copyright protection, however, UCD protection is short lived – lasting for only three years following the date on which the design is first made available to the public, published, exhibited etc in the European Union. This limited protection can nonetheless still provide very useful protection in some cases – in particular in fast-moving sectors in which there is a high turnover of new designs, such as fashion. Products/matter which might be protectable by designs in the life sciences sector will more often be at the other end of the spectrum, requiring a long lead time to market due to the obvious regulatory and other constraints. Nevertheless, UCD could still prove useful on occasion (eg, for protecting elements of product branding to the extent to which they are insufficiently distinctive to qualify for trade mark registration).

Registered Community Designs (RCDs) provide a more certain right of significantly longer duration. RCDs are protected initially for five years, renewable up to a maximum duration of 25 years. However, a key restriction on the ability to obtain RCD protection is that designs must be registered within 12 months of disclosure.

Unlike trade mark protection, which (primarily at least) provides protection in relation to specified goods and/or services, the protection given by an RCD is not limited to use of the design on/for any particular product, even though, for administrative purposes (eg, to assist searching), applications for RCD protection require an indication of the object to which the design is applied.

Apart from the obvious difference of duration and the need for registration in order for the RCD to exist, the essential differences between the two rights are twofold: first, the additional difficulty entailed in enforcing UCD due to the need to demonstrate its existence (whereas with an RCD the existence of the right is a matter of registration); and secondly, the fact that UCD (much like copyright) requires there to have been copying in order for the right to be infringed. In theory at least, therefore, it would be possible for a third party to come up with a design which does not create a different overall impression but which does not infringe the UCD, because it was created independently. In contrast, the RCD gives its proprietor an exclusive right to use the design in question and, to that end, to prevent third parties (without the proprietor's consent) from making, putting on the market etc products that incorporate a design which does not produce a different overall impression on the informed user.

It will be apparent from the scope of what may be protected by Community Designs that their application overlaps with signs which could potentially be protected as registered trade marks (eg, as CTMs). The two rights are capable of co-existing such that a logo, for example, might be the subject of both a CTM and an RCD. Whilst trade mark registrations have the ability to last indefinitely (subject to payment of renewal fees), there are some notable advantages of RCDs which make them a very useful addition to the armoury of rights owners generally, and particularly for life sciences businesses.

One key advantage is in relation to protection of product and packaging shapes. Such shapes are potentially protectable through trade mark registrations. In practice, however, the case law of the European courts in recent years has made the prospects for obtaining (and retaining) trade mark registrations for three-dimensional product and packaging shapes extremely limited. Such shapes will, in practice, only be registrable as trade marks if they are very unusual in the context of the prevailing norms for the types of product/packaging and the difference in shape is not related to function. Even in these limited circumstances, it will often only be possible to obtain registration on the basis of being able to demonstrate acquired distinctiveness – which is generally only possible once the product has been on the market (successfully) for a considerable time. In contrast, it is possible to obtain RCD registration of such product/packaging shapes, regardless of functionality, and without the need to show inherent or acquired distinctiveness. In these circumstances the RCD can provide protection in the absence of a trade mark registration, or at least on a temporary basis until such time as the product/packaging shape is able to present sufficient evidence to obtain a trade mark registration based on acquired distinctiveness.

Registrations of RCDs are much cheaper than for trade marks and the facility to register multiple designs within the same registration is particularly helpful in this regard. Unlike for trade marks, which are subject to substantive examination, at least on absolute grounds (eg, to assess distinctiveness etc), the RCD registration system is essentially a simple deposit system, such that registration is more or less instant provided that the form is correctly submitted and the requisite fees paid. This speed to obtain a registered right can be useful, not just so that the owner can publicise the registration as a deterrent to infringers (and indeed assert the registration against such infringers), but also for the purposes of recording the right with EU customs authorities to enable border detention of infringing articles imported into the Community by third parties (see 'Border Enforcement' at section 8 below).

A potential drawback of RCDs in comparison with trade mark registrations, which follows from the fact that there is no substantive examination of applications, is that many RCDs are susceptible to be invalidated on the basis that the designs are not new and possessed of individual character. Any review of the designs which have been registered as RCDs will reveal large numbers of banal, entirely commonplace designs which would undoubtedly be declared invalid if challenged (which, typically, happens whenever a design owner asserts it against a potential infringer). However, so long as applicants are conscious of when they are obtaining a registration which is vulnerable to invalidation and do not have unrealistic expectations in relation to such designs, then they can still play a useful role.

6.   Database rights

The role of some European intellectual property rights in the life sciences remains speculative at present. One notable area where this is true is in respect of the ability to protect data with intellectual property. This may prove to be an important issue in respect of the genetic and protein data that increasingly underlies much biotechnology research.

The working draft of the entire generic human genome was published in 2000 after more than a decade of work. Since then, the speed of gene sequencing technology has increased to such an extent that some now predict that with emerging sequencing technologies the entire human genomes of particular individuals (approximately six gigabytes of data each) could be sequenced in a matter of hours. The ability to apply powerful data mining and bioinformatics techniques to this information to find biomarkers of disease presents implications for the development of personalised medicine, with the prospect of a generation of drugs being designed to counter the genetic flaws of small groups of people, rather than the alleviation of symptoms that are particular to whole populations. The vast amount of data that this technology promises is therefore highly valuable. So what is the challenge to the scope of existing intellectual property rights? In particular, as regards patents, the need to satisfy industrial applicability requirements under the European Patent Convention raises questions about the extent to which information about genes and the proteins that they code for is protectable by patent law.[8] Consequently, some attention may shift to the extent to which raw data may be protected by rights under the EU Database Directive 96/9 (the Directive), which was intended to harmonise the law on the protection of databases throughout the member states of the European Union.

The Directive defines a database broadly as "a collection of independent works, data or other materials arranged in a systematic or methodical way and individually accessible by electronic or other means", and it creates two types of protection: copyright (Database Copyright) and a *sui generis* database right (Database Right). Until the implementation of the Directive in the member states, there were different approaches adopted across the European Union to the level of originality required for a database to enjoy copyright protection. In some member states it was only necessary to satisfy a lower 'sweat of the brow' test for copyright to subsist in a database, whereas in others a tougher 'intellectual creation' test was used. The Directive seeks to set the standard required for Database Copyright to the higher 'intellectual creation' level, whilst providing the Database Right protection for those databases that would formerly have satisfied the lower 'sweat of the brow' test. Under the Directive, Database Copyright is granted to databases that constitute the author's own intellectual creation, by reason of the selection or arrangement of their contents. The *sui generis* database right is granted to databases for which it can be demonstrated that a substantial investment in either the obtaining, verification or presentation of the contents has been made by the creator of the database – either qualitatively and/or quantitatively.

To the disappointment of many, the new Database Right protection suffered an early setback in 2004, when the ECJ gave its first interpretation of the scope of the new right in *The British Horseracing Board Ltd et al v William Hill*.[9] This took a narrower view of the scope of the protection afforded than had previously been thought; it is only the investment in 'obtaining', 'verifying' or 'presenting' the contents of the database that is determinative, and not the investment in resources used for the creation of the materials themselves. In other words, if time and money is put into seeking out, collecting together, structuring and verifying independent materials, then a database right may arise – but it will not protect data that necessarily arises out of the activity for which investment has been made. This would seem to place a limitation on the application of this right to those systematically sequencing DNA to produce volumes of their own data; according to *British Horseracing Board*, the investment activity here would seem likely to be deemed not in the creation of the database but in the sequencing activity. However, database right protection may become relevant where a database is created from investing in the collection together from diverging sources of data (eg, collecting information

---

8   See, for example, *Eli Lilly and Company v Human Genome Sciences Inc* [2010] EWCA Civ 33 and the subsequent Supreme Court judgment of November 2 2011.

9   Case C-203/02 [2004] ECR I-10415.

from databases around the world for comparisons of a particular human gene or other biomarker in different populations linked to a particular disease).

Once the difficulty of demonstrating that a database right subsists has been overcome, however, the protection against infringement that it affords is broad. The Database Right is infringed where there is an "extraction and/or re-utilisation of the whole or a substantial part evaluated quantitatively and/or qualitatively" of the database.[10] Extraction or re-utilisation by repeated and systematic means that amount to a substantial part is also an infringing activity.[11] Extraction is a defined term in the Directive, meaning "the permanent or temporary transfer of all or a substantial part of the contents of a database to another medium by any means or in any form".[12] There is also a definition for re-utilisation as "any form of making available to the public of all or a substantial part of the contents of a database by the distribution of copies, by renting, by on-line or other forms of transmission". The Directive also prohibits the "repeated and systematic extraction and/or re-utilisation of insubstantial parts of the contents of the database implying acts which conflict with a normal exploitation of that database or which unreasonably prejudice the legitimate interests of the maker of the database …".[13]

The ECJ decision in *Directmedia Publishing GmbH v Albert-Ludwigs-Universitat Freiburg*,[14] illustrates how potentially powerful the Database Right is. Here, Directmedia compiled a CD-ROM called '1000 poems everyone should know', 856 of which also appeared in a database prepared at Freiburg University called 'The 1100 most important poems in German literature between 1730 and 1900'. Directmedia had consulted this database before producing their CD-ROM and were held to have infringed it, even though they had used their own discretion about what they would and would not include in their own list, and the fact that they had not accessed their content from the Freiburg database but had instead obtained it from other sources using their own resources. This last point is an interesting departure from the copyright in which Database Right arguably has its origins; this is because it does not necessarily require actual 'copying' of the infringed material but will instead find infringement for recreation in certain circumstances. This resembles more closely a pure monopoly right. It leaves a question about whether someone who recreates a substantial part of a genome database, of which they have merely had sight, by virtue of their own sequencing infringes the original.

Furthermore, the ECJ has held that the extraction can be permanent or temporary and still infringe. It also does not matter whether the data is subsequently modified and placed in a database with a substantially different structure, and nor is the motivation of the person doing the extracting relevant to whether there had been an extraction.[15]

Database Copyright has been less well served by decisions of the CJEU than

---

10   Article 7(1).
11   Article 7(5).
12   Article 7(2).
13   Article 7(5).
14   Case C-304/07, ECJ, October 9 2008, [2008] ECR I-7565.
15   *Apis-Hristovich EOOD v Lakorda AD*, Case C-545/07, ECJ, March 5 2009, [2009] ECR I-1627.

Database Right. However, further light is expected to be thrown on this neglected right by the reference to the CJEU from the English Court of Appeal in *Football Dataco Ltd v Brittens Pools Ltd.*[16] The questions referred concern, among other matters, the meaning of the 'author's own intellectual creation', necessary to show that Database Copyright subsists. At the time of writing there is no indication of when a decision will appear in this case and so it remains to be seen whether this right could afford protection for databases of genomic and proteomic data in circumstances in which Database Right does not.

7.   The Enforcement Directive

Directive 2004/48/EC of the European Parliament and of the Council of April 29 2004 on the enforcement of intellectual property rights is referred to commonly as 'the Enforcement Directive' or sometimes as 'the IPR Enforcement Directive' or 'IPRED'. The Directive was issued with the aim of producing a level of harmonisation between European member states' rules, procedures and systems for enforcing intellectual property rights and in particular to ensure they provide at least a minimum set of measures to enable the effective enforcement of IP rights. The Directive incorporates certain civil law measures under the TRIPS Agreement.[17] The Enforcement Directive goes further than the minimum standard set out in the TRIPS Agreement, however, covering such matters as the required standing to bring enforcement proceedings, matters concerning evidence, interim/interlocutory measures, injunctions and seizure orders, damages and costs.

The deadline for member states to implement the Directive into national law was April 29 2006. However, many member states were very tardy in complying, with the final implementation (Luxembourg) being in June 2009. This delay in implementation and the fact that the extent of the changes to national laws and procedures necessitated by the Enforcement Directive differed significantly between member states, means that there is a limited basis for assessing the effects and effectiveness of the Directive. Notwithstanding this, Article 18 of the Directive provided for member states to submit reports on the implementation of the Directive three years after the required implementation date and on the basis of those reports for the Commission to draw up a report assessing the effectiveness of the Directive and proposals for its amendment. The Commission duly issued its report (Commission Report) on December 22 2010 and there was a consultation process by which interested parties could submit comments/feedback, the closing date for which was March 31 2011. The Commission has since published a synthesis of the comments received on the Commission Report. The Commission has also since held a public hearing on the Directive and the challenges posed by the digital environment.

From the specific perspective of businesses in the life sciences sector, the effectiveness of the Directive is obviously of general importance as its effects impact on any and all civil proceedings relating to the infringement of intellectual property

---

16   [2010] EWCA Civ 1380.
17   Agreement on Trade-Related Aspects of Intellectual Property Rights (TRIPS Agreement) of 1994.

rights across Europe. It therefore has relevance in the context of the fight against counterfeit medicines and other products as well as in the conduct of, and remedies available in, IP disputes between competitor life sciences businesses. However, the area in which the effectiveness (or otherwise) of enforcement of IP rights in Europe has been under most scrutiny is in relation to the internet and online trade. Whilst this area is obviously of significance and growing importance for the life sciences sector, it is notable that the majority of comments and commentary on the effects and effectiveness of the Enforcement Directive come from industries with a focus on the supply of digital content (eg, major publishers and media rights holders) and from groups representing users of such content. Not surprisingly, the submissions of those respective parties take highly polarised positions. There also appears to be a significant divergence of views among member states, between those who are keen for substantial amendments at an early stage and those who consider that it is inappropriate to make significant changes, at least until there has been the opportunity for more evidence to have been gathered and assessed on the impact of the Enforcement Directive in its current form. In view of this, at the time of writing, it is very difficult to assess what amendments are likely to be made to the European regime for enforcement of IP rights which might affect interests in the life sciences sector in particular. It is interesting that among the list of additional issues raised by respondents to the consultation exercise on the Commission Report on the Directive (ie, issues that had not specifically been mentioned in the Commission Report or the Commission Staff Working Document) was a proposal that the Commission should broker a memorandum of understanding on online pharmacies.

Some other issues which seem more likely candidates for amendment include:

- clarification of the scope of the Directive (eg, as to whether it applies to trade secrets);
- further provisions on the basis and measure of damages – with possible further focus on damages as a deterrent to IP infringers rather than compensatory for IP holders; and
- dealing with the interplay/conflict between rules on evidence and disclosure and member states'/Community protection for the right of privacy and data protection.

8.   Border enforcement

Regulation 1383/2003/EC, the Counterfeit Goods Regulation, sets out the procedures for customs authorities throughout the European Union to take action in respect of importation of goods believed to infringe particular IP rights. IP rights which the regulation covers which are of most relevance to businesses in the life sciences sector include trade marks, design rights, patents and SPCs. Pursuant to the Regulation, rights owners can record their rights with customs authorities for the purposes of having suspected counterfeit or pirated goods detained at EU ports of entry, and for the rights owners to be notified of any such detentions for the purposes of taking action. Customs authorities can also initiate detentions of suspected counterfeit/pirated goods on their own account and thereafter contact rights owners.

In practice, whilst the procedures provided for by the Regulation are capable of working quite effectively to enable the detention of counterfeit products (eg, counterfeit pharmaceuticals) entering the European Union, the system is very far from perfect. In particular, the effectiveness and efficiency with which customs authorities in the different EU member states operate the system varies very significantly. The current system is also subject to the limitation that parallel imported goods from outside the community are not able to be detained by Customs, notwithstanding that dealings in those goods (eg, trade-marked goods which have not been put on the market by the trade mark proprietor or with its consent in the European Union) would amount to an infringement of the trade mark proprietor's rights.

Following the European Commission's publication of its 'comprehensive' strategy for intellectual property rights, the regime for customs action against infringing IP rights (as with a number of the other matters dealt with in this chapter – eg, the Enforcement Directive) is in a state of flux. The Commission has published a new draft Regulation[18] which would replace the existing Regulation 1383/2003 if adopted.

Changes to the existing system relevant to the life sciences sector which the new Regulation would bring (if adopted in its current form) include, in particular, extending the regime to enable detention of infringing parallel imported goods (ie, imports from outside the European Economic Area), introducing procedures to enable rights owners to call for Customs to carry out automatic destruction of small consignments of infringing goods without the need to revert to the rights owner on each occasion (which action could potentially be very useful in view of the increasing use by infringers of multiple small consignments as a tactic to seek to minimise the risk of Customs detention) and changes to streamline procedures for destruction of goods in default of objection, as well as for the costs in relation to storage and destruction of small consignments of infringing goods to be borne by customs authorities rather than rights owners.

One issue which the new draft Regulation does not address is the issue of goods in transit (ie, goods which enter the European Union en route from one country outside the European Union to another such country). The state of European law at present appears, in general, to be that such goods do not amount to infringements of intellectual property rights within the European Union and that they may not, therefore, be detained by customs authorities, although a number of uncertainties on this issue remain to be clarified.[19]

9.   Parallel importation issues

The aim of this section is just to provide a brief overview of the underlying principles and concepts which are relevant to parallel importation issues as they affect life sciences businesses in Europe. This is an extremely broad topic, encompassing, as it

---

18   COM(711)285 dated May 26 2011.
19   At the time of writing (July 2011) a decision is awaited from the CJEU in joined cases C-446/09 *Koninklijke Philips Electronics NV v Lucheng Meijing Industrial Company Ltd, Far East Sourcing Ltd, Röhlig Hong Kong Ltd and Röhlig Belgium NV* and C-495/09 *Nokia Corporation v Her Majesty's Commissioners of Revenue and Customs* on this issue.

does, fundamental principles of the European Union and, specifically, both competition law and IP issues.

Free movement of goods is one of the key principles of the European Single Market – allowing goods which are put on the market in one member state to move freely into other member states without being restricted by national laws. This principle is laid out in Articles 28 and 29 of the EC Treaty, which prohibit quantitative restrictions on imports and exports and measures having equivalent effect as between member states. This prohibition is subject to an exception set out in Article 30 of the EC Treaty to the effect that Articles 28 and 29 shall not preclude prohibitions or restrictions on imports, exports or goods in transit justified on the grounds of (*inter alia*) the protection of industrial and commercial property (which includes intellectual property rights). However, the Article 30 exception is, in turn, restricted so as to ensure that intellectual property rights (bearing in mind that these include national IP rights in member states such as national trade mark registrations and are not limited to EU-wide unitary rights such as CTMs and RCDs) are not used to defeat the key principle of the free movement of goods. That restriction involves the application of the concepts of the 'specific subject matter' of the intellectual property rights and the 'exhaustion of rights'.

The specific subject matter in respect of trade marks (to which many of the major European cases relate and which, in particular, are generally the rights in focus in the context of the parallel importation of pharmaceuticals) may be summarised as the right for a trade mark owner to use/control the first marketing in the European Economic Area of products bearing the trade mark. The principle of exhaustion of rights, as set out under Article 7(1) of the Trade Marks Directive, is that a trade mark registration shall not entitle its proprietor to prohibit its use in relation to goods which have been put on the market in the Community under that trade mark by the proprietor or with his consent. As will be apparent, the intention of this provision is that once goods bearing a registered trade mark have been put on the market in the European Economic Area by or with the consent of the trade mark proprietor, then the proprietor's rights in respect of the trade mark have been 'exhausted' and the goods are, thereafter, able to move around within the European Economic Area freely without the trade mark proprietor being able to assert its trade mark rights to prevent parallel imports between EU member states. After a series of cases (the key ECJ decision being that in the *Davidoff v A&G Imports etc* joined cases C-414 to 416/99), it has been determined that a trade mark owner's rights are not exhausted by marketing of products bearing its trade mark outside the European Economic Area and that the proprietor's consent to the importation of such goods into the EEA cannot generally be implied (eg, by the absence of express prohibition), nor inferred from the silence of the proprietor on the issue. In summary, therefore, trade mark owners are generally able to object to parallel imports of trade-marked products from outside the European Economic Area, but may not object to parallel imports between member states once their rights to first-market the products in the Community have been exhausted.

The exhaustion principle set out in Article 7(1) of the Trade Marks Directive is, however, qualified by Article 7(2), which provides that Article 7(1) shall not apply where there are 'legitimate reasons' for the proprietor to oppose further commercialisation of the goods, especially where the condition of the products is changed or impaired after they have been put on the market. What exactly amounts to 'legitimate reasons' which allow a trade mark owner to oppose parallel importation into other member states of products which have already been put on the market in the European Economic Area by it, or with its consent, has been the focus on the majority of the (extensive) case law in this connection, in relation to pharmaceuticals in particular. A set of criteria (referred to as the BMS criteria) was set out in the *Bristol Myers Squibb v Paranova* case (joined cases C-427/93, C-429/93 and C-463/93) to specify the requirements with which a parallel importer of pharmaceuticals, which needs to repackage them in order to comply with regulatory requirements in the country in which they are to be marketed, must comply, in order that the trade mark owner will not have any legitimate reason to prevent their marketing, as follows:

- The repackaging/labelling must be objectively 'necessary';
- The original condition of the product must not be affected;
- The repackager must be identified on the new packaging;
- The presentation of the goods must not damage the reputation of the trade mark or its owner; and
- The parallel importer must have given prior warning to the trade mark owner of its intended repackaging and importation.

Each of the above requirements has been the subject of multiple disputes between research-based pharmaceutical companies (trade mark owners) and parallel importers, leading to a swathe of decisions of national courts, the EU General Court (formerly the CFI) and, notably, the CJEU (formerly the ECJ). Recently, for example, the CJEU issued a preliminary ruling in July 2011 that it is sufficient for the purposes of complying with the third requirement referred to above (identification of the repackager on the new packaging) that the packaging shows the name of the marketing authorisation holder which has arranged for the repackaging to be carried out and that it is not required that the identity of the entity actually doing the repackaging (if different) is stated: *Orifarm A/S v Merck Sharp & Dohme* – joined cases C-400/09 and C-207/10.

All the while that there exist price differentials for pharmaceuticals between different EU member states, fuelling the parallel import market, these issues will continue to be hotly contested. In the wake of the recent economic turmoil in certain Eurozone countries, it would appear that such differentials are at least as likely to grow as they are to reduce or disappear in the short to medium term and that, accordingly, there will be more of these cases to come.

10. The Commission sector inquiry

On January 15 2008 the European Commission launched an inquiry into competition in the pharmaceuticals sector. The stated objective behind the inquiry was to investigate the reasons behind the decrease in numbers of new medicines being brought to market and the delay of entry of generic medicines.

In particular, the inquiry was set up to examine whether agreements between pharmaceutical companies, such as settlements in patent disputes, have blocked or led to delays in market entry (eg, through the misuse of patent rights, vexatious litigation or other means).

The inquiry was essentially a fact-gathering exercise intended to allow the Commission or national competition authorities to focus any future action on the most serious competition concerns, and to identify remedies to resolve the specific competition problems in individual cases.

The Commission's Final Report of the Sector Inquiry, published in July 2009, (the Final Report) was more measured in its conclusions than its preliminary findings. In particular, the criticisms of the Final Report go beyond measures taken by patentees, extending also to: the impact of bottlenecks in the regulatory system on the delay between patent expiry and generic entry and the time taken for grant of pricing and reimbursement decisions. In addition the Final Report encourages member states to consider national measures to improve generic uptake (including compulsory generic substitution and more frequent adjustment of reimbursement levels) and endorses the need for the establishment of a EU-wide unitary patent and unified specialised patent litigation system.

That said, the Final Report maintains its attack on certain 'instruments' used by innovator companies and interventions before regulatory bodies that may delay generic entry.[20] Although the Final Report does not clarify how the competition rules in this area will be enforced, the Commission has indicated that it will apply greater scrutiny to the pharmaceutical sector in the future. The areas identified as being the likely subject of closer attention (eg, through targeted information requests and inspections) include 'vexatious' litigation campaigns, 'defensive' patenting practices, settlement agreements, and interventions before marketing authorisation and pricing and reimbursement bodies. The legal and evidential issues in these areas will not be straightforward from a competition law perspective.

As it promised, the Commission has maintained its scrutiny of the sector. Further 'dawn raids' have been conducted and on January 12 2010 the Commission confirmed that it had asked a large number of pharmaceutical companies to provide details of patent settlement agreements concluded between July 2008 and December 2009. Formal antitrust investigations against Les Laboratoires Servier (and others) and Lundbeck have also been opened.

Until these investigations are complete and final decisions have been reached (and perhaps throughout the years of appeals before the European courts) there will remain considerable legal uncertainty. For at least as long as such uncertainty remains, companies in the sector should pay particular attention to developments in the risk areas set out next.

10.1   Settlement agreements

One area of focus identified by the Commission is settlements "where an originator

---

20   When the Commission refers to 'delay' it appears that it refers to any delay to a generic drug entry after the expiry of the original 'new chemical entity' patent on a drug substance – that is, without regard to unchallenged development-phase patents or the impact of technical difficulties faced by the generic companies, regulatory delays or even the decision-making processes of generic companies.

company pays off a generic competitor in return for delayed market entry of a generic drug".

It is clear that the major sensitivity surrounds settlements in which there is a 'value transfer' from the patentee to the generic company in return for delayed market entry (so-called 'reverse payments' or 'pay for delay' settlements). This area has been the subject of significant antitrust litigation in the United States and whilst it would seem that the Commission has been influenced by the US situation, quite how far it will go in this regard remains to be seen. In the meantime the types of agreement to be treated with particular caution are:

- settlements that include reverse payments from patentee to generic company should be avoided. This includes where the 'value transfer' is not purely financial – the competition authorities are likely to want to test the justification for other 'side deals' such as a linked supply agreement or cross-licences; and
- settlements that include inappropriate limitations of generic entry should also be avoided (even without a reverse payment) – for example, where the settlement agreement imposes limitations outside the scope of the relevant patent, such as limitations in respect of non-infringing products, or beyond the term or territorial scope of the patent.

10.2 Disparaging generic product and interference with regulatory process

Another area of criticism by the Commission is the intervention by originator companies before marketing authorisation or pricing and reimbursement bodies (eg, raising concerns about the safety or efficacy of the generic product or arguing that the grant of a marketing authorisation or obtaining pricing and reimbursement status could violate patent rights). The low success rate of such cases and their potential to delay generic market entry was addressed in the Final Report.

10.3 Vexatious litigation and defensive patenting

Although the Final Report acknowledges that "enforcing patent rights in court is legitimate and a fundamental right guaranteed by the European Convention on Human Rights", it remains critical of 'vexatious' litigation campaigns and 'defensive patenting'. How, if at all, this will become an area for enforcement remains to be seen.

Chapter 3: An introduction to European regulatory rights

Sarah Bailey
Andrew Hutchinson
Marjan Noor
Alexandre Regniault
Simmons & Simmons LLP

1. Supplementary protection certificates

1.1 Introduction

A supplementary protection certificate (SPC) extends the duration of a patent covering a medicinal or veterinary product which has been granted a marketing authorisation. The rationale underlying the SPC regime is to compensate a patentee for its lost period of patent monopoly on a product arising from the lengthy procedure to obtain authorisation to sell the product.

The original SPC legislation, Council Regulation (EEC) No 1768/92 which came into force on January 2 1993, has subsequently been replaced and codified under Regulation (EC) No 469/2009 of May 6 2009 (the SPC Regulation), which entered into force on July 6 2009.

The SPC regime provides a patentee with a maximum 15-year period of patent monopoly, during which time the product is covered by an authorisation allowing its sale (and a further six months may be added to this period through the extension introduced by the Paediatric Regulation (EC) No 1901/2006, discussed in more detail in section 1.4 below). Subject to that maximum, the term of an SPC is dependent on the interval between the filing of a patent and the authorisation for the product – the greater that interval (in other words, the greater the length of time during the patent term taken to obtain authorisation), the greater the SPC term.

1.2 Requirements for an SPC

The applicant for the SPC must be the holder of a basic patent (Article 6). There is no express requirement that the applicant needs to be the holder of the marketing authorisation, which in some instances has resulted in patentees obtaining SPCs based on the marketing authorisation of a third-party competitor product. The application should be filed either within six months of grant of the relevant marketing authorisation (Article 7(1)) or, if the marketing authorisation is granted before the basic patent is granted, within six months of grant of the basic patent (Article 7(2)).

An SPC will be granted if:

(a) the product is protected by a basic patent in force;

(b) a valid authorisation to place the product on the market as a medicinal product has

been granted in accordance with Directive 2001/83/EC or Directive 2001/82/EC, as appropriate;

(c) the product has not already been the subject of a certificate;

(d) the authorisation referred to in point (b) is the first authorisation to place the product on the market as a medicinal product.

(Article 3)

The key definitions are reproduced below:

'Product': an active ingredient or combination of active ingredients of a medicinal product.

'Medicinal Product': any substance or combination of substances presented for treating or preventing disease in human beings or animals and any substance or combination of substances which may be administered to human beings or animals with a view to making a medical diagnosis or to restoring, correcting or modifying physiological functions in humans or in animals.

'Basic Patent': a patent which protects a product as such, a process to obtain a product or an application of a product, and which is designated by its holder for the purpose of the procedure for grant of a certificate.

The above requirements, which are set out in Article 3 and were first introduced in 1993, represented what was intended to be a 'simple, transparent system'. The SPC regime was intended to sit at the interface between the patent and regulatory regimes. A patentee should only obtain an extension where its patent monopoly has been eroded by more than five years on account of an authorisation process to sell a product. Therefore, Article 3 required a patent and marketing authorisation covering the product in question (Articles 3(a) and (b) respectively). Furthermore, the same patent holder was only entitled to an SPC once on the same product, so as to prevent successive SPCs for every minor modification of the product (hence Article 3(c)).

Indeed, the system works well for the simplest and most common scenario where a company: (i) owns a patent covering product A (satisfying Article 3(a)); (ii) holds an authorisation for product A (satisfying Article 3(b)); (iii) has never applied for an SPC for product A before because it is a new product (satisfying Article 3(c)); and (iv) has obtained an authorisation for product A for the first time in the country in which the application is being made (satisfying Article 3(d)).

1.3 Complex scenarios

However, the healthcare sector has increased in complexity since the original SPC regulation was introduced in 1993 and most would agree that the current SPC regime does not cater for this complexity. This is reflected in the high number of references made to the Court of Justice of the European Union (the CJEU) seeking clarification of the meaning of the provisions in the SPC Regulation. A number of these are summarised below:

- Article 3(c) does not prevent more than one patent holder obtaining an SPC on the same product (*AHP Manufacturing BV v Bureau voor de Industriële Eigendom* CJEU Case C-482/07).
- A new formulation of an active which is already the subject of an earlier

marketing authorisation or SPC is unlikely to be entitled to a further SPC because the CJEU has stated that the concept of 'active' needs to be interpreted strictly (Massachusetts Institute of Technology CJEU Case C-431/04).

- An active which has already been authorised for one use is to be treated as the same 'product' as an active authorised for another use (*Yissum Research and Development Company of the Hebrew University of Jerusalem* CJEU Case C-202/05). For example, an authorisation for product A for use in indication X is the first authorisation for the purposes of Article 3(d), even where the later authorisation cited in support of an SPC application covers product A for use in indication Y.

- An active which has already been authorised for use as a veterinary product is to be treated as the same 'product' as an active authorised for human use (*Pharmacia Italia SpA* CJEU Case C-31/03).

- It is not clear whether a marketing authorisation for a medicinal product containing A+B (or A+B+C) can be said to be an authorisation of product A (or A+B respectively) alone for the purposes of Article 3(b). This has relevance in the vaccine field, for example where a company applies for an SPC for product A based on its patent covering product A but where, because of government policy on vaccines, product A is required to be sold in combination with other vaccines in medicinal product A+B or A+B+C. The CJEU's decision in the case of *Medeva BV's SPC Applications* (Case C-322/10) on this issue is expected in late November 2011. The Advocate General delivered its Opinion on July 13 2011 and stated that a marketing authorisation for A+B+C was an authorisation covering product A within the meaning of Article 3(b).

- It is not clear whether a patent covering A can be said to protect the combination product A+B (or A+B+C, etc). In the vaccine example above, the applicant may instead apply for an SPC for product A+B+C such that its SPC application mirrors the marketing authorisation rather than the patent. The requirements of Article 3(b) would be satisfied as the marketing authorisation covers A+B+C. The remaining issue is whether the patent for A would satisfy the requirement of Article 3(a). There has been much disparity within Europe as to what should be the appropriate test under Article 3(a). Some countries have adopted a wider 'infringement' test such that if a product infringes a patent, then the patent protects it for the purposes of Article 3(a). Therefore, product A+B (or A+B+C, etc) would be protected by a patent on product A. Other countries have adopted a stricter 'subject matter' test, such as whether the product is part of the new and inventive subject matter of the patent, the approach adopted by the Dutch courts. Clarification on this issue was also sought from the CJEU in the case of *Medeva BV's SPC Applications* (see above). The Advocate General stated that the 'subject matter' rather than the infringement test should be applied on the basis of national rules, but provided no guidance as to its criteria (other than that the 'protective effect' of the patent is not one of them).

1.4 Duration of an SPC

The duration of an SPC is calculated by determining the interval between the patent filing date and the date of authorisation for the product less five years and up to a maximum of five years (Article 13). Therefore, a product authorised on January 1 2005 covered by a patent filed on 1 January 1998 would be entitled to an SPC of two years. Had the patent been filed on January 1 1995, the SPC duration would have been five years but a filing date of January 1 1993 would still provide an SPC duration of five years, that being the maximum available.

(a) Earlier authorisation which is non-EC compliant

Articles 13 and 2 have been the subject of two recent CJEU referrals from the UK courts (*Synthon BV v Merz Pharma Gmbh & Co KG* C-195/09 and *Generics (UK) Ltd v Synaptech Inc* C-427/09), which relate to whether an earlier marketing authorisation granted by a member state at a time when the relevant EC regulatory legislation (Directive 65/65) had not been implemented prevents an SPC applicant relying on a later, EC-compliant, authorisation.

On July 28 2011, the CJEU ruled that if a product has been placed on the market in the Community prior to obtaining an EC-compliant authorisation, it is not within the scope of the SPC Regulation under Article 2. Therefore, any later SPC granted for such a product based on an EC-compliant authorisation will be invalid.

(b) Paediatric extension

The Paediatric Regulation introduced a six-month extension to the duration of an SPC under Article 13(3) of the SPC Regulation.

An application for a paediatric extension can be made at the same time as an SPC application. For an SPC already granted, the transitional provisions allow an application for an extension to be made no later than two years before the expiry of the SPC. (Until January 26 2012, an application for a paediatric extension to an SPC already granted by January 26 2007 can be made no later than six months before the expiry of the SPC.)

Article 13(3) of the SPC regulation permits the extension if the product satisfies the requirements of the Paediatric Regulation, including compliance with an agreed paediatric investigation plan (a six-month extension is available even where the paediatric testing has not led to an authorised paediatric indication) and the product is authorised in all member states.

One issue that has arisen is whether a six-month extension is available where the interval between the authorisation and patent filing date is five years or less, which ordinarily would mean that an SPC is not available because its duration would be zero or negative. In this situation can an SPC extension be granted, bringing the zero or negative term to a positive term of six months or less?

Patent Offices in some countries, such as Germany and Portugal, have determined that SPCs should not be granted for zero or negative terms, whilst others such as the United Kingdom and Netherlands have ruled that that they should. A reference was made to the CJEU in March 2010 by the German Federal Patent Court (*Merck & Co Inc v Deutsches Patent und Markenamt*, Case 15 W (pat) 36/08, CJEU case

C-125/10). The CJEU judgment is pending; however, the Advocate General has concluded that it should be possible to obtain a negative or zero-term SPC.

2. Bringing drugs to the market

2.1 Obtaining authorisation of a new drug

(a) *Framework and legislation*

The regulation of medicinal products in Europe is governed by the following principal European legislation:

- Directive 2001/83/EC (as amended) on the Community Code relating to medicinal products for human use (the 'Community Code').
- Regulation 726/2004 laying down Community procedures for the authorisation and supervision of medicinal products for human and veterinary use and establishing a European Medicines Agency (the 'EMA Regulation'), which came into effect on November 20 2005 and which repealed and replaced Regulation 2309/93.

In addition to the European legislation, medicinal products are further regulated by the laws of each member state.

Guidelines issued by the Commission, whilst not binding, are persuasive on the determination of issues arising in connection with the authorisation and later regulation of medicines. Such guidelines include the Commission's Rules Governing Medicinal Products in the European Union, in particular Volume 2 – Notice to Applicants (published by the Committee for Medicinal Products for Human Use (CHMP), a committee within the EMA) and the EMA's Scientific Guidelines for Human Medicinal Products.

Each member state has established an authority to be responsible for regulating medicinal products at a national level – for example, in the United Kingdom this is the Medicines and Healthcare Products Regulatory Agency (MHRA). The EMA is an agency of the European Commission and is the responsible authority for pan-European authorisations.

(b) *Routes*

Subject to a few limited exceptions, medicinal products must obtain a marketing authorisation (MA) before they can be sold in Europe. There are four routes available for obtaining an MA: the national procedure; the mutual recognition procedure; the decentralised procedure; and the centralised procedure. In more detail:

- **National procedure:** The application under the national route is made to the relevant national authority of a member state. This route is not permitted where an authorisation is required in more than one member state and therefore it is of limited applicability in practice (Articles 17 and 18 of the Community Code).
- **The mutual recognition procedure:** The mutual recognition procedure is governed by Articles 27 to 32 of the Community Code. It is used where an

MA has already been granted by one member state (the Reference Member State (RMS)) and that authorisation is relied upon by the holder to obtain MAs in other member states (referred to as Concerned Member States (CMSs)). There is a set procedure with timelines leading up to grant and an arbitration procedure in case of disagreement between the member states.

The Coordination Group for Mutual Recognition and Decentralised Procedures, Human (CMD(h)), which consists of one representative per member state, considers points of disagreement between member states involved in a mutual recognition or decentralised application (described below).

- **The decentralised procedure:** The decentralised procedure is governed by Articles 27 to 32 of the Community Code. It was introduced through the amendments to the Community Code by Directive 2004/27. It is used where the applicant in question has not received an MA in any member state and so multiple applications are sent to all member states of interest, one of them being designated as the RMS. Under the decentralised procedure, the RMS carries out an initial assessment and then sends the draft assessment report together with other material to the CMSs. Again, there is a set procedure with timelines leading up to grant and an arbitration procedure in case of disagreement between the member states.

- **Centralised procedure:** The centralised procedure is governed by the EMA Regulation and allows a single application to be made to the EMA for an MA covering the entire European Community. At the end of the application procedure an opinion is issued by the EMA on whether an MA ought to be granted. In most cases the European Commission follows that opinion. The MA then granted is in the form of a Commission Decision. MAs granted in the Community under the centralised procedure are published on the Community Register of medicinal products for human use on the Commission's website.

Under Article 3(1) of the EMA Regulation, the centralised procedure is the mandatory route for specific types of MA application:

- products developed by one of a number of specified biotechnological processes;
- products containing a new active for treatment of acquired immune deficiency syndrome, cancer, neurodegenerative disorder, diabetes, auto-immune diseases and other immune dysfunctions and viral diseases; and
- medicinal products designated as orphan medicinal products pursuant to Regulation 141/2000 (discussed below).

The centralised procedure may, under Article 3(2), also be used for:

- a product containing a new active ingredient; and
- a product which is shown to constitute a significant therapeutic, scientific or technical innovation or where the granting of the authorisation is in the interests of patients at Community level.

Further guidance on the scope of the centralised procedure has been published by the Commission/EMEA and is available on their websites.

Although the legislation and guidelines specify timelines for the procedures, including clockstops during which the applicant addresses issues raised by the agencies, in practice an application for a standalone or full application will typically take between 12 and 18 months, although there is less predictability in the decentralised and mutual recognition systems where coordination between a number of agencies is often required.

(c)   *Particulars to accompany an application for a marketing authorisation*
The basic principles and requirements for obtaining an MA are substantively the same under all the above routes and an application is only successful if: (i) the medicinal product's risk–benefit balance is considered favourable; and (ii) the quality, safety and efficacy of the medicinal product is sufficiently guaranteed.

The scientific data required for a standalone/full application under Article 8(3) of the Community Code consists of:
- pharmaceutical (physico-chemical, biological or microbiological) tests;
- pre-clinical (toxicological and pharmacological tests); and
- clinical trials.

The full requirements of what must be contained within the MA application are set out in the Annex to the Community Code and the Notice to Applicants.

Once granted, the initial MA lasts for five years with the requirement to renew only once at least six months before the end of that initial term.

(d)   *The 'sunset clause'*
Article 24(4) of the Community Code states:
Any authorisation which within three years of its granting is not followed by the actual placing on the market of the authorised product in the authorising Member State shall cease to be valid.
Article 24(5) states:
When an authorised product previously placed on the market in the authorising Member State is no longer actually present on the market for a period of three consecutive years, the authorisation for that product shall cease to be valid.
Article 24(6) provides exemptions to the above in exceptional circumstances and on public health grounds.

Guidance has been issued by the EMA on the application of the so-called 'sunset clause', which states that the three-year period does not start until the expiry of any protection period – for example, third-party patents.

2.2   Data exclusivity

(a)   *Generic application route*
Article 10(1) of the Community Code states:
By way of derogation from Article 8(3)(i) … the applicant shall not be required to provide

the results of pre-clinical tests and of clinical trials if he can demonstrate that the medicinal product is a generic of a reference medicinal product which is or has been authorised under Article 6 for not less than eight years in a Member State or in the Community.

A generic medicinal product authorised pursuant to this provision shall not be placed on the market until ten years have elapsed from the initial authorisation of the reference product.

Accordingly, Article 10(1) (commonly known as the 'generic route') allows a later applicant (usually a generic company) to rely on the data of an earlier applicant (usually an innovator), provided (i) it can show it is a 'generic' of the reference product and (ii) the so-called data exclusivity period of eight years from the date of first authorisation in the Community has expired. Despite being an 'abridged' application, the time taken to grant is not dissimilar to a full application and can be in the region of one year.

A generic product is defined as:

a medicinal product which has the same qualitative and quantitative composition in active substances and the same pharmaceutical form as the reference medicinal product, and whose bioequivalence with the reference medicinal product has been demonstrated by appropriate bioavailability studies. The different salts, esters, ethers, isomers, mixtures of isomers, complexes or derivatives of an active substance shall be considered to be the same active substance, unless they differ significantly in properties with regard to safety and/or efficacy... (Article 10(2)(b))

Although the (generic) company can make an application after eight years, it cannot sell its product for a further two years. The 8+2-year period of data/market exclusivity also applies to products authorised through the centralised procedure (Article 14(11) of the EMA Regulation).

The 8+2-year data/market exclusivity period only applies in respect of reference medicinal products for which an application was made after October 30 2005 (if using the decentralised procedure or the mutual recognition procedure) or November 20 2005 (if using the centralised procedure).

For products in respect of which the applications were made before these dates, the period of data exclusivity under the Community Code which existed prior to the amendments made by Directive 2004/27 continue to apply. These are:

- 10 years where the product was authorised under the centralised procedure; and
- six or 10 years (depending on the member state in question) where the product was authorised under the national route or the mutual recognition procedure.

(b) **Extensions for new indications**

For products in respect of which an application was made after October 30 2005 (or November 20 2005 in the case of centralised authorisations), the 8+2-year period can be extended to a maximum of 11 years if:

during the first eight years of those ten years, the marketing authorisation holder obtains an authorisation for one or more new therapeutic indications which, during the scientific evaluation prior to their authorisation, are held to bring a significant clinical benefit in comparison with existing therapies.

(c)   ***Extensions for other development products – Global Marketing Authorisation***
Article 6 of the Community Code states:

When a medicinal product has been granted an initial marketing authorisation in accordance with the first sub-paragraph, any additional strengths, pharmaceutical forms, administration routes, presentations, as well as any variations and extensions shall also be granted an authorisation in accordance with the first sub-paragraph or be included in the initial marketing authorisation. All these marketing authorisations shall be considered as belonging to the same global marketing authorisation, in particular for the purpose of the application of Article 10(1).

The global marketing authorisation encompasses the original authorisation and later authorisations covering modifications of the type set out in Article 6. The data/marketing exclusivity period ends 10/11 years from the date of the first authorisation of the global marketing authorisation. Thus, the effect of the global marketing authorisation is to avoid developments to the same active substance resulting in successive periods of data/marketing exclusivity.

2.3   Paediatric testing

Regulation (EC) No 1901/2006 of the European Parliament and of the Council of December 12 2006 on medicinal products for paediatric use, as amended by Regulation (EC) No 1902/2006 (the Paediatric Regulation) came into force on January 26 2007.

Subject to a system of waivers and deferrals, the Paediatric Regulation introduced a requirement that applicants must conduct clinical trials on children before submitting an application for a marketing authorisation with respect to any drug. The trials must be developed following a paediatric investigation plan (PIP) (to be agreed by the Paediatric Committee within the EMA).

The Paediatric Regulation contains a system of rewards, including a six-month extension to the term of protection of a patent or SPC (subject to the terms of the SPC Regulation discussed in an earlier section). If the relevant drug has received orphan drug designation (see section 2.4 below), the Paediatric Regulation provides for a two-year extension to the data exclusivity period, increasing it from 10 to 12 years.

The Regulation also creates a new type of MA, the Paediatric Use Marketing Authorisation (PUMA), which is only granted for a product which is not covered by an SPC or a patent which is entitled to an SPC and which exclusively covers therapeutic indications for use in the paediatric population. The PUMA provides the 8+2-year period of data/market exclusivity associated with marketing authorisations obtained through the standalone route.

2.4   Orphan medicinal products

Orphan drugs (ie, those for the treatment of rare conditions) are governed by Regulation (EC) No 141/2000 of December 16 1999, as amended.

An applicant seeking an orphan drug designation must demonstrate that the drug in question:

•   is intended for the diagnosis, prevention or treatment of a condition affecting no more than five per 10,000 persons in Europe; and

- is intended for treating a serious or debilitating disease and it is unlikely that without incentives marketing it would generate sufficient return to justify the necessary investment.

The Committee for Orphan Medicinal Products (within the EMA) is responsible for assessing the applications for orphan designations. If successful, the relevant drug is entered on the Community Register of Orphan Medicinal Products.

Following orphan drug designation, applications to obtain an MA for the orphan drug must be made through the centralised procedure (described above). Where an MA is granted for an orphan drug, with certain exceptions and limitations, the holder will benefit from exclusive marketing rights for a 10-year period. More precisely, regulatory authorities cannot for a period of 10 years accept another application for an MA or grant an MA or accept an application to extend an existing MA for the same therapeutic indication in respect of a similar medicinal product (Article 8(1)). This period may be reduced to six years if, at the end of the fifth year, it is established that the criteria for orphan drug designation are no longer met (Article 8(2)). The exception most likely to be used in practice to circumvent the exclusivity is where the second application can establish that the second medicinal product, though similar to the orphan medicinal product already authorised, is safer, more effective or otherwise clinically superior. The definitions for 'similar medicinal product' and 'clinical superiority' are contained in Regulation (EC) 847/2000 of April 27 2000.

2.5   Biosimilar products

(a)   Legislation and guidelines

The generic route under Article 10(1) discussed above provides an exemption to the extensive scientific data required under the 'standalone' application. Recognising that biological products are unlikely to satisfy the 'generic' definition, Article 10(4) of Directive 2001/83 specifically provides a route for authorisation of such 'biosimilar' products.

Biosimilar guidelines issued by the EMA consist of a number of guidelines of general applicability adopted in the 2005/2006 timeframe:

- *Similar Biological Medicinal Products;*
- *Similar Biological Medicinal Products containing Biotechnology-Derived Proteins as Active Substance: Non-Clinical and Clinical Issues; and*
- *Similar Biological Medicinal Products Containing Biotechnology-Derived Proteins as Active Substance: Quality Issues.*

Following these, the EMA has adopted product-specific guidelines including ones for recombinant follicle stimulation hormone, recombinant interferon beta, recombinant erythropoietins, low-molecular-weight-heparins, recombinant interferon alpha, recombinant granulocyte-colony stimulating factor, Somatropin and recombinant human insulin.

(b)   *Requirements for authorisation*

An applicant for a biosimilar product must demonstrate to the regulating authority that its biosimilar drug shares the requisite similarities to the reference biologic drug.

However, given the nature of biological products, there are inherent differences between the biosimilar drug and the reference biologic drug arising, for example, from differences in amino acid sequence caused by natural mutation or from the manufacturing process. As a result, the EMA adopts a 'comparability' exercise to demonstrate similarity as opposed to the standard generic approach of demonstrating bioequivalence by appropriate bioavailability studies.

In practice, the data sufficient to establish comparability lies between the simple bioequivalence testing for generics and the full detailed data necessary for standalone applications. The success of a biosimilar application is very much dependent on whether a product-specific guideline, which sets out in detail the type of studies that should be carried out to demonstrate comparability, has been issued.

2.6   The 'Bolar' defence

Article 10(6) of the Community Code (introduced by Directive 2004/27) contains the so-called Bolar defence to patent infringement. It states:

> *Conducting the necessary studies and trials with a view to the application of paragraphs 1, 2, 3 and 4 and the consequential practical requirements shall not be regarded as contrary to patent rights or to supplementary protection certificates for medicinal products.*

Whilst some European countries have adopted more or less the exact wording of the Community Code, thus making the defence only available for studies leading to generic or biosimilar application and for EU marketing authorisation (eg, the United Kingdom and the Netherlands), the position across Europe is inconsistent and a number of member states (eg, Germany and Italy) have opted for a broader interpretation to cover testing, leading to non-EU marketing authorisations and originator applications.

3.   Drugs, 'non-drugs' and 'neutraceuticals'

3.1   The non-cumulation principle

The emergence of 'borderline products' – named due to their location at the borderline of the medicinal products market (and covering both foodstuffs and medical devices) – makes researching and defining the frontier between 'drugs' and 'non-drugs' a delicate matter.

Community law has addressed these issues, and little by little it has published specific definitions enabling, in theory, a distinction to be made between medicinal products and foodstuffs.

Indeed, medicinal products are defined at Article 1 of the Community Code as:

> *any substance or combination of substances which may be used in or administered to human beings either with a view to restoring, correcting or modifying physiological functions by exerting a pharmacological, immunological or metabolic action, or to making a medical diagnosis,*

or

any substance or combination of substances presented as having properties for treating or preventing disease in human beings.

'Foodstuffs' are specifically defined by Regulation 178/2002 of January 28 2002.[1] Article 2 of this text provides that foodstuffs are defined as:

any substance or product, whether processed, partially processed or unprocessed, intended to be, or reasonably expected to be ingested by humans.

However, practice is more complicated than theory, and qualifications can vary from state to state. This is the reason why a 'non-cumulation' principle was introduced into Community law at Article 2.2 of Directive 2004/27,[2] which stipulates:

In cases of doubt, where, taking into account all its characteristics, a product may fall within the definition of a 'medicinal product' and within the definition of a product covered by other Community legislation, the provisions of this Directive [ie, the Directive relating to medicinal products] shall apply.

Consequently, if there is the slightest doubt in a judge's mind when examining the nature of a product from a regulatory perspective, the qualification of a medicinal product shall prevail over the others, and the most stringent regulatory framework shall apply. This is, of course, a reflection of the need to safeguard public health.[3]

3.2 Dietary foods

Foodstuffs are specifically defined by Regulation 178/2008 of 28 January 2002.[4] As stated above, Article 2 of this text provides that foodstuffs are defined as:

any substance or product, whether processed, partially processed or unprocessed, intended to be, or reasonably expected to be ingested by humans.

The notion of 'ingestion' is at the heart of the definition of foodstuffs. A product with some nutritive functions but which is not ingested could not be considered as a foodstuff.

Within this general definition, it is important specifically to identify dietary foods or foodstuffs intended for particular uses. Indeed, due to their specific composition or manufacturing processes, these products are clearly distinguishable from general foodstuffs for standard consumption. These specific products are commercialised and produced in order to meet a specific nutritional purpose, in accordance with Directive 89/398 of May 3 1989.[5]

In this context, the criteria for identifying these foodstuffs reside in the recognition of a particular diet, which must be defined by reference to the specific nutritional needs of:

---

1   Regulation EC N° 178/2002 of the European Parliament and of the Council of January 28 2002 laying down the general principles and requirements of food law, establishing the European Food Safety Authority and laying down procedures in matters of food safety, *OJ* L 31/1, February 1 2002.

2   *OJ* L 136, April 30 2004, pp 0034 to 0057.

3   CJEU, March 21 1991, case C-269/88, Delattre, Rec CJEU I, p 1487. CJEU, November 30 1983, case C-227/82, Van Bennekom, Rec CJEU, p 3883, for an extensive interpretation of the definition of 'medicinal products'.

4   Regulation EC N° 178/2002 of the European Parliament and of the Council of January 28 2002 laying down the general principles and requirements of food law, establishing the European Food Safety Authority and laying down procedures in matters of food safety, *OJ* L 31/1, February 1 2002.

5   Directive 89/398 of May 3 1989 on the approximation of the laws of the member states relating to foodstuffs intended for particular nutritional uses.

- certain categories of persons of whom the digestive processes or metabolism are impaired; or
- certain categories of persons in a particular physiological condition and who are therefore able to obtain specific benefits from controlled consumption of certain substances in foodstuffs; or
- infants or young children in good health.

These dietary products thus fall within a specific legal framework, distinct from that applicable to medicinal products, although more stringent than the framework applicable to foodstuffs for standard consumption.

Dietary foods for special medical purposes are also regulated by Commission Directive 1999/21/EC of March 25 1999,[6] which defines those products intended to meet particular nutritional requirements of persons whose ability to consume, digest, absorb or metabolise ordinary foodstuffs is impaired.

Dietary foods for special medical purposes are classified in the following three categories:

- nutritionally complete foods with a standard nutrient formulation which, used in accordance with the manufacturer's instructions, may constitute the sole source of nourishment for the persons for whom they are intended;
- nutritionally complete foods with a nutrient-adapted formulation specific for a disease, disorder or medical condition which, used in accordance with the manufacturer's instructions, may constitute the sole source of nourishment for the persons for whom they are intended; and
- nutritionally incomplete foods with a standard formulation or a nutrient-adapted formulation specific for a disease, disorder or medical condition which are not suitable to be used as the sole source of nourishment.

3.3   **Vitamins and minerals as supplements or as vitamin and mineral food additives**
Food supplements may be defined as:

> ... *foodstuffs the purpose of which is to supplement the normal diet and which are concentrated sources of nutrients or other substances with a nutritional or physiological effect, alone or in combination, marketed in dose form, namely forms such as capsules, pastilles, tablets, pills and other similar forms, sachets of powder, ampoules of liquids, drop dispensing bottles, and other similar forms of liquids and powders designed to be taken in measured small unit quantities ...*

These food supplements, which are subject to a specific legal framework since Directive 2002/46 of July 10 2002[7] on the approximation of the laws of the member states relating to food supplements, must not have therapeutic effects; otherwise they may be regarded as medicinal products (either by function or by presentation).

It should be noted, however, that at present this text only concerns a list of vitamins and minerals. The drawing-up of minimum and maximum vitamin content

---

6   Commission Directive 1999/21/EC of March 25 1999 on dietary foods for special purpose, *OJ* L 091, July 4 1999, pp 0029 to 0036.
7   Directive 2002/46/EC of the European Parliament and of the Council of July 10 2002 on the approximation of the laws of the member states relating to food supplements, *OJ* K 183/51, July 12 2002.

provided for by this text has been postponed to a later date.

The use of therapeutic claims is restricted to medicinal products, as they refer to a disease and present the product in question as possessing curative and preventive properties.

Nutritional claims imply that a foodstuff has been shown to have a beneficial nutritional effect (either because of the energy that it provides, or because of the presence of nutrients or other specific substances).

Health claims, defined by Regulation (EC) 1924/2006 of December 20 2006 as "any claim, that states, suggests or implies that a relationship exists between a food category, a food or one of its constituents and health",[8] present the products as beneficial for health in general. These claims are found in borderline products, such as food supplements.

Drawing the frontier between a foodstuff and a medicinal product is sometimes difficult: for example, the mere reference to the fact that a product facilitates the treatment or the prevention of illnesses, in a general way, will not necessarily result in the product being treated as a medicinal product by presentation, even though the claim relates to the curative or preventive properties of the product.

Finally, Regulation EC 1925/2006 on the addition of vitamins and minerals and of certain other substances to foodstuffs[9] now also covers the addition of vitamins and minerals (with the exception of food supplements) and of substances having a nutritional or physiological role (applicable to any type of foodstuff).

Under this Regulation, the addition of vitamins and minerals to foodstuffs (other than unprocessed foodstuffs and beverages containing a certain volume of alcohol) is now lawful provided that:

- only vitamins and/or minerals listed in the Annex of the EC Regulation may be added to foodstuffs;
- the purity criteria provided for by Community legislation or recommended by international organisations is respected; and
- when a vitamin or mineral is added to foods, the minimum and maximum amounts (not established yet by law) are respected.

3.4 Genetically modified foods

Genetically modified organisms are defined as organisms in which the DNA has been altered in a way that does not occur naturally:

> It allows selected individual genes to be transferred from one organism into another, also between non-related species. Such methods are used to create genetically modified plants – which are then used to grow genetically modified food crops.[10]

---

8   Regulation (EC) 1924/2006 of the European Parliament and of the Council of December 20 2006 on nutrition and health claims made on foods, completed by Commission Regulation (EU) 665/2011 of July 11 2011 on the authorisation and refusal of authorisation of certain health claims made on foods and referring to the reduction of disease risk (text with EEA relevance), and by Regulation (EU) 666/2011 of July 11 2011 refusing to authorise certain health claims made on foods, other than those referring to the reduction of disease risk and to children's development and health (text with EEA relevance).

9   Regulation EC n° 1925/2006 of the European Parliament and of the Council of December 20 2006 on the addition of vitamins and minerals and of certain other substances to foods, OJ L 404/26, December 30 2006.

10 World Health Organization, "20 questions on genetically modified food".

Genetically modified foods are developed and marketed because of the advantage either to the producer or to the consumer of these foods (lower price, greater benefit in terms of durability or nutritional value or both).[11]

Genetically modified foods are the subject of extensive Community legislation which has been in place since 1990, in order to ensure maximum product safety. One of the key texts is this area is the Directive 2001/18/EC of March 12 2011,[12] which has codified the principles regarding:

- the assessment of environmental risks;
- post-marketed monitoring requirements;
- the obligation to inform consumers;
- the obligation for member states to ensure labelling and traceability;
- information to identify and detect genetically modified organisms in order to facilitate inspection and monitoring post-marketing;
- the obligation to consult scientific committees;
- the obligation to consult the European Parliament about decisions to authorise the release of genetically modified organisms; and
- the possibility for the Council to adopt or reject with a qualified majority a proposal from the Commission concerning the authorisation of a genetically modified organism.

Furthermore, Regulation 1829/2003 EC[13] on genetically modified food and feed provides a general framework for regulating genetically modified food and feed in the European Union, in order to ensure a high level of protection.

This Regulation provides a single authorisation procedure for food products containing genetically modified organisms. It also provides that food and feed products containing genetically modified organisms must be labelled as such. The words 'genetically modified' or 'produced from genetically modified [*name of the organism*]' must be clearly visible on the labelling of these products.

Food and feed products which contain a proportion of genetically modified organisms of less than 0.9% of each ingredient are not required to be labelled as genetically modified organisms provided that the presence of the genetically modified organism is technically unavoidable.

Regulation 1830/2003 EC of September 22 2003[14] provides that all products approved in accordance with its provision are subject to compulsory labelling; consumers will therefore be better informed about genetically modified products, whether for human or animal consumption.

---

11   *Ibid.*
12   Directive 2001/18/EC of the Parliament and of the Council of March 12 2011 on the deliberate release into the environment of genetically modified organisms and repealing Council Directive 90/220/EEC, modified by Directive 2008/27/EC of the European Parliament and of the Council of March 11 2008 amending Directive 2001/18/EC on the deliberate release into the environment of genetically modified organisms, as regards the implementing powers conferred on the Commission.
13   Regulation 1829/2003 EC of the European Parliament and of the Council of September 22 2003 on genetically modified food and feed.
14   Regulation 1830/2003 EC of the European Parliament and of the Council of September 22 2003 concerning the traceability and labelling of genetically modified organisms and the traceability of food and feed products produced from genetically modified organisms and amending Directive 2001/18/EC.

In this context, the main objectives of the regulations on genetically modified food and feed are:

- to protect human and animal health by introducing a safety assessment of the highest possible standard at EU level before any genetically modified food and feed is placed on the market;
- to have in place harmonised procedures for risk assessment and authorisation of genetically modified food and feed that are efficient, time-limited and transparent;
- to ensure clear labelling of genetically modified food and feed in order to respond to consumers' concerns and enable them to make an informed choice, and to avoid misleading consumers; and
- to set labelling requirements for genetically modified feed which provide farmers with accurate information on the composition and properties of feed, thereby enabling them to make an informed choice.

3.5   Regulating medical devices

Medicinal products and medical devices are products which are likely to have the same end purpose, or at least function within a framework of prevention and/or treatment of disease. The 2007/47/EC Directive of the European Parliament and the European Council of September 5 2007[15] harmonised the numerous definitions for medical devices as follows:

> *any instrument, apparatus, appliance, material or other article, whether used alone or in combination, including the software necessary for its proper application, intended by the manufacturer to be used for human beings for the purpose of:*
>
> *– diagnosis, prevention, monitoring, treatment or alleviation of disease,*
> *– diagnosis, monitoring, treatment or alleviation or compensation for an injury or handicap,*
> *– investigation, replacement or modification of the anatomy or of a physiological process,*
> *– control of conception,*
>
> *and which does not achieve its principal intended action in or on the human body by pharmacological, immunological or metabolic means, but which may be assisted in its function by such means.*

In order to ascertain whether a product is governed by the definition of a medicinal product or a medical device, the purpose of the product must be taken into account (ie, the use for which the device is intended according to the indications provided by the manufacturer in the labelling, the instructions and/or the promotional materials).

In this context, the principal action of the product can also be taken into consideration: medical devices generally act through physical and mechanical means, whereas medicinal products act through pharmacological, immunological or metabolic means.

The qualification of 'medical device' determines the applicable legal framework.

---

15   Directive 2007/47 EC of the European Parliament and of the Council of September 5 2007 amending Council Directive 90/285/EEC on the approximation of the laws of the member states relating to active implantable medical devices, Council Directive 93/42/EEC concerning medical devices and Directive 98/8/EC concerning the placing of biocidal products on the market, *OJ* L 247/21, September 21 2007.

It is not possible in the context of this chapter to present the specificities of each set of rules in a comprehensive manner. However, it is interesting to highlight a few significant differences.

The legal framework specific to medical devices is based on the CE marking systems. A medical device may not be put on the market without a CE marking which certifies compliance with the mandatory requirements of health and safety defined in the relevant harmonised European directives, transposed by the member states into their national laws. In particular, and by contrast to medicinal products, the competent health authority intervenes after commercialisation of a medical device. Medical devices are, in addition, subject to a classification (I, IIa, III), which is specific to them.

In the field of medical devices, Directive 2007/49/EC has introduced many developments, in particular: the conformity assessment, including design documentation and design review; the sufficiency and adequacy of clinical data for all classes of devices; post mark and surveillance; notified bodies; increased transparency to the general public in relation to the approval of devices; and modification of the AIMDD (Active Implantable Medical Devices Directive) in order to align it with the other framework directives on medical devices.

4. Protecting individuals and communities

Against the strict regulatory backdrop for pharmaceuticals, there are certain areas where individual rights and freedoms take precedence over intellectual property rights, effectively limiting the reach of patent monopolies in favour of public order considerations, human dignity and morality. One of the most controversial areas is, of course, the issue of stem cells but traditional or indigenous knowledge also raises important questions regarding how to protect long-standing traditions and practices from commercial monopolies. Data privacy also raises important issues, where it is necessary to weigh the rights and freedom of patients against the needs of medical research.

4.1 Stem cells

Stem cell research whilst potentially highly promising in many therapeutic fields, remains controversial, raising many complex, and often diverging, moral and ethical issues across the European Union and further afield. The principal concern regarding this type of research is due to the fact that the most effective stem cells can only be sourced from early human embryos.

The significance of human stem cells is that they can develop into specialised cells and be used to repair or replace damaged tissue. They are considered by many to hold 'considerable promise' in relation to metabolic, degenerative and inflammatory diseases.

Whilst adult cells are currently used in relation to certain pathologies, they are not able to develop into all types of cells, limiting their R&D potential. Stem cells obtained from embryos are, however, able to do just that and are referred to as 'totipotent' cells and 'pluripotent' cells, by opposition with 'multipotent' adult stem cells. Totipotent cells develop into human beings, whereas pluripotent cells can develop into almost all type of cells and tissues (but not a human being).

It is the origin of pluripotent cells which causes debate as it raises the difficult question of where society draws the line between, on the one hand, respect for human life and, on the other, the needs of medical research and the positive impact that it can have on serious illness. This debate resulted in US President George W Bush imposing restrictions on the federal funding of embryonic stem cell research during his mandate – restrictions which have subsequently been lifted by President Obama.

The use of stem cells is, as one would expect, regulated both at EU and national level,[16] with legislators attempting to keep up with scientific developments. The issue which remains to be determined in the patent field is the extent to which inventions based on stem cell research should be capable of patent protection in Europe.

Common principles in relation to biotechnological inventions were introduced in Europe in the form of Directive 98/44/EC of July 6 1998 on the Legal Protection of Biotechnological Inventions (the Biotech Directive) which member states have now implemented into their national legislation.

Article 5(2) of this Directive provides a general statement of principle to the effect that:

an element isolated from the human body or otherwise produced by means of a technical process, including the sequence or partial sequence of a gene, may constitute a patentable invention, even if the structure of that element is identical to that of a natural element.

Based on this text, a process involving stem cells may, in principle at least, constitute a patentable invention provided that it meets the standard requirements of patentability (ie, novelty, inventive step and industrial application).

However, the Directive goes on to stipulate at Article 6(2) that inventions whose commercial exploitation would be contrary to public order or morality cannot be considered as patentable. Article 6(2)(c) further specifies that this prohibition applies to "uses of human embryos for industrial or commercial purposes". This prohibition is mirrored in both the national patent systems across Europe and in the European Patent Convention.

The position of the European Patent Office on this question was expressed in the WARF decision,[17] where it decided that inventions which necessarily involve the destruction of human embryos could not qualify for patent protection. It considered that the intention behind Article 6 of the Directive was to prevent the destruction of embryos and preserve human dignity.

The WARF decision does not, however, give guidance on what should be considered to constitute a human embryo. This question has recently been examined by the Court of Justice of the European Union in the *Brüstle v Greenpeace* case[18] which relates to pluripotent cells. The Court was asked to examine in a reference for a preliminary ruling whether the exclusion of the human embryo from patentability covers all stages of life, from fertilisation of the ovum, or whether other requirements, such as the attainment of a certain stage of development, must be satisfied in order to be treated as an embryo.

---

16   In the United Kingdom: Human Fertilisation and Embryology Act 1990, as amended by the Human Fertilisation and Embryology Act 2008, Human Tissue Act 2004.
17   *Wisconsin Alumni Research Foundation*, EP patent app EP96903521.1.
18   C-34/10.

The Grand Chamber of the Court of Justice has given a very wide definition of 'human embryos', drawing on the fact that "the context and aim of the Biotech Directive show that the European Union legislature intended to exclude any possibility of patentability where the respect of human dignity could thereby be affected". Consequently, the Court found that "any human ovum must, as soon as is it fertilised, be regarded as a human embryo". The Court applies the same reasoning to non-fertilised human ova where they are capable of developing into human beings as a result of techniques applied to them.

Unlike its Advocate General, the Court does not rule out the possibility that pluripotent cells may also meet the definition of an embryo. Instead, the Court considers that it is for the referring court to ascertain whether a stem cell obtained at the blastocyst stage constitutes a human embryo. If the national court considers that this type of stem cell constitutes a human embryo, it will fall under the prohibition as to patentability at Article 6(2).

However – and this is the point in the Advocate General's opinion that generated much debate and concern – the Court considers that an invention based on embryonic pluripotent stem cells (not embryos) should not be patentable where the application of the technical process for which the patent is filed necessitates the prior destruction of human embryos or their use as base material at whatever stage it takes place, and regardless of whether this is specified in the claims of the invention. The only circumstances where this would be acceptable under the terms of the Directive, according to the Court, relate to inventions having therapeutic or diagnostic purposes which are applied to the human embryo and are useful to it (eg, to correct a malformation).

Much of the criticism centred on the Advocate General's initial Opinion and the Court's ruling springs from the fact that embryonic stem cells are derived using surplus *in vitro* fertilised eggs that are grown in culture. Indeed, there exist today more than 100 stem cell lines available via national and international cell banks, which have never formed part of an embryo. The concern remains that, as a result of the Court's ruling regarding the irrelevancy of when the destruction takes place, research into stem cells is likely no longer to be able to benefit from patent protection in Europe. This could well result in European researchers, and private funding, leaving the continent of Europe for countries which welcome scientific research on this subject.

4.2 Traditional or indigenous knowledge

Traditional knowledge, also often referred to as 'TK', is "a form of know-how resulting from intellectual activity in a traditional context. This includes agricultural, environmental and medicinal knowledge, and knowledge associated with genetic resources".[19]

Whilst there is increasing recognition of the need to protect TK from unjust exploitation by others, it does not currently enjoy any specific intellectual property protection despite the efforts of members of the World Intellectual Property Organization and the EC to try to organise international legal instruments to ensure

---

19   www.wipo.int/tk/en/law/index.html.

such protection. WIPO has set up an Intergovernmental Committee in Intellectual Property and Genetic Resources, Traditional Knowledge and Folklore.[20] This Committee is now looking at how to regulate the interface between intellectual property rights and the genetic resources on which TK is based.

Indeed, traditional knowledge represents a tremendous value and is therefore the object of all sorts of misappropriations and claims, known as 'bio-piracy'). It is for this reason that the Indian government has provided the EPO access to its Traditional Knowledge Digital Library (TKDL) to help examiners to evaluate whether new patent applications should be rejected because they are merely trying to patent traditional knowledge. The TKDL provides a description of medical treatments and the curative properties of plants, which were contained in ancient texts and languages.

In this respect, the EPO recently rejected a patent application describing a method for producing an extract from a plant.[21] The plant had traditionally been used as a drug in southern Africa and, arguably, belonged to 'traditional knowledge'. It was submitted that the patent was against public order and morality within the meaning of Article 53(a) of the EPC, although the patent was actually revoked on the basis of 'lack of inventive step'.

TK is not only a question raised in developing countries outside the European Union, but it is also crucial in certain European countries such as France which has developed a form of protection of TK constituted by 'indications of origin'. These indications of origin cover not only wine and cheese, but also local food products which claim this form of IP protection.

These examples show how countries are seeking to protect traditional knowledge from being fraudulently used or appropriated by third parties and that their arms for achieving this result are diverse but still relatively weak. At a time when there is clearly a return to traditional values, the risk of misappropriation of traditional knowledge is all the more present and the need for protection at international level urgent.

4.3   Data privacy

The legislative framework for data privacy in the European Union is based on Directive 95/46/EC of the European Parliament and of the Council of October 24 1995 on the protection of individuals with regard to the processing of personal data and on the free movement of such data (the Data Privacy Directive).[22] This text which is currently under review, has formed the basis for a harmonised approach to data privacy issues within the European Union and is the legislative reflection of Article 8 of the Charter of Fundamental Rights of the European Union which expressly recognises the fundamental right to the protection of personal data.

Personal data in the pharmaceutical sphere is both vital and, at the same time, highly sensitive. Indeed, pursuant to Article 8 of the Data Privacy Directive, "member states shall prohibit... the processing of data concerning health". This prohibition is, of course, subject to exceptions in favour of the provision of healthcare and medical

---

20   See www.wipo.int/tk/en/
21   EPO – Opposition Division – April 10 2010.
22   *Official Journal* L 281, November 23 1995 pages 0031–0050.

research. As a result of the discretion that member states enjoy when implementing a directive, experience shows that national regulatory authorities do not always adopt the same approach to the issues raised.

As a general rule of thumb, the data controller (ie, the person responsible for determining the purposes and means of the processing of personal data) must obtain the patient's consent to the processing of his personal data or render the data anonymous.

Both of these solutions pose practical problems which have a significant financial and ethical impact for the life sciences industry. These issues are exacerbated by globalisation of personal data processing in the medical field, with the creation of biobanks for biological data, electronic health records and a move towards genetic medicine. Indeed, it is currently not clear whether genetic data falls within the category of 'sensitive' health data; accordingly, the manner in which its processing is regulated varies across member states.

In addition, the technical capacity to collect this type of data and transfer it effectively has increased immeasurably since the Data Privacy Directive was introduced in 1995. This also raises significant data privacy concerns where the data is being transferred outside the European Union or the European Economic Area. Indeed, pursuant to Article 25 of the Directive, a data controller is effectively prohibited from transferring data to countries outside these areas where the data privacy legislation of the country where the recipient is located does not offer an 'adequate level of protection'. The adequacy of protection is determined by the EU Commission, which publishes specific Opinions in which it sets out the circumstances in which the relevant foreign legislation offers an equivalent level of protection to that available in the European Union under the terms of the Directive.[23]

Given the leeway that the Directive provides members states when implementing these provisions, this effectively results in an additional layer of regulatory requirements, which can make the processing of data in the context of cross-border data sharing very challenging.

The aim behind these provisions is clearly to protect individuals from the inappropriate processing of their personal data – the unauthorised use or disclosure of health data to employers, insurers or banks clearly poses serious risks for individuals and their personal freedoms. Against this background, it remains true that the sharing of health data amongst health professionals and researchers is recognised as beneficial. The challenge today, when the legal framework of the data protection system within Europe is under review,[24] is to ensure that the rights of individuals retain their central position in the modified regime, whilst ensuring that the conditions for processing this type of data are both clarified and harmonised across the European Union.

---

23   Opinions have so far been given for Andorra, Argentina, Australia, Canada, Switzerland, Faroe Islands, Guernsey, the State of Israel, the Isle of Man, Jersey and the United States (see http://ec.europa.eu/justice/policies/privacy/thridcountries/ index_en.htm#countries).
24   See http://ec.europa.eu/justice/policies/privacy/review/index_en.htm

Australia

Wayne Condon
Griffith Hack

1.   Small molecules

1.1   Product and process claims

In Australia product, process and method claims are all potentially patentable. In addition to base chemical or compound claims, other common types of pharmaceutical claims are:

- salts, esters and solvates;
- enantiomers;
- polymorphs;
- combinations of APIs;
- formulations;
- methods of use;
- methods of manufacture or process claims;
- product-by-process claims;
- release profiles; and
- methods of medical treatment.

1.2   Scope of protection of claims and Markush formulae

New synthetically manufactured compounds are usually delineated in the claims by a generalised structural formula. This is known as a Markush claim. The scope of such a claim depends upon which compounds may be created by combining the different alternatives mentioned in the different positions in the formula. That type of claim makes it possible to define a very large number of compounds concisely. Markush claims are common in Australia and their scope is determined in accordance with the usual principles of patent claim construction set out below.

The orthodox approach to claim construction was set out by the Full Court of the Federal Court in *PAC Mining Pty Ltd v Esco Corporation* (2009) 80 IPR 1; [2009] FCAFC 18 at [27] to [29] (referring to the earlier Full Court decision in *Pfizer Overseas Pharmaceuticals v Eli Lilly & Co* (2005) 225 ALR 416; [2005] FCAFC 224).

The Full Court noted, apparently with approval, the 10 principles of construction set out by Sheppard J in *Decor Corp Pty Ltd v Dart Industries Inc* (1988) 13 IPR 385 at 400; [2005] FCAFC 224; 225 ALR 416 at 468 [249]:

- The claims define the invention which is the subject of the patent. These must be construed according to their terms upon ordinary principles. Any purely verbal or grammatical question that can be answered according to ordinary rules for the construction of written documents is to be resolved accordingly.

- It is not legitimate to confine the scope of the claims by reference to limitations which may be found in the body of the specification but are not expressly or by proper inference reproduced in the claims themselves. To put it another way, it is not legitimate to narrow or expand the boundaries of monopoly as fixed by the words of a claim by adding to those words glosses drawn from other parts of the specification.

- Nevertheless, in approaching the task of construction one must read the specification as a whole.

- In some cases the meaning of the words used in the claims may be qualified or defined by what is said in the body of the specification.

- If a claim is clear, it is not to be made obscure because obscurities can be found in particular sentences in other parts of the document. But if an expression is not clear or is ambiguous, it is permissible to resort to the body of the specification to define or clarify the meaning of words used in the claim.

- A patent specification should be given a purposive construction rather than a purely literal one.

- In construing the specification, the court is not construing a written instrument operating *inter partes*, but a public instrument which must define a monopoly in such a way that it is not reasonably capable of being misunderstood.

- The body, apart from the preamble, is there to instruct those skilled in the art concerned in the carrying out of the invention; provided it is comprehensible to, and does not mislead, a skilled reader, the language used is seldom of importance.

- Nevertheless, the claims, since they define the monopoly, will be scrutinised with as much care as is used in construing other documents defining a legal right.

- If it is impossible to ascertain what the invention is from a fair reading of the specification as a whole, it will be invalid. But the specification must be construed in the light of the common knowledge in the art before the priority date.

In *Pfizer*, the Full Court also noted that the "clear words of a claim are not to be distorted by importing passages from the specification" and that "when a claim is clear it is not to be glossed or obscured by reference to the specification".

The correct approach to the construction of claims has been referred to in two recent judgments of the Full Court. In *Insta Image Pty Ltd v KD Kanopy Australasia Pty Ltd* [2008] FCAFC 139 at [82], the Full Court said (citations excluded):

> *Construction of claims is a matter for the Court ... When claims are being construed, the specification is to be read as a whole in order that the necessary background to the invention may be known and words or expressions used in the claims may be understood ... However, where a claim is clear and unambiguous, it is not permissible to resort to the specification in order to qualify it ... It may, however, be possible to resolve any*

ambiguity or lack of clarity by reference to the body of the specification ...

In *Kinabalu Investments Pty Ltd v Barron and Rawson Pty Ltd* [2008] FCAFC 178 at [44] to [45], the Full Court said:

The principles of construction applicable were not in dispute. When determining the nature and extent of the monopoly claimed, the specification must be read as a whole. But as a whole it is made up of several parts which have different functions. The claims mark out the legal limits of the monopoly granted. The specification describes how to carry out the process claimed and the best method known to the patentee of doing that. Although the claims are construed in the context of the specification as a whole, it is not legitimate to narrow or expand the boundaries of monopoly as fixed by the words of a claim, by adding to those words glosses drawn from other parts of the specification. If a claim is clear and unambiguous, it is not to be varied, qualified or made obscure by statements found in other parts of the document. It is legitimate, however, to refer to the rest of the specification to explain the background of the claims, to ascertain the meaning of technical terms and resolve ambiguities in the construction of the claims.

Other more specific principles of construction collected in *Flexible Steel Lacing Co v Beltreco Ltd* [2000] FCA 890; (2000) 49 IPR 331 at [73] to [75] are as follows:

- A specification should be given a purposive construction rather than a purely literal one.
- The hypothetical addressee of the specification is the non-inventive person skilled in the art before the priority date.
- The words used in a specification are to be given the meaning the hypothetical addressee would attach to them, both in the light of the addressee's own general knowledge and in the light of what is disclosed in the body of the specification.
- As a general rule, the terms of the specification should be accorded their ordinary English meaning.
- Evidence can be given by experts on the meaning those skilled in the art would give to technical or scientific terms and phrases, and on unusual or special meanings given by such persons to words which might otherwise bear their ordinary meaning.
- The construction of the specification is for the court, however, not for the expert. In so far as a view expressed by an expert depends upon a reading of the patent, it cannot carry the day unless the court reads the patent in the same way.

1.3   Metabolites

An active metabolite, if novel and inventive and of utility, can be patented separately to the base compound – which then gives rise to the question of whether sales of the original drug substance infringe the patent claiming the active metabolite.

That was the question considered by the House of Lords in the United Kingdom in *Merrell Dow Pharmaceuticals Inc v HN Norton and Co Ltd* [1996] 33 IPR 1. In that case the House of Lords accepted the argument that, although some members of the public had been making the active metabolite by swallowing terfenadine before the date of the metabolite patent, they had not been aware of that fact and thus the prior use was not novelty destroying. However, the disclosure of the terfenadine patent specification itself, by teaching that it could be used as a pharmaceutical if injected, was held to anticipate the claim to the acid metabolite because the specification communicated to the public "information which enables it to do an act having the inevitable consequence of making the acid metabolite".

Although it has not been the subject of final judicial determination in Australia, it is strongly arguable that the reasoning applied by the House of Lords in *Merrell Dow* also holds good in Australia.

2. Second-generation inventions

2.1 Combinations

Combinations, including combinations of active pharmaceutical ingredients, are inherently patentable if they are novel, inventive and useful.

Under Section 50(1)(b) of the Patents Act 1990 a patent application may be refused on the ground that the specification claims as an invention a substance which is capable of being used as a medicine and is a mere mixture of known ingredients, or a process producing such a substance by mere admixture.

'Mixtures' embrace not only powders or granules, either loosely or in compacted form, but also mixtures of liquids or gases and include suspensions and solutions.

By 'mere mixture of known ingredients' is meant a mixture exhibiting only the aggregate of the known properties of the ingredients. If the result achieved by the invention is more than might be expected from a mere mixture (ie, synergism), the invention is patentable.

2.2 Enantiomers

Enantiomers are inherently patentable in Australia provided that they satisfy the basic requirements for patentability – novelty, inventiveness and utility.

In the last few years there have been two important decisions, dealing with clopidogrel and escitalopram respectively, which have set out the principles relating to the novelty and inventiveness of enantiomer patents.

In the *Lundbeck*[1] and *Apotex*[2] cases, it was held that the mere disclosure of a compound by exact naming in the prior art was sufficient, of itself, to constitute anticipation. Those were not the facts in the *Lundbeck* case, but it was in *Apotex* in which the court held that the disclosure of a particular enantiomer in the prior art anticipated the later patent which claimed that the enantiomer, notwithstanding the compound consisting only of that enantiomer, had never been made.

The alternative line of authority in Australia, to the effect that a prior art disclosure must be 'enabling' in order to anticipate a later patent, is based on the decision in *Hill v Evans* [1862] ER 365 and of the decision of the Australian High Court in *Olin Corporation v Super Cartridge Co Pty Ltd* [1977] 180 CLR 236. In *Hill v Evans* it had been held that:

---

1   *H Lundbeck A/S v Alphapharm Pty Ltd* [2009] 177 FCR 151.
2   *Apotex Pty Ltd v Sanofi-Aventis* [2009] 82 IPR 416.

the prior knowledge of an invention to avoid a patent must be knowledge equal to that required to be given by a specification, namely, such knowledge as will enable the public to perceive the very discovery, and to carry the invention into practical use.

The very issue of which strand of authority in relation to anticipation applied in the case of an enantiomer patent was recently considered in the decision of *Albany Molecular Research Inc (AMR) v Alphapharm Pty Ltd* [2001] FCA 120 in February 2011.

Faced with the apparently conflicting line of authority, the court in that case followed the *Lundbeck* and *Apotex* authorities to the effect that any prior disclosure of the compound anticipates, and it rejected the argument that only an enabling prior disclosure of the compound anticipates. The judge in that case was, however, clearly uncomfortable in the conclusion which His Honour felt bound to reach.

The judge considered the evidence of eminent organic chemists and concluded that the prior-art disclosure relied on did not disclose an effective means of preparing substantially pure fexofenadine, being the compound at issue in that case, and stated:

if I am wrong about the law as established in Lundbeck *and* Apotex, *I would hold that the invention, so far as claimed in the claims which are presently relevant, was not anticipated by [the Carr patents].*

It will come as no surprise that the *AMR* decision has been appealed and it will now be for the Full Court of the Federal Court of Australia on appeal to consider which of the conflicting lines of authority in relation to the law of anticipation apply in Australia as a matter of principle.

In assessment of the inventiveness of a patent for an enantiomer, the so-called 'worthwhile to try' test has recently been applied to determine obviousness. This is so because it is often the case, when dealing with an enantiomer patent, that the racemate forms part of the common general knowledge. In *Hässle v Alphapharm Pty Ltd* [2002] 212 CLR 411, the Australian High Court, by majority, approved the restatement of the so-called 'Cripps question' of Graham J in *Olin Mathison v Biorex*[3] which can be expressed in general terms as follows:

Would the notional research group, at the relevant date, in all the circumstances, directly be led as a matter of course, to try [the invention] in the expectation it might well produce a useful result?

The most recent application of that general approach is the *AMR* case referred to above. In that case the judge, holding the invention not to be obvious said:

I do not accept that ... a research group would have been directly led, as a matter of course, to try the [successful process] in the expectation that it might well provide a link in the synthetic chain of the production of fexofenadine ... Rather I accept the submission made on behalf of AMR ... that example was much closer to what has been described as trying to 'mark' each of numerous possible choices until one possibly arrived at a successful result, where the prior art gave either no indication of which parameters were critical or no direction as to which of many possible choices is likely to be successful.

---

3   [1970] RPC 157.

2.3   Selection inventions

In Australia, currently, the law in relation to selection patents is unsettled.

In the *Apotex* case referred to above, the trial judge expressed some doubt about the applicability in Australia of the concept of selection patents but nevertheless proceeded to consider the criteria for such a patent as set out in *IG Farbenindustrie* as referred to by the Full Court of the Federal Court of Australia in *Ranbaxy Australia Pty Ltd v Warner-Lambert Co LLC* [2008] 778 IPR 449. At first instance in the *Apotex* case the trial judge said that, in his view, and after considering various authorities, the concept of a selection patent does not recognise a special class of patents at all but is merely "a convenient shorthand to pick up the relevant principles concerning anticipation".

The Full Court in *Ranbaxy* set out the principles by which a claimed invention constitutes a selection such that it is not anticipated by the disclosure of a class of compounds of which it is a member. The claimed member of the class must satisfy the following requirements:

- There must be some substantial advantage to be secured by the use of the selected members.
- The whole of the selected members must possess the advantage in question.
- The selection must be in respect of a quality of a special character, which can fairly be said to be peculiar to the group.
- The advantage possessed by the selected members must be clearly disclosed in the specification.

In the *Apotex* case, at first instance, the judge found that, if a selection were available, it would be satisfied by the selection of the d-enantiomer from the class comprising the d-enantiomer, the l-enantiomer and the racemate. The judge concluded that there was a substantial advantage to be secured by the use of the d-enantiomer, as compared with the use of either the racemic mixture or the l-enantiomer. The advantage was that the d-enantiomer was better tolerated or less toxic and as effective as the racemate at a lesser concentration. The trial judge also considered the second and fourth criteria set out immediately above to have been satisfied in that case.

On appeal, the Full Court of the Federal Court of Australia in *Apotex* avoided the issue of selection patents and in so doing stated:

> It is not necessary to decide whether or not there is a special category of selection patents which, if they satisfy the test in IG Farbenindustrie, may overcome a lack of a claim of lack of novelty. Any such category was not, in our view, intended to exclude from the requirement of novelty a compound (here the d-enantiomer) that was previously disclosed and claimed as one of a class of inventive compounds that demonstrated, or were predicted to demonstrate, particular activity and tolerance at various levels, and the compound was then shown to demonstrate that same activity at a high level, with high tolerance.

Consequently, currently in Australia, as a matter of law, it is not clear that the courts will recognise the concept of a 'selection patent', whether in a pharmaceutical context or otherwise. The comments of the trial judge in the *Apotex* case would suggest, however, that in the modern context in Australia the concept of a 'selection patent' may no longer find favour – at least as a separate and discernable category of patent but that, rather, considerations of the *IG Farbenindustrie* kind will be taken into account in any consideration of novelty.

2.4 Methods of use and secondary indications

Patents for new methods of use and new indications can be granted in Australia, provided the basic test for patentability is satisfied – in particular the novelty, inventiveness and utility requirements.

In Australia there is currently also a threshold test of patentability, separate to novelty and inventiveness. In order to be patentable, an invention must also be a 'manner of manufacture' within the concept of Section 6 of the Statute of Monopolies 1623 – an ancient English statute given currency in Australia by reason of its adoption into the Australian patent law pursuant to Section 18 of the Patents Act 1990. In *National Research and Development Corporation v Commissioner of Patents* (1959) 102 CLR 252, the High Court of Australia considered the expression 'manner of manufacture' to require an invention to give rise to an artificially created state of affairs in a field of economic endeavour for it to be patentable.

In the context of pharmaceuticals it is not unusual for an invention to be centred on the concept of a new use or application of a known product. It is commonplace for new uses that are found for known things to be the subject of the claims of a patent. Claims of that type, however, can only be valid where the new use arises from some previously unknown property of the known thing or in some adaptation of the thing to suit the new purpose or in other cases where the adaptation of an old process to a new use involves an inventive step. It was held in *Commissioner of Patents v Microcell Limited* [1959] 102 CLR 232 by the Australian High Court that a use of a known thing that is an exploitation of a known property of that thing is not patentable, notwithstanding that the use may be a new one.

2.5 Methods of treatment

Unlike Europe, for example, methods of medical treatment are directly patentable in Australia. Technically, the courts which have made that pronouncement have done so *obiter* such that it could be suggested that the issue has still not received final judicial pronouncement in Australia. However, in the two cases which have considered this issue directly, six of the eight judges who considered it were of the opinion that methods of medical treatment could be the proper subject of a patent in Australia.

While it is also the practice in the Australian Patent Office for Swiss-style claims to be accepted, they have not been judicially considered in detail in Australia and their validity therefore remains an open question. It has been observed by a delegate of the Commissioner of Patents that Swiss-style claims should be interpreted in Australia in the same way that they are interpreted in Australia and elsewhere – see *Smith Kline Beecham plc v Lek Pharmaceutical and Chemical Company* [2004] 61 IPR 626.

2.6 Formulations and physical forms

A patent is available in Australia for a new, inventive and useful formulation of a pharmaceutical compound. Polymorph inventions are similarly patentable.

2.7 Reach-through claims

In some cases an invention will reside in a method of testing or identifying properties of a particular entity but the claims will be directed to the entity itself, unlimited to that testing environment or any field of use. Such claims are known as 'reach-through' claims.

These reach-through claims are rarely supported over their full scope but, currently in Australia, so long as there is an exemplification of at least one preferred embodiments the Patents Office will not be able to object to a reach-through claim set.

Such claims may, however, lack fair basis. The concept of a 'reach-through' claim contemplates that the underlying entity already existed prior to the testing process and was generated without recourse to it. Invention does not reside in the entity itself but in the consequential uses of the entity (based on the properties ascertained from the inventive method). Claims to the entity *per se*, without further limitation, therefore may be argued to claim matter beyond the subject matter of the invention and if they do that they are not fairly based on the disclosure, in contravention of Section 40(3) of the Patents Act 1990.

3. DNA, biologicals and personalised medicine

3.1 Discoveries

The orthodox position in Australia is that mere discoveries, fundamental concepts and principles, which include mathematical algorithms, formulas, calculations, directions for use, or some other form of intellectual information, are precluded from patentability.[4] However, if the application of knowledge in a particular scenario creates an 'artificial state of affairs', then it will be eligible for a patent (supposing it satisfies other criteria). For example, with regard to DNA, the current practice is that a claim to a naturally occurring DNA sequence is a mere discovery, but an isolated and purified sequence is proper subject matter for a patent.[5]

It is worth noting, that there is currently a bill before the Australian federal parliament which will exclude biological material that is 'structural and functionally identical to that which exists in nature.[6] Whether or not this becomes law, and what its limits on patentability will be though, is unknown.

3.2 Gene patents and industrial application

A valid Australian patent is required to pass a test of 'utility'.[7] It is akin to 'industrial

---

4   'Patentable Subject Matter' (Final Report, Australian Council on Intellectual Property) 6. Also, see generally, *National Research Development Corporation v Commissioner of Patents* (1959) 102 CLR 252; *Grant v Commissioner of Patents* (2006) 154 FCR 62.

5   *Kirin-Amgen Inc v Board of Regents of University of Washington* (1996) 33 IPR 557, 569.

6   Patent Amendment (Human Genes and Biological Materials) Bill 2010 (Cth).

7   Patents Act 1990 (Cth), Section 40(2)(a).

application' but differs in function. The test requires that the claims, as defined and properly interpreted, are capable of achieving the promise of the patent.[8] Unlike European law, utility is not concerned with the question of whether the result claimed by the patentee in the patent specification would be commercially viable.[9]

There is currently a Bill before the Australia federal parliament to modify the test of utility to include a US-modelled 'specific, substantial and credible' use test.[10] This is colloquially referred to as the 'Raising the Bar' Bill. However, whether it will be passed into law and how it will be interpreted by an Australian court is not presently clear.

3.3   Stem cells, organic tissue and *ordre public*

Although there is no case in point, the Australian patent examiner's manual reflects the Australian Patents Office practice which permits patents in respect of organic tissue, including totipotent (but not pluripotent/multipotent) stem cells. No objection can be taken to a claim to a new organism on the ground that it is something living;[11] however, 'human beings and the biological processes for their generation' are not patentable.

Although Australia is a signatory to TRIPS, Australian patent law does not, at the moment, expressly include an *ordre public* or other type of morality exclusion. There is discussion that an amendment to Australian patent law should be made to provide for a European-modelled *ordre public* exclusion;[12] but even if this was enacted, Australian courts are very wary of ethical and social policy concerns, having previously stated that it is an area in which they have no special expertise.[13]

3.4   Bioinformatics systems

Bioinformatics systems are made up of various components and each can attract different intellectual property protection. In Australia, copyright law will protect lines of code, as well as compilations of data.[14] User interfaces and programs can receive patent protection.[15] And for other aspects such as algorithms, which may not otherwise receive intellectual property protection, if it is only known to a small number of people, then it may be protectable as a trade secret.[16]

3.5   Copyright and sequence information

The Australian judiciary has not considered whether nucleotide or peptide sequence information acquires copyright protection. However, in light of recent decisions, sequence information would probably be unlikely to acquire copyright protection.[17]

---

8   *Wimmera Industrial Minerals Pty Ltd v RGC Mineral Sands Ltd* (1996) 66 FCR 41, 59.
9   *Rehm Pty Ltd v Websters Security Systems (International) Pty Ltd* (1988) 81 ALR 79, 95 to 96.
10   Intellectual Property Laws Amendment (Raising the Bar) Bill 2011 (Cth) 4; Explanatory Memorandum, Intellectual Property Laws Amendment (Raising the Bar) Bill 2011 (Cth) 43 to 45.
11   Australian Patent Office, *Manual of Practice and Procedure* (February 22 2006) 2.7.1 and 2.9.2.14.
12   "Patentable Subject Matter" (Final Report, Australian Council on Intellectual Property) pt C-4.
13   *Anaesthetic Supplies Pty Ltd v Rescare Ltd* 50 FCR 1, 45.
14   Copyright Act 1968 (Cth), Section 10(1) (definition of 'literary work').
15   *International Business Machines Corp v Commissioner of Patents* (1991) 33 FCR 218.
16   *Saltman Engineering Co Ltd v Campbell Engineering Co Ltd* [1963] 3 All ER 413n, 415; *Moorgate Tobacco Co Ltd v Philip Morris Ltd [No 2]* (1984) 156 CLR 414, 437.

Although skill and labour is involved in sequence production, copyright protection is only afforded to effort that is directed to expression itself.[18] Since the expression of sequence information is primarily dictated by the nature of sequences itself, it is unlikely that sufficient originality would be found to acquire copyright protection.

3.6 Sufficiency issues

The requirement of a patent specification to describe an invention fully is the Australian equivalent of sufficiency.[19] What this item requires is that the specification makes it plain, to a hypothetical skilled person in the art, how to create or perform an invention.[20] This requirement is met if the specification discloses how to produce *one embodiment* of an invention,[21] even though the patent may claim multiple embodiments of the invention which are not described. As a result of this, the Raising the Bar Bill includes amendments to Australian 'sufficiency' law aimed at aligning the Australian law with that of the European approach to sufficiency.[22] Whether this will give rise to *Biogen*[23]-like sufficiency issues in the future remains to be seen.

4. Acts of patent infringement

4.1 Infringement

There is no definition of infringement in the Patents Act 1990. However, a patent gives the patentee the exclusive rights, during the term of the patent, to 'exploit' the invention and to authorise another person to exploit the invention.[24]

The rights given are:

- where the invention is a product – to make, hire, sell or otherwise dispose of the product, offer to make, sell, hire or otherwise dispose of it, use or import it, or to keep it for the purpose of doing any of those things; or
- where the invention is a method or process – to use the method or process or do any act mentioned in the bullet point above in respect of a product resulting from such use.

The patentee (or any exclusive licensee) may take legal action to prevent infringement of the exclusive rights granted pursuant to a patent. Patent infringement may be either direct or contributory. The infringement is direct if a person, without licence or authorisation, exercises any of the exclusive rights of the patent holder.

---

17   See generally, *IceTV Pty Ltd v Nine Network Australia Pty Ltd* 239 CLR 458; *Acohs Pty Ltd v Ucorp Pty Ltd* 86 IPR 492; *Telstra Corporation Ltd v Phone Directories Company Pty Ltd*, 264 ALR 617.
18   *IceTV Pty Ltd v Nine Network Australia Pty Ltd* 239 CLR 458, 481.
19   Patents Act 1990 (Cth), Section 40(2)(a).
20   *Patent Gesellschaft AG v Saudi Livestock Transport and Trading Company* (1997) 37 IPR 523, 530.
21   *Lockwood Security Products Pty Ltd v Doric Products Pty Ltd* (2004) 217 CLR 274, 297.
22   Intellectual Property Laws Amendment (Raising the Bar) Bill 2011 (Cth) 5; Explanatory Memorandum, Intellectual Property Laws Amendment (Raising the Bar) Bill 2011 (Cth) 46 to 48.
23   *Biogen Inc v Medeva plc* [1997] RPC 1.
24   Section 13 of the Act.

4.2 Contributory infringement

By virtue of Section 117 of the Patents Act, contributory infringement occurs where there is a supply of a product the use of which would constitute infringement of the patent.

The supply of a product will only amount to contributory infringement if one of the following conditions is satisfied:

- where the product is capable of only one reasonable use, having regard to its nature or design – supply for that use; or
- where the product is not a staple commercial product – supply for any use of the product, if the supplier had reason to believe that the person would put it to that use; or
- in any case – supply for the use of the product in accordance with any instructions for the use of the product, or any inducement to use the product, given to the person by the supplier or contained in an advertisement published by or with the authority of the supplier.

4.3 'Bolar'-type provisions

In Australia, Section 119A of the Patents Act provides a Bolar-type defence to infringement.

This section was added to the Patents Act 1990 in 2006 and provides an exemption from infringement of a 'pharmaceutical patent', at any time during the term of such a patent, if the acts done are solely for obtaining regulatory approval in Australia or elsewhere. (A 'pharmaceutical patent' is defined as one which claims a pharmaceutical substance or a method, use or product relating to one. A pharmaceutical substance is, in turn, defined, in effect, as a substance for therapeutic use.)

The exemption does not apply to permit export of goods from Australia, unless the term of the patent has been extended and the goods contain a pharmaceutical substance *per se* (or have been produced by a process that involves the use of a recombinant DNA technology) (Section 119A(2)).

4.4 Experimental-use exemptions

The Act currently contains no express exemption from infringement in relation to acts done purely for private non-commercial or experimental purposes.

However, an implied experimental-use defence may be available under common-law principles. See, for instance, the 19th-century English case of *Frearson v Loe*, which suggested that acts may not constitute 'use' of an invention if there is no commercial purpose.[25]

Concerns that the absence of an explicit experimental-use defence inhibits research in Australia have led both the Australian Law Reform Commission and the Australian Advisory Council on Intellectual Property to recommend amendments to the Patents Act.

Specifically, a Bill currently before the Australian Parliament proposes[26] changes

---

25   See the judgment of Jessel MR in *Frearson v Loe* (1878) 9 Ch D 48 at 66 to 67.
26   Intellectual Property Laws Amendment (Raising the Bar) Bill 2011, Section 119C.

to the Patents Act to ensure that acts done for 'experimental purposes' do not infringe a patent. 'Experimental purposes' will be defined to include, but are not limited to, the following:

- determining the properties of the invention;
- determining the scope of a claim relating to the invention;
- improving or modifying the invention;
- determining the validity of the patent or of a claim relating to the invention; and
- determining whether the patent for the invention would be, or has been, infringed by the doing of an act.

4.5 Submitting authorisations and offers to supply

The mere filing of an application to list a therapeutic product on the Australian Register of Therapeutic Goods (ARTG) does not amount to an infringement of a patent relating to that product.

In Australia there is no linkage between the obtaining of marketing approval of pharmaceutical products and patents akin to the US Orange Book procedure. However, as a result of the Australia–US Free Trade Agreement, which took effect on January 1 2005, an applicant applying for the listing or registration of a therapeutic good on the ARTG must provide a certificate to the Australian Therapeutic Goods Administration (TGA) in relation to certain patent matters. Therapeutic goods are entered in the ARTG as either listed goods or requested goods, depending on their ingredients and intended purpose.

The listing or registration applicant must certify to the effect that:

- either acting in good faith, it believes on reasonable grounds that it is not marketing, and does not propose to market, the therapeutic goods in a manner or in circumstances that would infringe a valid claim of a patent that has been granted in relation to the therapeutic goods (a Section 26B(1)(a) certificate); or
- a patent has been granted in relation to the therapeutic goods and the listing or registration applicant proposes to market the therapeutic goods before the end of the term of the patent and that the applicant has given to the patentee notice of the application for registration or listing of the therapeutic goods concerned (a Section 26B(1)(b) certificate).

A present offer to supply is almost certainly likely to infringe a relevant patent. However, an offer to supply only after patent expiry will not – see *Gerber v Lectra* [1995] RPC 383, at 411.

4.6 Summaries of product characteristics (SmPCs)

Statements in a product information document required in respect of all therapeutic products will be relevant to considerations of infringement.

5. Patent enforcement

5.1 Obtaining information on the infringer and the infringement

The usual way in which a patentee will first become aware of a potential launch of a generic pharmaceutical product which may infringe one or more of the patentee's patents is when the generic product is registered on the ARTG. The ARTG is a public document available electronically at the website of the TGA.

The information on the ARTG will include information as to the product sponsor and the identity of the compound concerned.

It is likely that the patentee will require more information about the generic compound and about the planned date of launch into the Australian marketplace. Commonly, the patentee will seek that information by letter from the generic applicant. If the generic applicant does not provide that information voluntarily, then the patentee can apply to the Federal Court of Australia, pursuant to Order 15A, Rule 6 of the Federal Court Rules, seeking discovery from the prospective respondent (infringer). Where the patentee has a reasonable cause to believe that it would be entitled to relief for patent infringement from the respondent and after making all reasonable enquiries it has been unable to obtain sufficient information to enable it to make a decision as to whether to commence an infringement proceeding or not, a discovery order can be sought against the prospective respondent requiring it to provide any document relating to the question whether the patentee has the right to obtain relief for patent infringement. Usually documents relating to the generic compound and its manufacture (eg, extracts from the relevant Drug Master File) will be sought so as to arm the patentee with the information necessary for it to make a decision as to whether or not to commence patent infringement proceedings.

5.2 Interim relief

In Australia, preliminary injunctions are commonly referred to as interlocutory injunctions.

The Australian courts apply the traditional analysis to the grant of interlocutory injunctions, namely:

- Is there a serious question to be tried such that, if the evidence remains the same at trial, the patentee would be entitled to relief?
- Will the patentee suffer irreparable harm for which damages will not be adequate compensation unless an injunction is granted?
- Does the balance of justice favour the grant of an injunction?

Whether or not there is a serious question to be tried depends on the nature of the rights which the patentee asserts. In the case of a pharmaceutical patents it will be incumbent upon the patentee to set out in the form of an affidavit from an expert witness why it is said that the alleged infringer exploits or intends to exploit the relevant claims of the patent. The onus is on the patentee to establish an arguable case of infringement. Similarly, if invalidity is raised by the alleged infringer as a cross-claim, then the onus is on the alleged infringer to establish that the case for invalidity is a triable issue. However, unless the issue of invalidity is so strong as to qualify the patentee's serious question on infringement, the court will go on to consider the adequacy of damages and the balance of justice. It is very rare in Australia that the decision of a court as to whether or not to grant an interlocutory injunction is determined on the basis of whether or not there is a serious question to be tried. In most circumstances that requirement will have been satisfied.

The second issue which the court will consider in determining whether or not to grant an interlocutory injunction is the adequacy of damages as a remedy. In the case of pharmaceuticals in Australia, this issue is complicated by reason of the Pharmaceutical Benefits Scheme (PBS) and price disclosure.

The listing on the PBS of a new generic drug will in most cases trigger an automatic reduction in the price of the drug and/or price disclosure requirements which are likely to have similar effect on the price of the drug. In the case of interlocutory injunction applications, therefore, it is usual for the brand company to argue that the entry of a generic product will automatically lead to immediate and ongoing price reductions for which the brand company would never be compensated. There is, for example, no precedent for a readjustment of the PBS price to the brand name company if it is later shown that an injunction ought to have been granted. Some generic pharmaceutical companies have attempted to avoid the consequence of this argument by providing an undertaking to the brand company that it will compensate the company for any loss which it might suffer because of the triggered PBS price reductions; however, such a strategy is heavily risk laden and the undertaking may well expose the generic company to a potentially huge compensation claim.

There are other issues as to irreparable harm which an innovator or brand company will often raise, such as the 'floodgates' argument which is to the effect that if one generic entrant comes into the marketplace then this may encourage numerous others to do so thereby flooding the market and making it impossible for the brand company ever to regain its market share. The brand companies also often argue that other more subtle forms of loss may be occasioned if an interlocutory injunction is not granted, which are not easily compensable by any damages award – for example, changes to doctors' prescribing practices.

In considering whether the balance of justice favours the grant of an interlocutory injunction, the Federal Court gives considerable emphasis to the *status quo*. In most instances the *status quo* will favour the patentee as it is the brand company that will have been in the market for some time and the generic entrant will be looking to break into that market.

It has recently been observed that the obligation on a generic pharmaceutical company to 'clear the path' of patents before launching its product is not a universal principle or rule of law. Further, any delay on the part of the patentee in bringing an application for an interlocutory injunction will tend to favour the refusal of the application.

5.3   Springboarding/post-patent expiry injunctions
The is no reported example of a 'springboard' or post-expiry injunction being granted in Australia to prevent an infringer from obtaining a post-expiry advantage

from conduct carried out prior to expiry. However, nor is there any express prohibition on such an injunction being granted in an appropriate case.

5.4 Unjustified threats

In Australia declarations, injunctions and damages are available for an unjustified threat of patent infringement (Patents Act 1990, Section 128).

The court may grant relief unless the respondent (patentee) satisfies the court that the acts about which the threats were made infringed a claim (or would infringe a claim in a published specification, once granted) that was not shown to be invalid.[27]

The mere notification of the existence of a patent, or an application for a patent, does not constitute a threat.[28] However, a veiled threat is still a threat. The question is whether or not the communication, in the circumstances in which it was made, would convey to a reasonable man that there was an intention to bring infringement proceedings.

Once a threat has been established, it is *prima facie* unjustifiable unless the person making the threat establishes that it was justified because the relevant conduct infringes or would infringe the patent.[29] However, the relief the court will grant an applicant is discretionary.[30]

5.5 Remedies

The primary relief granted for infringement of a patent is an injunction against the infringing party and either damages or an account of profits.[31] The judge may provide for an additional award of damages where the circumstances call for it.[32] Delivery-up or destruction of infringing articles can also be ordered.

(a) Injunction

When infringement is confirmed by the court, the key form of relief that may be supplied is the grant of a final injunction restraining the respondent from continuing its infringement.[33]

(b) Damages

Damages are assessed with the intention of placing the patentee in the position it would have been had the infringement not occurred. Subject to Section 122(1A), as to which see above, damages are compensatory only and cannot be punitive; they are designed to repay the applicant for the loss suffered due to the respondent's wrongful acts.

---

27 See Section 129 of the Act.
28 See Section 130 of the Act.
29 Crennan J in *JMVB Enterprises Pty Ltd v Camoflag Pty Ltd* (2005) 67 IPR 68 at [210].
30 See the discussion by Dowsett J in *Occupational and Medical Innovations v Retractable Technologies* [2007] FCA 1364.
31 Patents Act 1990, Section 122(1).
32 Patents Act 1990, Section 122(1A).
33 *Ibid.*

(c)   *Account of profits*
As an alternative to damages, a patentee may elect to pursue against the respondent an account of the profits the infringer has made as a result of its infringement.

(d)   *Delivery-up or destruction*
Where the respondent still has in its possession infringing goods, an order for either the delivery-up of these goods to the patentee, or their destruction on oath may be made. These orders are also used at the discretion of the judge and can be combined with the other forms of relief.

6.   Compulsory licensing
Pursuant to Section 133(1) of the Patents Act, a person may apply to a prescribed court for a compulsory licence to work a patent after a period of three years has elapsed since the patent was granted. The court may make the order if it is satisfied that:

- the 'reasonable requirements of the public' with respect to the patented invention have not been satisfied;
- the patent holder has not given a satisfactory reason for failing to exploit the patent; and
- the applicant has attempted for a reasonable period to obtain a licence on reasonable terms and conditions, but without success.

According to Section 135, the "reasonable requirements of the public with respect to a patented invention are to be taken not to have been satisfied if":

- an existing trade or industry in Australia, or the establishment of a new trade or industry in Australia, is unfairly prejudiced, or the demand in Australia for the patented product, or for a product resulting from the patented process, is not reasonably met, because of the patentee's failure:
  - to manufacture the patented product to an adequate extent, and supply it on reasonable terms; or
  - to manufacture, to an adequate extent, a part of the patented product that is necessary for the efficient working of the product, and supply the part on reasonable terms; or
  - to carry on the patented process to a reasonable extent; or
  - to grant licences on reasonable terms; or
- a trade or industry in Australia is unfairly prejudiced by the conditions attached by the patentee (whether before or after the commencing day) to the purchase, hire or use of the patented product, or the use or working of the patented process; or
- the patented invention is not being worked in Australia on a commercial scale, but is capable of being worked in Australia.

A compulsory licence must not grant the exclusive right to work the patented invention. Therefore, the patent holder is entitled to be paid for use of the patent. Moreover, a compulsory licence may be revoked where the circumstances that justified its grant cease to exist and are unlikely to recur, and the legitimate interests of the licensee are not likely to be adversely affected by the revocation.

A compulsory licence of a patent may also be granted by a court if the patentee has engaged in conduct, in connection with the patent which contravenes the Australian Competition and Consumer Law.

7. Ownership, inventors and compensation

In Australia, a patent may only validly be granted to a person who is the inventor or a person who derives entitlement to the invention from the inventor (Section 15). Although the Patents Act 1990 (Cth) does not define the term 'inventor', it is the person who 'performed the intellectual and practical work involved in the development of the invention' (see *Stack v Davies Shephard* (2001) 108 FCR 422 at 428).

Pursuant to Section 16 of the Act, co-ownership of a patent is possible and can result from joint inventorship. Section 31 of the Act provides that two or more persons may make a joint application for a patent. Where there are two or more patentees, each of them is entitled to an equal undivided share in the patent (Section 16(1)(a)) and each is entitled to work the patent without accounting to the others (Section 16(1)(b)). However, none of them may grant a licence or assign an interest in the patent to another without the consent of all patentees (Section 16(1)(c)).

The Australian Full Federal Court in *Polwood Pty Ltd v Foxworth Pty Ltd* considered the issue of joint authorship and held at paragraphs [33] and [34] that:

> The entitlement to the grant of a patent as the inventor is not determined by quantitative contribution. The role of joint inventors does not have to have been equal; it is qualitative rather than quantitative. It may involve joint contribution or independent contributions. The issue is whether the contribution was to the invention. What constitutes the invention can be determined from the particular patent specification which includes the claims … One criterion for inventorship may be to determine whether the person's contribution had a material effect on the final invention.

In circumstances where the inventor is an employee and makes the invention in the course of his ordinary duties as an employee, the employer is usually the 'person entitled to have the patent assigned to the person' (see Section 15(1)(b)) by the inventor and therefore the person entitled to the grant of the patent. On this basis, the contractual relationship between the employer and employee plays a key role in determining the ownership of inventions and the right to seek patents.

7.1 Compensation

There is no provision in the Australian Patents Act akin to the statutory compensation procedure provided for employees who have made inventions such as those in the UK Patents Act 1977.

8. Branding and designs

8.1 Registrability issues specific to pharmaceuticals

In the 1950s the World Health Organization began a programme of identifying each pharmaceutical substance by a unique, universally recognisable name to be known as an International Non-proprietary Name (INN). The INN list, which is regularly updated, is included on the electronic database operated by IP Australia under the heading 'Search for Pharmaceutical Names'. When examining applications for registration of trade marks in class 5 covering pharmaceutical preparations and substances, veterinary substances, or pesticides, trade mark examiners in Australia check the trade mark against the INN list using the Search for Pharmaceutical Names and the list of INN stems. If a trade mark for a pharmaceutical preparation or substance or a veterinary substance or a pesticide contains or consists of an INN or an INN stem, this would provide a ground for rejection of the application.

Trade marks in Australia have a limited life span if they are not being used. If a trade mark has not been used continuously and in good faith for three years, an application can be made to have it removed from the trade marks register for non-use. This can be problematic in the pharmaceuticals sector where trade marks are often filed early and well before regulatory approval permitting product launch. To address that issue, pharmaceutical companies often seek trade mark registration for goods under the broad heading of 'pharmaceutical preparations' so as to capture all possible indications of the substance which is the subject of the trade mark.

Branding and trade marks are important to the pharmaceuticals sector, notwithstanding that direct-to-consumer advertising of prescription pharmaceuticals in Australia is not permitted.

8.2 Enforcement

A registered trade mark allows the owner to prevent the use of that mark as a trade mark by a third party and to obtain relief for any unauthorised use. To establish that trade mark infringement has occurred it is necessary to establish that:

- the infringing mark was used by the alleged infringer as a trade mark to indicate a connection between the goods or services and the trader or service provider;
- the infringing mark is either substantially identical or deceptively similar to the registered trade mark; and
- the goods or services in relation to which the infringing mark is used are the same or closely related to the goods or services for which the trade mark is registered.

8.3 Use as a trade mark

Section 120 of the Australian Trade Mark Act specifies that, in order for infringement to be established, the alleged infringer must use the sign as a trade mark.

There is use as a trade mark where the alleged infringer uses the sign to distinguish its goods (or services) from those of others – that is, as a badge of origin. Use of a mark in a descriptive way will not amount to use of a trade mark sufficient to ground an infringement allegation.

An illustration of the concept of 'use as a trade mark' occurs in the so-called Caplets case.[34] There the word 'caplets' was used by the defendant, Johnson &

---

34   *Johnson & Johnson Australia Pty Ltd v Sterling Pharmaceutical Pty Ltd* (1991) 30 FCR 326 at 350 to 351.

Johnson, in phrases such as 'Tylenol Paracetamol Caplets', 'each caplet contains paracetamol 500mg' and '24 Caplets'. The Full Court, overruling the decision of the trial judge, found that Johnson & Johnson had used the word solely to indicate the shape and dosage. This decision emphasises the significance of the context of the defendant's use. Gummow J said:

> It is undoubtedly true, as counsel for the respondent [plaintiff] urged upon us, that a term such as 'caplets' might readily be susceptible of use on packaging in a trade mark sense so as to indicate a particular form or method of dosage emanating from Johnson & Johnson whilst 'Tylenol' was used concurrently and to indicate a different connection in the course of trade, namely the trade origin not of the dosage form but of the active ingredient, paracetamol 500mg.

Nonetheless, in the setting of the package, Gummow J did not consider that the word 'caplet' appeared as a mark for distinguishing the product from other pain-killing products in the course of trade, with the consequence that there was no use of the word in a trade mark sense and therefore no infringement.

8.4 Substantially identical

A sign which is identical to the registered trade mark is substantially identical to it. A side-by-side comparison of the registered mark and the alleged infringing mark is required.

8.5 Deceptive similarity

The question of deceptive similarity is not to be judged by a side-by-side comparison. Instead, what is involved is a comparison, on the one hand, of the impression based on recollection of an applicant's mark that persons of ordinary intelligence would have and the impression such persons would get from the respondent's mark.[35] In that regard it is important to consider the 'idea of the mark'.

In assessing deceptive similarity, questions of aural impression may be important. The risk of deception must be tangible, but it is enough if an ordinary person entertains a reasonable doubt.

Allowance must be made for imperfect recollections in considering whether a mark so nearly resembles another mark that it is likely to cause confusion or deception.

That principle extends even to marks which are not just invented words. If a registered trade mark includes words which can be regarded as an essential feature of the mark, another mark that incorporates those words may well infringe the registered trade mark.

8.6 Defences to infringement

A number of defences exist to an action for trade mark infringement, the most relevant being:

- use of a sign in good faith to indicate the kind, quality, quantity, intended purpose, value, geographic origin, or some other characteristics of the goods or services;

---

35   *Ibid.*

- use of sign in good faith to indicate the intended purpose of goods/services;
- use of a trade mark for the purposes of comparative advertising; and
- where a court finds that the alleged infringer would obtain registration of the alleged infringing trade mark if the person alleged to have infringed were to apply for it.

8.7 Unfair competition

Although there is no tort of unfair competition in Australia, Section 18 of Schedule 2 of the Competition and Consumer Act 2010 comes close to providing an equivalent cause of action. Section 18 states that a corporation shall not, in trade or commerce, engage in conduct that is misleading or deceptive or is likely to mislead or deceive. This provision not only protects consumers from misleading and deceptive conduct but has also become a valuable weapon for business competitors.

There is an overlap between causes of action arising under Section 18 of the above-mentioned Act and the tort of passing off, but each is distinct and not necessarily coextensive. Section 18 gives wider protection than passing off.

Some examples of the operation of Section 18 are as follows:

- *Janssen Pharmaceutical Pty Ltd v Pfizer Pty Ltd:*[36] the applicant succeeded in establishing that the respondent's television advertisement for its drug treating worms in humans was misleading because it stated that the drug treated three types of worms, two of which were of little significance, implying that they were all significant. The implication was that the respondent's drug was more effective than the applicant's.
- *Sterling Winthrop Pty Ltd v Boots Company (Australia) Pty Ltd:*[37] an interlocutory injunction was granted to prevent publication of advertisements for a pain reliever in trade magazines, because the comparisons used in the advertisement omitted relevant information necessary to make the advertisement fair.
- *Bristol-Myers Squibb Australia Pty Ltd v Astra Pharmaceuticals Pty Ltd:*[38] the court granted an interlocutory injunction restraining a pharmaceutical company making certain advertising claims about a new pharmaceutical product that was compared with the applicant's product.
- *AstraZeneca Pty Ltd v GlaxoSmithKline Australia Pty Ltd:*[39] this case involved a flyer for a new drug sent to general practitioners, which was alleged to be misleading and deceptive. The court determined that the relevant group in relation to whom the flyers were to be tested were ordinary or reasonable members of the community of general practitioners throughout Australia.
- *Schwabe Pharma (Aust) Pty Ltd v AusPharm.Net.Au Pty Ltd:*[40] the applicant succeeded in obtaining an interlocutory injunction restraining the respondent making certain allegedly misleading and deceptive claims about the therapeutic effect of ginkgo in the relief of tinnitus.

---

36   (1985) 6 IPR 227.
37   (1995) 32 IPR 361.
38   (1999) 45 IPR 144.
39   (2006) ATPR 42-106.
40   [2006] FCA 868.

9. Protecting valuable information

Confidential information and trade secrets are protected in Australia under common law and equity via an action for breach of confidence. An action for breach of confidence in relation to a trade secret arises where a person who has a duty of confidence not to disclose or utilise the information provided discloses or uses that information for an unauthorised purpose. To obtain relief for a breach of confidence the owner of the trade secret must establish the following (*Coco v AN Clarke (Engineers) Ltd* [1969] RPC 41 at 47 to 48):

- The information had the necessary quality of confidence. That is, the information had been kept secret and not published or placed in the public domain.
- The person receiving the information must know that it is confidential or likely to be confidential (unless it was published or supplied in breach of a pre-existing obligation of confidence) (*O Mustad & Son v Allcock & Co Ltd & Dosen* [1963] 3 All ER 416).
- The information must have commercial importance or value.
- The information must go beyond mere employee 'know-how'.
- There has been actual or threatened use of the information to the detriment of the owner.

Reliance on such an action to protect bioscience inventions is rare given that the primary means of protecting innovation in this sector is by means of patents. However, surrounding technical know-how and information is often disclosed under conditions of confidence. Trade secrets may be a preferred option for the protection of an invention if the nature of the invention is not disclosed or capable of being reverse-engineered from a product in the marketplace which embodies the invention.

9.1 Copyright, product information leaflets and packaging

Recent amendments to the Australian Copyright Act 1968 (Cth) by the Therapeutic Goods Legislation Amendment (Copyright) Act 2011 now mean that a Product Information (PI) document issued for a generic version of a registered medicine can be substantially identical to the already approved PI for the medicine. This is intended to prevent claims of copyright infringement based on similarities between the brand company's PI documents and that of the generic.

New sub-section 44BA(1) of the Copyright Act 1968 (Cth) offers relief to generic pharmaceutical companies by broadly providing that any act done "in respect of PI under the Therapeutic Goods Act 1989 (Cth) relating to an application for the registration of a medicine for which draft PI must be submitted, or an application to vary approved PI, is not an infringement of any copyright subsisting in the approved PI for the medicine".

Further, new sub-sections 44BA(2) and (3) exempt from infringement all those involved in making approved PI documents publicly available for safety and efficacy purposes in respect of a medicine.

10. Counterfeiting

The Therapeutic Goods Administration (TGA), with cooperation from the Australian state and territory governments, closely monitors the supply chain of therapeutic goods and devices in Australia to prevent counterfeit medicines and medical devices from entering the market.

Pursuant to Section 42E of the Therapeutic Goods Act, it is an offence intentionally to manufacture, supply, import or export counterfeit therapeutic goods. Those found guilty face a maximum term of imprisonment of up to seven years and a fine of up to A$220,000. There are also alternative civil penalties available in respect of counterfeiting conduct.

If counterfeit medical products are detected, the TGA regulatory compliance unit may investigate the allegation with a view to initiating legal action against the importer, supplier or manufacturer of the products in either criminal or civil law courts. If the counterfeiting activity is pursued civilly, then civil penalties apply rather than the criminal penalty provided for.

Further, the Department of Health has the power to notify the Australian Customs Service of an import or an export of counterfeit therapeutic goods, in which event the goods are deemed forfeit to the Crown as prohibited imports/exports.

11. Collaborative models – 'open innovation'

Assuming that the concept of 'open innovation' encompasses the use of external and internal ideas to develop technology, this concept is not new in Australia, particularly in the pharmaceutical industry. The pharmaceutical and life sciences sector generally recognises the need for collaboration to develop products and that the development of products solely in-house is becoming rare. In Australia, we are continuing to see more pharmaceutical companies (including global companies) in-licensing products from biotechnology companies.

Such collaborative or 'open innovation' models pose similar issues to any other licensing models, including the need for clarity on IP ownership (both background and foreground IP), methods of protection of new IP (including by the use of the patent system or by trade secrets), commercialisation rights and obligations, access to commercialisation revenue for the research partners, and decisions on whether or not to continue with the research. None of these issues are peculiar to Australia.

Australia continues to utilise publicly funded collaborative models, including through the Co-operative Research Centres (CRC) programme and other joint-venture models, many of which involve research in the area of human health. Many of these collaborative research models are by their nature examples of open innovation.

12. Hot topics

The Intellectual Property Laws Amendment (Raising the Bar) Bill 2011 has been introduced into the Australian parliament. As the name suggests, the Bill will see a number of amendments to Australia's IP laws, and in particular to the Patents Act, which are intended to raise the quality of granted patents.

The Bill seeks to amend the Patents Act to address four key areas of patentability:

- to raise the standard set for inventive step to that which is more consistent with Australia's major trading partners;
- to strengthen the utility requirement to prevent the grant of patents for speculative inventions;
- to raise the standard set of disclosure of an invention by introducing a requirement of sufficient disclosure across the full scope of each claim; and
- to increase the grounds of invalidity which can be considered at the time of grant to be consistent with those that can be considered by a court, and to apply a consistent standard of proof so that the Patents Office is not obliged to grant patents which would not pass scrutiny in a court challenge.

There has been much discussion surrounding the bill, which is likely to be passed into law in late 2011 or early 2012. The new laws will apply to pending patent applications unless a request for examination has been made as at the commencement date. The Bill will not have retrospective effect.

Canada

Greg Beach
Neil Belmore
Afif Hamid
Lindsay Neidrauer
Marian Wolanski
Belmore Neidrauer LLP
Andrew McIntosh
Bereskin & Parr
Robert Shapiro
Lawyer

1. Small molecules

Small molecule compounds are litigated in Canada in traditional patent infringement actions, and under a pharmaceutical regulatory framework in which the court is asked to determine whether a generic manufacturer is prohibited from receiving authorisation to market the drug. Canadian law regarding small molecules in both infringement actions and in the regulatory context is discussed below.

1.1 Product and process claims

The Canadian pharmaceutical regulatory regime allows certain patent claims to be listed against a drug. Only patent claims that are listed against a drug need be addressed by a generic challenger, which must prove allegations of non-infringement or invalidity in order to gain regulatory approval to market the drug. The following claims are eligible for listing: a claim for the medicinal ingredient; a claim for the formulation that contains the medicinal ingredient; a claim for the dosage form; and a claim for the use of the medicinal ingredient.[1] Claims that do not fall within these categories are not eligible to be listed and thus would not have to be addressed by a generic manufacturer seeking marketing approval. Product claims fall within the definition of a 'claim for a medicinal ingredient' and are covered by the regulations. Process claims do not fall within any of the four categories, so are not covered by the regulations (which are focused on approval of drug products rather than manufacturing processes).

Outside the pharmaceutical regulatory regime, both products and processes are within the definition of 'invention' in the Patent Act.[2] A patent for a process must describe the process clearly and concisely in exact terms, and must explain the necessary sequence (if any) of the various steps so as to distinguish the invention from other inventions.[3] In an infringement action for a patented process for obtaining a new product, the product is considered to have been produced by the patented process.[4]

---

1   Patented Medicines (Notice of Compliance) Regulations, SOR/93-133, Sections 2, 4.
2   Patent Act, RSC 1985, c P-4, Section 2.
3   *Ibid.*, Sections 27(3)(b) and (d).
4   *Ibid.*, Section 55.1.

1.2   Scope of protection of claims and Markush formulae

Patent claims (generally, including claims for small molecules) are to be construed on the basis of a purposive construction, identifying essential and inessential elements by taking into account the state of the art and the inventor's intent as expressed or inferred from the claims.[5]

Markush claims cover selected members of a genus rather than all members of the genus, so as to exclude inoperative members. Markush groupings are considered to cover one invention when all members of the group have a common basic structure and/or when a common property or activity is present. Where a common property or activity is present, all members are expected to behave in the same way in the context of the claimed invention.[6] The group of subject matter could be claimed in separate applications, but may be claimed in a single application because it incorporates a single inventive concept.[7]

The Federal Court has held that the key to a Markush claim is to claim a homemade genus, all the members of which can be used interchangeably. In one case, the Markush claim claimed a number of solvents, selected from the group consisting of "acetone… and methanol-ethanol". The Court held that the claim was a claim to a group of solvents, rather than each solvent in the alternative.[8]

1.3   Metabolites

In the pharmaceutical regulatory regime, patents claiming only metabolites cannot be listed against a drug product. In other words, such patents are not an obstacle to a generic drug manufacturer seeking marketing approval for an innovative drug.[9,10] This aside, metabolites are patentable in Canada.

2.   Second-generation inventions

In addition to protecting a pharmaceutical compound or device, there are other areas in Canadian patent law where a secondary invention could provide additional patent protection. These include obtaining patents on combinations, an enantiomer (if one enantiomer is superior to the other), a selection from a group of compounds, a use of the invention, a method of treatment, claims to formulations and physical forms (such as crystals and salts), and, finally, using reach-through claims to claim an unknown product made by a particular method.

2.1   Combinations

Canadian patent law differentiates between a new combination (which is patentable) and an aggregation of elements (which is not patentable). A new combination exists when the elements of the invention cooperate and interact together to create a

---

5   *Free World Trust v Électro Santé Inc* 2000 SCC 66 at paragraph 31.
6   *Manual of Patent Office Practice*, Canadian Intellectual Property Office, Section 14.03.02.
7   Ibid., Section 14.03.
8   *Abbott Laboratories v The Minister of Health* 2005 FC 1095 at paragraphs 26 to 27.
9   *Merck Frosst Canada v The Minister of Health* 2001 FCA 136 at paragraphs 5 to 9.
10   See also "Regulatory Impact Analysis Statement for October 2006 Amendments to the Patented Medicines (Notice of Compliance) Regulations, SOR/2006-242", page 1517.

unitary result.[11] In contrast, an aggregation is a sum of known parts without any synergy between the parts.[12]

To determine the patentability, a court will look at the functionality of the invention,[13] and whether the patent discloses the aggregation.[14] Once it is determined that the invention is a combination, and therefore patentable, the next step is to consider whether that combination is obvious and therefore not patentable on that basis.

2.2   Enantiomers

Enantiomer patents are valid so long as the state of the art did not teach the relevant person to pursue enantiomers. For example, in two cases that considered enantiomer patents with priority dates of 1985 and 1987, an enantiomer patent was valid when the degree of difference between the enantiomers was unknown,[15] or when there was no disclosure in the prior art that one isomer had beneficial properties over the other.[16]

A court considering the obviousness of selecting an enantiomer in 1991 held that it was known that the pharmacokinetic properties of a racemate could well differ from those of its racemate and would have been important to evaluate those properties.[17] The state of the art had changed, such that by 1991 it was obvious to pursue enantiomers.[18]

2.3   Selection inventions

Following the UK jurisprudence,[19] the Supreme Court of Canada upheld the validity of selection patents, so long as there is (i) a substantial advantage to be secured; (ii) the whole of the selected members possess the advantage in question; and (iii) the selection must be in respect of a quality of a special character peculiar to the selected group.[20]

A selection patent cannot be challenged as an 'invalid selection' *per se* but can be attacked on the basis of any ground set out in the Patent Act[21] (eg, lack of novelty or lack of utility).[22]

2.4   Methods of use and secondary indications

The Supreme Court of Canada has held that claims directed to a new and unobvious use of a known compound are allowable, as long as that new use is included in the

---

11   *Whirlpool Corp v Camco Inc* (1997) 76 CPR (3d) 150 at 180 to 181 (FCTD); aff'd (1999) 85 CPR (3d) 129 (FCA); aff'd (2000) 9 CPR (4th) 129 (SCC).
12   *Hershkovitz v Tyco Safety Products Canada Ltd* 2009 FC 256 at 148; aff'd 2010 FCA 190.
13   *Ibid.*
14   *Domtar Ltd v McMillan Bloedel Packaging Ltd* (1977) 33 CPR (2d) 182 at 191 (FCTD); affirmed (1978) 41 CPR (2d) 182 (FCA).
15   *Janssen-Ortho Inc v Novopharm Ltd* 2006 FC 1234 at paragraphs 114 to 115; aff'd 2007 FCA 217; leave to appeal to SCC dismissed 2007 CanLII 66767 (SCC).
16   *Apotex Inc v Sanofi-Synthelabo Canada Inc* 2008 SCC 61 at paragraphs 74 to 93.
17   *Novo Nordisk Canada Inc v Cobalt Pharmaceuticals Inc* 2010 FC 746 at paragraphs 318 to 323.
18   *Apotex v Sanofi-Synthelabo*, above note 16, at paragraph 90.
19   *In re I G Farbenindustrie A G's Patents* (1930) 47 RPC 289 at 303 (Ch D).
20   *Apotex v Sanofi-Synthelabo*, above note 16, at paragraph 10.
21   Patent Act, above note 2, Sections 2, 27, 28, 53.
22   *Eli Lilly Canada Inc v Novopharm Limited* 2010 FCA 197 at paragraph 39.

claims.[23] Claims to a new property or advantage of a known article or method are not patentable.[24] Claims must cover a practical solution to a practical problem – something more than a mere discovery.[25]

2.5 Methods of treatment

A patent cannot claim a method that provides a practical therapeutic benefit, such as cures, prevents or ameliorates an ailment or pathological condition, or treats a physical abnormality or deformity such as by physiotherapy or surgery.[26] In those cases, the claims can be drafted in Swiss-style *use* claims.[27]

2.6 Formulations and physical forms

Placing a known compound in a known formulation is novel if there are numerous choices with respect to the delivery system, excipients and processing, such that it is not self-evident that the formulation will work.[28] Claims to crystal forms are patentable in Canada, and the validity of crystal form patents has been upheld by the court.[29] In addition, claims to salt forms of known compounds are patentable and the court has upheld the validity of salt form patents.[30]

2.7 Reach-through claims

A reach-through claim is a claim that defines a method and then claims the products that will be discovered using that method, even though those products have not yet been identified (typically early-stage discoveries).[31] There has not been any jurisprudence in Canada on the validity of reach-through claims. However, the decision of the United States Court of Appeals for the Federal Circuit, which invalidated the claims to the unknown products on the basis that the claims failed adequately to describe the invention, would be a useful starting point in determining the patentability of these claims in Canada.[32] Therefore, it is expected that these claims would be susceptible to attack on the grounds of lack of utility, claims broader than the invention made or disclosed, and insufficient disclosure.[33]

3. DNA, biologicals and personalised medicine

3.1 Discoveries

Discoveries must fit the definition of 'invention' in the Patent Act,[34] namely "any

---

23   *Shell Oil v Commissioner of Patents* (1982) 67 CPR (2d) 1 at 11 to 13 (SCC); *Manual of Patent Office Practice*, above note 6, Section 11.10.02.
24   *Calgon Carbon Corporation v North Bay, City* 2005 FCA 410 at paragraph 16.
25   *Ibid.*, at paragraphs 16 to 19.
26   *Tennessee Eastman Co et al. v Commissioner of Patents* [1974] SCR 111 at 118 to 119; *Manual of Patent Office Practice*, above note 6, section 12.05.02.
27   *Merck & Co, Inc v Pharmascience Inc* 2010 FC 510 at paragraphs 109 to 110, 114.
28   *Purdue Pharma v Pharmascience Inc* 2009 FC 726 at paragraphs 92 to 100.
29   *Abbott Laboratories v Canada (Minister of Health)* 2004 FC 1349 at paragraphs 117 to 122, aff'd 2005 FCA 250.
30   *AB Hassle v Apotex Inc* 2003 FCT 771 at paragraphs 82 to 97; aff'd 2004 FCA 369.
31   *Manual of Patent Office Practice*, above note 6, Section 17.07.03.
32   *University of Rochester v GD Searle & Co* 358 F.3d 916 at paragraphs 54 to 56 (Fed Cir 2004).
33   Patent Act, Sections 2 and 27(3).

new and useful art, process, machine, manufacture or composition of matter, or any new and useful improvement" thereof,[35] to be patentable. An invention cannot be a disembodied idea; it must have a method of practical application.[36]

A mere scientific principle or theorem[37] and laws of nature are not patentable.[38] The following discussion highlights some additional limitations on the scope of patentable subject matter in the area of biotechnology and medicine.

(a)   **Gene patents**

According to the Canadian Intellectual Property Office (CIPO), nucleic acid sequences are patentable,[39] provided they have, *inter alia*, utility, such as encoding therapeutic proteins or diagnostic probes. The patentability of genes does not necessarily extend to the entire organism since lower,[40] but not higher,[41] life forms are patentable. In *Harvard*, the Supreme Court of Canada held that a transgenic mouse, genetically modified to increase its susceptibility to cancer, was not patentable subject matter.[42] In a later decision, the Supreme Court held that claims to a plant gene and a modified plant cell were valid even without a proviso that would limit them to the genes and cells only when in an isolated laboratory form.[43]

(b)   **Stem cells**

CIPO's position is that fertilised eggs and totipotent stem cells (stem cells that have the inherent ability to develop into animals) fall within the Supreme Court's definition of higher life forms and do not constitute statutory subject matter.[44] Embryonic, multipotent and pluripotent stem cells, which do not have the inherent ability to develop into an animal, are considered to be lower (patentable) life forms.[45]

(c)   **Organic tissue**

Organic tissue is generally not considered by CIPO to be patentable subject matter.[46]

---

34   Patent Act, above note 2.

35   *Ibid.*, Section 2. A claimed invention must also be non-obvious pursuant to Section 28.3 of the Patent Act.

36   *Shell Oil Co v Canada (Commissioner of Patents)* [1982] 2 SCR 536 at 554. Also see *Riello Canada Inc v Lamber* (1986) 9 CPR (3d) 324 (FCTD) at 335.

37   Patent Act, above note 2, Section 27(8).

38   *Pioneer Hi-Bred Ltd v Canada (Commissioner of Patents)* [1989] 1 SCR 1623, 25 CPR (3d) 257 at 264 to 265 [*Hi-Bred* cited to CPR].

39   *Manual of Patent Office Practice*, above note 6, Section 17.02.04. The *Manual of Patent Office Practice* serves as a guide on the application of the Patent Act and its associated rules.

40   In *Re Application of Abitibi Co* (1982) 62 CPR (2d) 81 at 88 to 89 and 91 (PAB), lower life forms were held to be patentable so long as the organism claimed has not existed previously in nature, has utility beyond being a starting material for further research and its creation involved inventive ingenuity.

41   *Harvard College v Canada (Commissioner of Patents)* [2002] 4 SCR 45, 2002 SCC 76 at paragraph 155. See also *Manual of Patent Office Practice*, above note 6, at Section 17.02.01a, where CIPO defines 'lower' and 'higher' life forms to mean, in general, whether the life form is unicellular (lower) or multicellular (higher).

42   *Ibid.*

43   *Monsanto Canada Inc v Schmeiser* [2004] 1 SCR 902, 2004 SCC 34 at paragraphs 17 to 24.

44   *Manual of Patent Office Practice*, above note 6, section 17.02.01a. See also CIPO, "Office Practice Regarding Fertilized Eggs, Stem Cells, Organs and Tissues" (2006) available at www.cipo.ic.gc.ca/epic/site/cipointernet-internetopic.nsf/en/wr00295e.html (last accessed May 1 2011).

45   *Manual of Patent Office Practice*, ibid., Section 17.02.01a.

46   *Ibid.*, Section 17.02.01b.

Conversely, artificial organ- or tissue-like structures, created by combining cellular and/or inert components, may, in some cases, be considered to be statutory subject matter.[47] However, methods that involve excision of tissue, organ, or tumour samples from the body cannot be claimed.[48]

(d)   Bioinformatics systems
Despite the fact that information and mathematical formulae[49] are not patentable *per se*, even when a computer is used, CIPO has recognised that methods involving computer models can be used in certain inventions, such as in *in silico* screening methods.[50]

(e)   Copyright and sequence information
The Copyright Act[51] prescribes that copyright subsists in "original literary, dramatic, musical and artistic work"[52] and defines 'literary work' to include "tables, computer programs, and compilations of literary works".[53] Whether or not copyright can subsist in biological sequences under the Copyright Act is unknown and has yet to be considered by Canadian courts.

3.2   Disclosure requirement
A patent specification must contain a disclosure of the invention,[54] which must, on its own, be sufficiently detailed so as to enable a person skilled in the art to make and use the claimed invention.[55] For some biotechnological inventions, there are specific disclosure requirements.

3.3   Sequence listing
Where a nucleotide or an amino acid sequence is disclosed, other than a sequence identified as forming a part of the prior art, a 'sequence listing' in electronic form must be included.[56] The sequence listing is a detailed disclosure of the nucleotide and/or amino acid sequence(s) and other available information.[57]

---

47   *Ibid.*
48   *Ibid.*, Section 17.02.03.
49   *Schlumberger Canada Ltd v Commissioner of Patents* (1981) 56 CPR (2d) 204 at 205 to 206 (CA).
50   *Manual of Patent Office Practice*, above note 6, Section 17.02.04.
51   Copyright Act, RSC 1985, c C-42.
52   *Ibid.*, Section 5(1).
53   *Ibid.*, Section 2.
54   Patent Act, above note 2, Section 27(3).
55   *Hi-Bred*, above note 38 at 268; *Consolboard Inc v Macmillan Bloedel (Saskatchewan) Ltd* [1981] 1 SCR 504 at 520.
56   Patent Rules, SOR/96-423, Section 111(1) as amended by SOR/2007-90, Section 24 [Rules]. The 'sequence listing' must be compliant with the 'PCT sequence listing standard', defined in Section 2 of the Rules to mean the Standard for the Presentation of Nucleotide and Amino Acid Sequence Listings in International Patent Applications under the PCT provided for in the Administrative Instructions under the Patent Cooperation Treaty. This standard is available online at www.wipo.int/pct/en/texts/ ai/annex_c.html (last accessed May 1 2011). Also see *Manual of Patent Office Practice*, above note 6, Section 17.04.01 for an overview of the Patent Rules as they relate to sequence listings.
57   Administrative Instructions under the Patent Cooperation Treaty, *ibid.* at Section 2(i).

(a) **Biological materials**

For some inventions, a deposit of biological material with an International Depository Authority will be required to satisfy the disclosure requirement.[58] Where a specification refers to a deposit of biological material, the deposit forms part of the specification,[59] but the fact that a biological deposit has been made does not necessarily mean that an invention has been adequately described.[60]

(b) **Inclusion of examples**

Although there is no absolute requirement in the Patent Act for a patent specification to include examples, for some biotechnological inventions it may not be possible sufficiently to describe what the invention is without exemplary support and the necessity for such support is assessed on a case-by-case basis.[61]

4. Acts of patent infringement

The starting point for liability for patent infringement in Canada is Section 55(1) of the Patent Act.[62] While there is no statutory definition provided, the Supreme Court of Canada has adopted a purposive definition of patent infringement as "any act that interferes with the full enjoyment of the monopoly granted to the patentee".[63] Infringement is linked to the concept of 'use', which Section 42 of the Patent Act defines as one of the exclusive rights granted to the patentee, in addition to making, constructing and selling. When dealing with infringement by use, the relevant inquiry becomes whether the defendant's activity deprives the inventor in whole or in part, directly or indirectly, of full enjoyment of the monopoly conferred by law.[64]

Unlike in other jurisdictions, Canada's Patent Act does not enumerate the various heads of patent infringement. To date, two broad categories of patent infringement have been recognised in the case law. The first is the case of direct patent infringement, where the alleged infringer completes all of the acts of infringement. Secondly, the Federal Court of Appeal has explicitly referred to the following two cases of indirect patent infringement, as exceptions to the rule that selling an article which does not itself infringe the patent does not constitute infringement:

- if the vendor alone or in association with another person, sells all of the components of the invention to a purchaser in order that they be assembled by him; and
- if the vendor, knowingly and for his own ends and benefit induces or procures the purchaser to infringe the patent.[65]

The second scenario is referred to as inducing infringement, and it has the

---

58   *Manual of Patent Office Practice*, above note 6, Section 17.04.02. Sections 103 to 110 of the Rules govern the procedure for depositing biological materials.
59   Patent Act, above note 2, Section 38.1(1).
60   *Hi-Bred*, above note 38 at page 271.
61   *Manual of Patent Office Practice*, above note 6, Section 17.04.03.
62   Patent Act, above note 2, Section 55.
63   *Monsanto*, above note 43, at paragraph 34.
64   *Ibid.*, at paragraph 35.
65   *Valmet Oy v Beloit Canada Ltd* (1988) 20 CPR (3d) 1 at 14 (FCA).

following three elements:

- an act of infringement was completed by the direct infringer;
- the act of infringement was influenced by the inducing party to the point that, without said influence, infringement would not take place; and
- the inducing party must know that its influence would result in the completion of the act of infringement.[66]

Statements made by generic drug companies in their product monographs that bear on the intended use have been held to constitute inducing infringement.[67]

The theory of contributory infringement, where the acts of influence fall short of being the sole cause but nevertheless materially contribute to the infringement, has been rejected by the Federal Court in the past.[68] However, it has recently been argued before the court that contributory infringement is a valid cause of action, as the inventor is deprived in whole or in part, directly or indirectly, of full enjoyment of the monopoly conferred. The issue is currently before the court.[69]

The Patent Act[70] provides an exception to patent infringement, by permitting a generic company to take required steps to be able to submit an application for approval to a governmental authority. However, based on the data protection regulations subsumed in the Food and Drug Regulations,[71] a generic company may not file a drug submission in respect of a new drug for six years, and the minister may not issue a notice of compliance to the generic company for eight years, from the time the notice of compliance was issued to the patentee. Furthermore, a generic company may not advertise a drug for sale before having submitted a drug submission, labels and product brochures and a statement setting out the proposed date on which they will first be used, and the minister has granted a notice of compliance.[72]

There is also an exception for use, manufacture, construction or sale solely for the purpose of experiments related to the subject matter of the patent.[73]

5. Patent enforcement

An action for infringement can be brought in either the Superior Court of the province where infringement is alleged to have occurred, or the Federal Court.[74] In terms of interim relief, a patentee may seek an interlocutory injunction.[75] To succeed, the plaintiff must demonstrate that there is a serious question to be tried, and that irreparable harm will result, after which the court will then assess the balance of convenience.[76] An interlocutory injunction can also be sought in the case of

---

66   *Solvay Pharma Inc v Apotex Inc* 2008 FC 308 at paragraph 136.
67   *AB Hassle v Genpharm Inc* 2003 FC 1443 at paragraph 155, aff'd 2004 FCA 413 at paragraphs 22 to 23; *Abbott Laboratory Ltd v Canada (Minister of Health)* 2006 FC 1411 at paragraph 59, aff'd 2007 FCA 251; *Procter and Gamble Pharmaceuticals Canada v Canada (Minister of Health)* 2002 FCA 290 at paragraph 34.
68   *Faurecia Automotive Seating Ltd v Lear Corporation Canada Ltd* 2004 FC 421 at paragraph 50.
69   Federal Court Files No T-1786-08 and T-368-08.
70   Patent Act, above note 2, Section 55.2(1).
71   Food and Drug Regulations, CRC, c. 870, Section C.08.004.1.
72   *Ibid.* Section C.08.002(1).
73   Patent Act, above note 2, Section 55.2(6).
74   *Ibid.* Section 54.
75   Federal Courts Rules, SOR/98-106, rule 373.
76   *RJR-Macdonald Inc v Canada (AG)* [1994] 1 SCR 311, 54 CPR (3d) 114 at 140 to 141.

springboarding, but it can only be obtained up until the expiry of the patent, if that occurs before the trial date.[77] Other types of interim relief include Mareva injunctions, the right to freeze exigible assets when found within the jurisdiction,[78] and Anton Piller orders, to preserve property that is, or will be, the subject matter of proceeding or as to which a question may arise therein.[79]

In the pharmaceutical drug context, the patentee's rights are governed by the Patented Medicines (Notice of Compliance) Regulations.[80] The patentees can enforce their patent rights by submitting a list of patents to be included on the patent register maintained by the Minister of Health. This must be done after the patentee has submitted a new drug submission,[81] and the patent(s) will not be added to the register until the minister has issued a notice of compliance.[82] Any generic company that wishes to enter the market must then address the patent(s) registered against the drug. The patentee can then bring an application under the Patented Medicines (Notice of Compliance) Regulations, which institutes a statutory stay for the duration of the proceeding, for up to 24 months.[83]

If a patentee makes unjustified threats against another, the offended party has recourse under Section 7(a) of the Trade-marks Act. This provision is invoked when the patentee alleges that another party's wares infringe its patent, and threatens that party's customers with infringement proceedings if they continue to deal with that other party. The representation as to the alleged infringement, which is later found to be false, is actionable under Section 7(a).[84] These situations have been referred to as "patent abuse cases".[85]

Successful patentees, or persons claiming under them, can seek damages for conduct after the grant of the patent,[86] and for reasonable compensation for infringement occurring prior to the issuance of the patents but after publication.[87] No remedy may be awarded for an act of infringement committed more than six years before the commencement of the action for infringement.[88] The plaintiff is entitled to its lost profits. Where the infringer's sales do not damage the plaintiff's sales, the plaintiff can receive a reasonable royalty. As an alternative, a successful plaintiff can elect an accounting of the defendant's profits that exceed the ones that would have been made using the best non-infringing option.[89] Other relief includes permanent injunctions, delivery-up and destruction of infringing goods, punitive damages and an award of costs.

---

77 *China Ceramic Proppant Ltd v Carbo Ceramics Inc* 2004 FCA 283 at paragraph 13.
78 *Aetna Financial Services v Feigelman* [1985] 1 SCR 2 at paragraph 26.
79 Federal Courts Rules, above note 75, rule 377.
80 PMNOC Regulations, above note 1.
81 *Ibid.*, Section 4(1).
82 *Ibid.*, Section 3(7).
83 *Ibid.*, Section 7(1)(e).
84 *S & S Industries Inc v Rowell* [1966] SCR 419, 48 CPR 193; *Riello Canada, Inc v Lambert* (1986) 9 CPR (3d) 324 (FCTD).
85 *Canadian Copyright Licensing Agency v Business Depot Ltd* 2008 FC 737 at paragraphs 31 to 32.
86 Patent Act, above note 2, Section 55(1).
87 *Ibid.*, Section 55(2).
88 *Ibid.*, Section 55.01.
89 *Monsanto*, above note 43, at paragraphs 98 to 105.

6.   Compulsory licensing

Three compulsory licensing regimes are set out in the Patent Act. They deal with abuse of patent rights, international humanitarian use, and government use of patents.

6.1   Abuse of patent rights

The Patent Act sets out a legislative framework for remedies to be granted as a result of the abuse of rights under a patent.[90] A patentee may not hold its patent unworked for the purpose of blocking trade.[91] The Patent Act states that an interested person can, three years after the grant of the patent, apply to the Commissioner of Patents (Commissioner) seeking relief on the basis that there has been an abuse of rights under the patent.[92]

The Federal Court has found that this scheme is based on a positive obligation on the patentee to work its invention after having been granted a monopoly. The failure to do so, either directly or by licence, may amount to an abuse of the exclusive rights under the patent.[93] Abuse can be found where demand for a patented article (including an article made by a patented process) is not being adequately met or if certain forms of prejudice can be made out.[94] Section 65(2) of the Patent Act states that the Commissioner can make a finding of abuse:

[(a) and (b) have been repealed]

(c) if the demand for the patented article in Canada is not adequately being met;

(d) if by refusing to grant a license, the trade or industry of Canada or the trade of any person trading in Canada, or the establishment of any new trade or industry in Canada is prejudiced and it is in the public interest that a license be granted;

(e) if any trade or industry in Canada is unfairly prejudiced by the conditions on which the patented article or process is being purchased, hired, licensed or used; or

(f) if it is shown that the existence of the patent, being a patent covering a process involving the use of materials not covered by the patent or for an invention covering a substance produced by such a process, has been used by the patentee to unfairly prejudice the manufacture, use or sale of any materials in Canada.[95]

Under Section 65(2)(c), the first step for the Commissioner is to determine what the patented article is, and then to determine whether demand for the article is adequately being met.[96] 'Demand' has been interpreted as the demand of the marketplace (ie, the demand of the public at large rather than a single trader).[97] The demand must be an existing rather than a potential, anticipated or future demand, and it does not include a potential demand for a cheaper version of the article being sold.[98] A request for a licence solely for the purpose of manufacturing and exporting from Canada may not be covered by these provisions.[99]

---

90   Patent Act, above note 2, Sections 65 to 71.
91   *Brantford Chemicals Inc v Canada (Commissioner of Patents)* 2006 FC 1341 at paragraphs 68 to 69.
92   Patent Act, above note 2, Section 65(1).
93   *Brantford Chemicals,* above note 91, at paragraph 68.
94   Patent Act, above note 2, Sections 65(2) and 65(5).
95   *Ibid.,* Sections 65(2)(c) to (f).
96   *Brantford Chemicals* above note 91, at paragraph 72.
97   *Ibid.,* at paragraph 86.
98   *Ibid.,* at paragraphs 96 to 98, 103.
99   *Torpharm Inc v Commissioner of Patents* 2004 FC 673 at paragraphs 29 to 30.

Under Section 65(2)(d), the Federal Court has held that three things must be established in order to find abuse of exclusive rights granted under a patent:

- there must be refusal by the patentee to grant a licence on reasonable terms;
- there must be prejudice to a trade or industry in Canada, the trade of any persons trading in Canada, or the establishment of a new trade or industry in Canada; and
- it must be in the public interest that the licence be granted.[100]

When abuse has been found, the Commissioner may grant a licence to the applicant and to its customers on terms deemed to be expedient.[101] The Commissioner is to secure the widest possible use of the invention by one or more licensees, and ensure the patentee derives an advantage from his patent rights.[102] A licensee is entitled to call on the patentee to commence proceedings to prevent infringement of the patent. If the patentee refuses, the licensee may commence proceedings in his own name.[103] The Commissioner may alternatively order that the patent be revoked if the grant of a licence would not rectify the abuse.[104]

6.2 International humanitarian use – Canada's access to medicine regime

The Patent Act also sets out a compulsory licensing framework for providing drugs for humanitarian purposes in developing countries. The government refers to this scheme as Canada's Access to Medicine Regime (CAMR),[105] which came into force in 2005. Its specified goal was to facilitate "access to pharmaceutical products to address public health problems afflicting many developing and least-developed countries, especially those resulting from HIV/AIDS, tuberculosis, malaria and other epidemics".[106] For specified drugs and specified developing countries, a drug manufacturer may apply to the Commissioner for a two-year compulsory licence (renewable once for an additional two-year period) to sell the drug in the developing country.[107] To obtain a compulsory licence if the developing country is on a prescribed list of World Trade Organization members, the drug must either not be patented in the country or the country must grant a compulsory licence for the drug domestically.[108]

A royalty amount may be determined by the Federal Court based on humanitarian and non-commercial factors.[109] If the drug is sold for more than 25% of the average price in Canada, the patentee may request that the Federal Court terminate the licence or order increased royalties.[110]

---

100   *Brantford Chemicals,* above note 91, at paragraph 111.
101   Patent Act, above note 2, Sections 66(1)(a) to (e).
102   *Ibid.,* Section 66(4).
103   *Ibid.,* Section 66(2).
104   *Ibid.,* Section 66(1)(d).
105   Government of Canada, "Canada's Access to Medicines Regime", online: www.camr-rcam.gc.ca/index-eng.php
106   Patent Act, above note 2, Section 21.01.
107   *Ibid.,* Sections 21.04(1), 21.09 and 21.12.
108   *Ibid.,* Section 21.04(3).
109   *Ibid.,* Section 21.08.
110   *Ibid.,* Sections 21.17(1) and 21.17(3).

Only one company has used the CAMR, for a single shipment to a single country.[111] There have been no court decisions on the CAMR.

6.3   Government use

The Patent Act also allows the federal or provincial government to apply to the Commissioner for authorisation to use a patented invention.[112] The federal or provincial government must establish that it has made efforts to obtain a licence on reasonable terms and that a licence has not been granted within a reasonable period. Note that there are exceptions for national emergencies, extreme urgency, or where the authorisation sought is for a public, non-commercial use.[113] The Commissioner shall consider the following principles when determining the terms of use:

- the scope and duration of the use shall be limited to the authorised purpose;
- the use shall be non-exclusive; and
- any use shall be authorised predominantly to supply the Canadian market.[114]

Note there is no demand requirement.

The applicant must pay the patentee an amount considered by the Commissioner to be adequate in the circumstances, who must take into account the value of the authorisation.[115] There have been no court decisions on these provisions of the Patent Act.

7.   Ownership, inventors and compensation

7.1   Inventorship

The Patent Act does not define 'inventor'. To qualify as an inventor, a person must have performed or assisted in the creative act of invention;[116] it is insufficient for him to merely have an idea that "floated through his brain".[117] He must be both the person who first conceives a new idea and the person who sets that conception into a definite and practical shape.[118] While inventors may seek the assistance of others in testing or implementing an invention, those others are not joint inventors unless they contribute to the conception, as opposed to verification, of the invention.[119] Where the invention is a combination, the question is who conceived the combination as a whole, and not who contributed each element to the combination.[120]

---

111   Canada News Wire, "Canada's Access to Medicines Regime Must Be Fixed", online: www.cnw.ca/en/releases/archive/March2011/08/c8711.html.
112   Patent Act, above note 2, Section 19.
113   *Ibid.*, Sections 19.1(1) and 19.1(2).
114   *Ibid.*, Section 19(2).
115   *Ibid.*, Section 19(4).
116   *Apotex Inc v Wellcome Foundation Ltd* (2002) 21 CPR (4th) 499 at paragraph 96 (SCC).
117   *Christiani v Rice* [1930] SCR 443 at 454, aff'd [1931] 4 DLR 273 (PC).
118   *Apotex Inc v Wellcome Foundation Ltd* (2000) 10 CPR (4th) 65 at paragraph 30 (FCA), aff'd (2002), 21 CPR (4th) 499 (SCC); *Weatherford Canada Ltd v Corlac Inc* (2010) 84 CPR (4th) 237 at paragraph 239 (FC).
119   *Apotex v Wellcome Foundation*, above note 116, at paragraph 97 (SCC).
120   *Ibid*, at paragraph 98; *Weatherford*, above note 118, at paragraph 240 (FC).

7.2 Ownership

The general rule is that the inventor is the first owner of an invention, even where it is made during the course of employment.[121] Exceptions exist where there is an express contract to the contrary, or where the employee is employed for the purpose of inventing and the invention is made in the ordinary course of his employment.[122] The court will look to a variety of factors including whether:

- the employee was hired for the purpose of inventing;
- at the time he was hired the employee had previously made inventions;
- the employer had incentive plans encouraging innovation;
- the employee's conduct suggested ownership was held by employer;
- the invention is the product of a problem the employee was instructed to solve;
- the employee sought help from others in the company;
- the employee was dealing with highly confidential information; and
- it was a term of the employee's employment that he could not use the ideas which he developed to his own advantage.[123]

A patent may be owned by more than one entity, and absent an agreement to the contrary, each co-owner is entitled to exploit the invention without consent from, or accounting to, other co-owners.[124] A co-owner may also assign its full interest in a patent, without consent from other co-owners.[125] However, a co-owner may not license its interest in the patent, hire an independent contractor to work the invention, or otherwise dilute a co-owner's interest.[126]

8. Branding and designs

8.1 Trade marks in life sciences

Trade mark protection can in principle extend to non-traditional trade marks such as the colour, shape and size of pills or medical devices such as inhalers.[127] However, the resulting mark and protection will generally be weak.[128] The difficulty in securing protection is establishing that the features sought to be protected distinguish the pills or devices from those of others.[129] This is at least partly because physicians and pharmacists typically differentiate pharmaceutical products on the basis of features other than their colour, shape or size.[130]

---

121   *Comstock Canada et al. v Electec Ltd et al.* (1991) 38 CPR (3d) 29 at pages 51 to 52 (FCTD).
122   *Ibid.*, at page 53.
123   *Ibid.*, at page 53.
124   *Forget v Specialty Tools of Canada Inc* (1995) 62 CPR (3d) 537 at page 14 (BCCA).
125   *Ibid.*, at page 10.
126   *Ibid.*
127   *Glaxo Group Limited v Apotex Inc* 2010 FCA 313.
128   *Novopharm Ltd v Bayer Inc et al.* (1999) 3 CPR (4th) 305 at paragraph 77 (FCTD); aff'd 2000, 9 CPR (4th) 304 (FCA).
129   *Ibid.*, at paragraphs 71 to 79.
130   *Ibid.*, at paragraph 79; *Eli Lilly and Company v Novopharm Ltd* 2006 FC 843 at paragraphs 89 to 94.

8.2 Extending protection using trade marks and designs

A trade mark cannot protect a functional structure or utilitarian feature of a product, regardless of whether it is distinctive.[131] The shape, configuration, pattern or ornament of a pill or device may be protected through industrial design registration, provided that the design was not published more than one year prior to filing the application, is not confoundingly similar to any design already registered, and is not dictated solely by utilitarian function.[132] Industrial design registrations have a term of 10 years.[133] Registration grants the owner the exclusive right to make and sell articles embodying the design, and designs not differing substantially therefrom.[134]

8.3 Comparative advertising

Comparative advertising is permissible but must not contain false or misleading claims, statements, illustrations or representations.[135] In addition to criminal and civil liability under the Competition Act, the Trade-marks Act prohibits false or misleading statements tending to discredit the business, wares or services of a competitor.[136] To be actionable, such statements must have a direct nexus to an intellectual property right.[137]

The Trade-marks Act also prohibits use of a trade mark registered by another person in a manner that is likely to depreciate the goodwill attaching thereto.[138] To succeed in an action, a plaintiff must establish that:

- its registered trade mark was used by the defendant;
- its registered trade mark is sufficiently well known to have goodwill attached to it;
- the trade mark was used in a manner likely to have an effect on that goodwill; and
- the likely effect would be to depreciate the value of its goodwill.[139]

It has often proved difficult to succeed in a claim under this section for comparative advertising of wares, because 'use' in association with wares requires the trade mark to be, at the time of transfer of the wares, marked on the wares themselves or on their packaging in which they are distributed, or in any other manner so associated with the wares that notice of the association is given to the person to whom the property is transferred.[140] The defendant in most comparative advertising cases does not use the plaintiff's trade mark at the time it sells its wares, and the first prong of the test is not met. In contrast, 'use' of a trade mark in association with

---

131 Trade-marks Act, RSC, 1985, c. T-13, Section 13(2); *Kirkbi AG et al. v Ritvik Holdings Inc et al.* 2005 SCC 65 at paragraphs 42 to 46.
132 Industrial Design Act, RSC, 1985, c. I-9, Sections 5.1, 6.
133 *Ibid.*, Section 10; Industrial Design Regulations, SOR/99-460, Section 18.
134 Industrial Design Act, above note 132, Section 11.
135 Competition Act, RSC, 1985, c. C-34, Section 36, 52.
136 *Ibid.* Section 52 and 36; Trade-marks Act, above note 131, Section 7(a).
137 *The Canadian Copyright Licencing Authority v The Business Depot Ltd* (2008) 330 FTR 133 at paragraphs 26 to 33 (FC); *Anheuser-Bush, Inc v Brick Brewing Co Ltd* (October 20 2009) T-1440-09 (FC).
138 Trade-marks Act, above note 131, Section 22.
139 *Veuve Clicquot Ponsardin v Boutiques Cliquot Ltée* 2006 SCC 23 at paragraphs 38 and 46.
140 Trade-marks Act, above note 131, Section 4; *Clairol International Corp et al. v Thomas Supply & Equipment Co Ltd et al.* (1968) 55 CPR 176 at pages 195 to 196 (Ex CR); *Cie Generale des Etablissements Michelin-Michelin & Cie v CAW – Canada et al.* (1996) 71 CPR (3d) 348 at pages 357, 362–365 (FCTD)

services occurs if it is displayed or used in connection with the performance or advertising of those services.[141] Although the distinction has been referred to as 'bizarre',[142] the scope of protection afforded to services is broader than wares.

8.4   Direct-to-consumer advertising

Direct-to-consumer advertising of prescription drugs is restricted in Canada to the brand name, common name, quantity and price of the drug.[143] Two types of advertising are currently permitted:

- 'reminder' ads, which include only a product's brand name and no health claims or other indications as to the use of the drug; and
- 'disease oriented' ads, which do not mention a specific brand but discuss a disease and recommend that consumers seek advice from their physician, without mentioning any specific treatment.

The two types of activities, where independently permissible, may not be combined as part of a larger campaign where a consumer could easily link the two and the linkage would contravene the Food and Drug Regulations.[144]

9.   Protecting valuable information

9.1   Preserving confidentiality

To succeed in an action for breach of confidence, the plaintiff must establish:

- that the information used by the defendant was confidential;
- the plaintiff disclosed the information to the defendant in confidence; and
- the defendant misused the information.[145]

(a)   The information must be confidential

While in certain circumstances information that is partly or even completely publicly available may qualify as confidential information,[146] if the information is generally known to the trade[147] or the information concerns a type of product which is commonly known and widely available,[148] the information will not usually constitute confidential information.[149] While publication to a limited number of persons may not destroy the confidentially of the information,[150] disclosure to a large number of people or to the public probably will.

---

141   Trade-marks Act, above note 131, Section 4.
142   *Eye Masters Ltd v Ross King Holdings Ltd* (1992) 44 CPR (3d) 459 at page 463 (FCTD).
143   Food and Drug Regulations, CRC c 870, Section C.01.044(1).
144   Health Canada Policy Statement – Advertising Campaigns of Branded and Unbranded Messages (November 1 2000).
145   *Coco v AN Clark (Engineers) Ltd* [1969] RPC 41 (Ch D); *Lac Minerals Ltd v International Corona Resources Ltd* [1989] 2 SCR 574.
146   *Schering Chemicals Ltd v Falkman Ltd* [1981] All ER 321 at 338 (CA).
147   *R L Crain Ltd v Ashton* [1950] 1 DLR 601 at 609 (Ont CA).
148   *Cooperheat of Canada Ltd v Slater* (1973) 13 CPR 25 (Ont HCJ).
149   *Saltman Engineering Co v Campbell Engineering Co* [1963] 3 All ER 413 (CA).
150   *Franchi v Franchi* [1967] RPC 149; and *Data General Corp v Digital Computer Controls Inc* 188 USPQ 276 (Del Cir Ct, 1975) at 179, as quoted in R Brait, "The Unauthorized Use of Confidential Information" (1990) 18 *CBLJ* at 353.

(b)   *The information must be given in circumstances imparting a duty of confidence*

Whether circumstances of disclosure impart an obligation of confidence is an objective issue: would the reasonable person believe the information was being imparted to him in confidence, even if the confider does not state as much?[151] Disclosure may be in writing or made orally, but if the latter it must be identified or specific.[152] Courts have placed significant weight on whether it would be industry practice to consider the information as being disclosed in confidence in the circumstances.[153]

(c)   *Unauthorised use*

Before attracting liability, the confidential information must be misused such that the recipient obtains an unfair advantage,[154] such as a springboard to obtain a start over others.[155] Further, when confidential information subsequently becomes public knowledge, the recipient, in order not to breach confidentiality obligations, must obtain the information from public sources before using it.[156]

9.2   Know-how and ex-employees: confidential information and trade secrets

In most situations a trade secret and confidential information are synonymous.[157] However, an ex-employee may not use or disclose a trade secret such as a secret manufacturing process but is able to use 'confidential information' such as customer names, addresses and times of delivery, and prices charged, so long as this information is a product of memory and not documents taken from the former employer.[158] In other words, the former employee is required to keep and not use trade secrets, but may use 'confidential information' – which, had the employee revealed it during his former employment, would have constituted a breach of the duty of good faith.[159]

(a)   *Fiduciary duty*

Where one party has an obligation to act for the benefit of another, and that obligation carries with it a discretionary power to affect the beneficiary's interest, and the beneficiary is vulnerable to the person holding the discretion, the party thus empowered becomes a fiduciary.[160] Equity then supervises the relationship by

---

151   *Coco*, above note 145, at page 48; See also *Tree Savers International v Savoy* (1992) 40 CPR (3d) 173 at 178 (Alta CA).

152   *Terrapin Ltd v Builders Supply Co (Hayes) Ltd* [1967] RPC 375 at 391.

153   *LAC Minerals Ltd v International Corona Resources Ltd* [1989] 2 SCR 574 at pages 659 to 62.

154   *Saltman,* above note 146. Also see *Chevron Standard Ltd v Home Oil Co* [1980] 5 WWR 624 (Alta QB) at 667, in which the Court held that "[t]here must be more than a mere possibility of misuse to settle liability on a party. Suspicion of misuse of confidential information is insufficient – there must be real evidence".

155   *LAC Minerals*, above note 153, at pages 613 to 615.

156   *Coco*, above note 145 at 47.

157   Vaver, "What is a Trade Secret?" in RT Hughes, ed, MG Horan and ID Werkler, "Trade Secrets, Confidential Information and Employment Relationship" in RT Hughes, ed, *Trade Secrets* (Toronto: Law Society of Upper Canada) 71, at 18 to 19 and 73 to 74. See also *Elsley v Collins Insurance Agencies* [1987] 2 SCR 916 at 923.

158   *Faccenda Chicken Ltd v Fowler* [1986] 1 All ER 617 (CA).

159   For further discussion, see Horan and Werkler, "Trade Secrets, Confidential Information and Employment Relationship" in RT Hughes, ed, *Trade Secrets* (Toronto: Law Society of Upper Canada) 71, at 73 to 76.

holding him to the fiduciary's strict standards of conduct.[161]

Unlike a fiduciary duty, a duty of confidence does not require an agreement or understanding between the parties. A fiduciary has a positive obligation to act for the benefit of the recipient. Under a duty of confidence, no such positive duty exists; there exists only the negative duty not to misuse the information.

Senior employees are held to the same standard of fiduciary duty as directors of a corporation,[162] but Canadian courts have not extended fiduciary duties to employees who rank below the level of senior management.[163] Non-senior employees who are not bound by an enforceable confidentiality agreement are free to continue to use business and trade information, including customer information acquired from their former employer, provided that they do not use physical records and provided that the information does not constitute a trade secret.[164]

10. Counterfeit goods

10.1 Remedies against counterfeiters

Rights holders have a variety of remedies against counterfeiters. Civil remedies include interlocutory and permanent injunctions, damages, profit disgorgement, and delivery-up of infringing product.[165] The Copyright Act[166] and the Criminal Code[167] also provide criminal liability for copyright and trade-mark infringement, with maximum penalties available under the Copyright Act of a fine not exceeding C$1 million, imprisonment for a term not exceeding five years, or both.[168]

(a) Anton Piller Orders

Anton Piller orders[169] are one of the most powerful remedies available against counterfeiters. These orders authorise the holder to attend a counterfeiter's premises, without notice, and request entry to search for and seize documents and objects relevant to the proceedings.[170] The Supreme Court has called Anton Piller orders an 'exceptional remedy' to be used only when there is a "real possibility that relevant evidence will be destroyed or otherwise made to disappear",[171] but it has also noted that such orders "are now fairly routinely issued in ordinary civil disputes".[172]

To obtain an order, the applicant must demonstrate:

---

160   *LAC Minerals*, above note 153, at pages 601 to 602; *Guerin v The Queen* [1984] 2 SCR 335 at 383 to 384.
161   *LAC Minerals*, above note 153, at page 602.
162   *Canadian Aero Services Ltd v O'Malle* [1974] SCR 592 at 606 to 607; see also *Genesis Canada Inc v Hill* (1994) 56 CPR (3d) 419 (Ont Ct (Gen Div)).
163   *Genesis Canada*, above note 162. But see *Canadian Industrial Distributors Inc v Dargue* (1994) CPR (3d) 22 (Ont Ct (Gen Div)), in which a non-fiduciary employee was held liable for participation in a fiduciary's breach.
164   *Faccenda Chicken*, above note 158.
165   Copyright Act, above note 51, Section 34; Trade-marks Act, above note 131, Section 53.2.
166   Copyright Act, above note 51, Sections 42 to 43.
167   Criminal Code, RSC 1985, c C-46, Sections 406 to 412.
168   Copyright Act, above note 51, Section 42.
169   *Anton Piller KG v Manufacturing Processes Ltd* [1976] 1 Ch 55 (CA).
170   *Celanese Canada Inc v Muray Demolition Corp* 2006 SCC 36 at paragraphs 1, 28 to 29; *British Columbia (Attorney General) v Malik* 2011 SCC 18 at paragraph 29.
171   *Ibid.*, at paragraph 1.
172   *Ibid.*, at paragraph 29.

- a strong *prima facie* case;
- the alleged damage is very serious;
- the counterfeiter has incriminating documents or things in its possession; and
- there is a real possibility that the counterfeiter may destroy the material before the discovery process.[173]

The court has also considered:
- whether the inspection would do harm to the counterfeiter or its case; and
- whether the interests of justice would be brought into disrepute by allowing the seizure order.[174]

Before granting an order, the court requires an undertaking from the rights holder to pay the defendant damages it suffers if it is ultimately determined that there is no infringement.[175]

Where the counterfeiters are unknown and transient, rolling 'John Doe' Anton Piller orders, against persons unknown, are available.[176] There is a higher threshold to obtain this type of order, with the rights holder needing to establish the nature and extent (both number of instances and geographic scope) of the problem.[177]

(b)   Border controls

Canadian law provides limited protection against counterfeit products entering Canada. Under both the Copyright Act[178] and the Trade-marks Act,[179] the Canada Border Services Agency, which oversees the importation of products, may seize counterfeit products destined for import. The enabling provisions require the rights holder to apply to the court for relief with appropriate evidence and potentially post security for damages, either before the counterfeit goods have been imported, or before they are released by Customs.[180]

Customs officials, acting as public officers, may also seize goods based on intelligence and directives received from the a police and in furtherance of police criminal counterfeit investigations.[181] However, absent such police directive or a civil court order, customs officials are not mandated, and have no authority, to independently search for or seize counterfeit goods.

11.   Collaborative models – 'open innovation'

There are many types of collaborative innovation models, including participation in

---

173   *Ibid.*, at paragraph 35.
174   *Vinod Chopra Films Private Limited and Reliance Mediaworks (USA) Inc v John Doe* 2010 FC 387 at paragraph 24.
175   *Club Monaco Inc v Woody World Discounts* (1994) 2 CPR (4th) 436 at paragraph 7 (FCTD).
176   *Ibid.*, at paragraph 6.
177   *Ibid.*, at paragraph 7.
178   Copyright Act, above note 51, Sections 44.1 to 44.3.
179   Trade-marks Act, above note 131, Sections 53.1 to 53.3.
180   Copyright Act, above note 51, Sections 44.1 to 44.3; Trade-marks Act, above note 131, Sections 53.1 to 53.3.
181   Criminal Code, above note 167, Section 489(2).

research consortia, strategic alliances, and collaboration with research-based institutions. While each raises its own unique intellectual property and competition issues, several common issues can arise, including patent pooling and data privacy. These are discussed next.

11.1 Patent pooling

Neither Canadian courts, nor the Competition Bureau appears to have considered a case involving patent pooling. Canada's Competition Bureau Intellectual Property Enforcement Guidelines acknowledge that patent pooling can be pro-competitive in certain circumstances such as cross-licensing to avoid litigation, reducing transaction costs by offering single-source licensing, or to clear blocking patents.[182] However, they also indicate that the Competition Bureau may view a patent pool as unnecessary and anticompetitive, and contrary to the Competition Act if such an arrangement is not required to permit a new technology to enter into the market.[183] This appears to be so notwithstanding that there may be other, legitimate, pro-competitive reasons for the patent pool.

Despite the lack of jurisprudence, it is noteworthy that patent pools have operated in Canada (eg, relating to the telecommunications industry) without apparent scrutiny from the Competition Bureau. It seems unlikely that patent pools offering a single licence to patents that are essential to the practice of a standard, that are owned by different parties, and that are offered to third parties on a fair, reasonable and non-discriminatory basis would attract the attention of the Competition Bureau.

11.2 Data privacy

Canada's Personal Information Protection and Electronic Documents Act[184] governs the collection, use and disclosure of personal information by commercial organisations. 'Personal information' means information about an identifiable individual but does not include the name, title or business address or telephone number of an employee of an organisation.[185] The purpose for which personal information is collected must be identified and documented, and should be disclosed to the individual before or at the time it is collected.[186] With a limited number of exceptions, personal information may only be collected, used and disclosed for the purpose for which it was specifically collected,[187] except with the consent of the individual or as otherwise required by law.[188] Personal information that is no longer required for the purpose for which it was collected should be destroyed or otherwise made anonymous.[189] Companies are required to create and follow policies and

---

182 Intellectual Property Enforcement Guidelines: Competition Bureau Canada (Ottawa: September 2000) (Commissioner: K von Finckenstein) at page 22.
183 *Ibid.*
184 Personal Information Protection and Electronic Documents Act, SC 2000, c 5.
185 *Ibid.*, Section 2.
186 *Ibid.*, Schedule 1, clause 4.2.
187 *Ibid.*, Section 5, Schedule 1 clauses 4.2, 4.3.
188 *Ibid.*, Section 5, Schedule 1 clause 4.5.
189 *Ibid.*, Section 5, Schedule 1 clause 4.5.3.

procedures that govern the collection, use, disclosure and destruction of personal information.[190]

The three major public scientific research funding organisations in Canada[191] have developed a joint policy statement[192] which includes comprehensive provisions relating to privacy and data protection, and to which institutions must adhere in order to receive funding.[193] The policy statement requires researchers to provide their research ethics boards with detailed proposed measures for safeguarding information for the full life cycle of information.[194] Research data transmitted over the internet may require encryption or use of special denominalisation software to prevent interception by unauthorised individuals. Identifiable data obtained through research that is kept on a computer and connected to the internet should be encrypted.[195]

12.   Hot topics

12.1   'Good faith' in patent prosecution

The Patent Act requires applicants to respond in 'good faith' to any requisition made by an examiner during prosecution of the application, failing which the application is deemed abandoned.[196] There was no similar language in the former Act,[197] and courts held that an applicant had no overriding duty of candour and good faith when dealing with the Patent Office.[198] Recently, the court confirmed that the 'good faith' requirement of the current Act places a duty on patent applicants and their counsel fully and fairly to describe the prior art when answering the examiner's requisitions.[199]

In *Lundbeck*, in response to an examiner's requisition, the applicant disclosed the existence of a prior-art reference (later determined by the court to be the most relevant to inventiveness) but did not provide a copy.[200] In a subsequent requisition, the examiner rejected the application on the grounds of obviousness. In responding, the applicant referred to four less relevant prior-art references that 'taught away' from the invention, but failed to comment on the most relevant reference that came to the opposite conclusion. The court found the response was not in good faith, and that the application had therefore been abandoned.[201] While there remains a paucity of jurisprudence on what the 'good faith' requirement entails, lack-of-good-faith

---

190   *Ibid.*, Section 5, Schedule 1 clause 4.1.4.
191   Canadian Institutes of Health Research, Natural Sciences and Engineering Research Council of Canada and the Social Sciences and Humanities Research Council of Canada.
192   Canadian Institutes of Health Research, Natural Sciences and Engineering Research Council of Canada and Social Sciences and Humanities Research Council of Canada, "Tri-Council Policy Statement: Ethical Conduct for Research Involving Humans", December 2010.
193   *Ibid.*, at page 5.
194   *Ibid.*, at page 60.
195   *Ibid.*, at page 61.
196   Patent Act, above note 2, Section 73(1)(a).
197   *Ibid.*, Section 30.
198   *Flexi-Coil Ltd v Bourgault Industries Ltd* (1999) 86 CPR 221 at paragraph 30 (FCA); *Janssen-Ortho Inc et al. v Apotex Inc et al.* 2008 FC 744 at paragraph 201,
199   *Lundbeck Canada Inc et al. v Ratiopharm Inc et al.* 2009 FC 1102.
200   *Ibid.*
201   *Ibid.*, at paragraph 352.

arguments are becoming more commonplace by those seeking to invalidate a patent. Until the law is more fully developed, applicants should be careful to present a full and fair description of the scope of the entire prior art when making submissions to the Patent Office.

12.2   Sufficiency issues for patents claiming antibodies

In light of the maturation in monoclonal antibody production technology, patents claiming monoclonal antibodies, without working examples, may now satisfy the disclosure requirements in the Patent Act.[202] In *Re Immunex Corp Application No 583,988*,[203] the Patent Appeal Board accepted the principle established in foreign jurisdictions that monoclonal antibodies that are immunoreactive with a novel polypeptide can be claimed without the antibody actually having been made or deposited, provided that the polypeptide target has been fully characterised. This reverses an earlier decision holding that describing monoclonal antibodies by reference to their antigens does not sufficiently enable the invention without considerable and protracted experimentation.[204]

In *Immunex*, the board found that claims to monoclonal antibodies of a distinct polypeptide receptor type met the disclosure requirements since the specification sufficiently described the receptor, taught how it could be expressed recombinantly, and disclosed the isolation of one of its domains.[205] However, this relaxed approach could not be extended to unqualified claims encompassing monoclonal antibodies to a class of polypeptides broader than those actually characterised.[206]

---

202   Patent Act, above note 2, Section 27(3).
203   *Re Immunex Corp Application No 583,988* (2010) 89 CPR (4th) 34 (PAB) at paragraph 68.
204   *Re Institut Pasteur Application* (1995) 76 CPR (3d) 206 (PAB).
205   *Immunex*, above note 200, at paragraphs 54 to 55.
206   *Ibid.*, at paragraphs 28, 44, 51 and 65.

China

Lewis Ho
Simmons & Simmons LLP

Overview

IP protection and enforcement has been a hot topic in the People's Republic of China (PRC). The issue has become more important, as the PRC government promulgated 'National IP Policy' (the Policy) in 2008. This seeks to shift the business model from 'OEM' (original equipment manufacturer) to 'ODM' (original design manufacturer). The Policy emphasises the importance of R&D activities in China and as a result the PRC government has recently amended the patent law, technology transfer regulations and patent examination guidelines and has issued a judicial interpretation on patent infringement actions. The amendments and the Policy have a significant impact on the ways in which to protect IP, which is key to the development of the life sciences sector in China.

Introduction to the PRC legal system

Chinese intellectual property laws

The main legislation governing Chinese intellectual property law can be found in the following legal instruments: the Patent Law, the Patent Law Implementing Regulations, the Patent Examination Guidelines; the Trade Mark Law, the Trade Mark Implementing Rules; the Anti-unfair Competition Law; and various other specific regulations. Detailed aspects of some of these are discussed in the sections that follow.

China has become the most litigious country in the world for IP enforcement, and there are now more IP infringement civil actions filed in China than in the United States per year.

The PRC legal system

The PRC is a civil law jurisdiction and its litigation system has more in common with Germany than, for example, England or the United States. Proceedings in PRC courts are inquisitorial, not adversarial. This means that judges are obliged to, and take an active role in, investigating the case, although each party must still present the facts and evidence in support. Other notable differences are that there are no general discovery obligations, documentary evidence takes precedence over oral testimony, and there is seldom cross-examination of witnesses.

Basic principles of patent litigation

Court system
China's judicial system consists of four levels of courts:
- the Basic People's Court (county and city district);
- the Intermediate People's Court (regional);
- the Higher People's Court (provincial); and
- the Supreme People's Court.

The Basic People's Court is only a trial court, while the others are both trial and appellate courts. Each court consists of several chambers, including civil, administrative and criminal, with IP cases usually heard by the civil chamber. The Supreme People's Court (SPC), located in Beijing, is the highest court in China. It handles appeals from the Higher People's Courts and also serves an administrative role, in which it issues judicial interpretations of PRC laws that are legally binding upon lower courts.

IP courts
Because of the complexity of IP cases (including patents, trade marks, copyright and trade secrets), the Supreme People's Court has designated approximately 80 courts around the country (mostly Intermediate People's Courts) as first-instance courts for adjudication of IP infringement claims. The Supreme People's Court also has its own special panel for handling IP-related cases.

Panels hearing patent infringement cases will usually consist of three judges, sometimes assisted by a court-appointed technical expert.

No precedent
The Chinese legal system does not have a case law system which follows a rule of binding precedent. Chinese courts generally do not have to follow earlier decisions of other courts, even higher courts. However, decisions and opinions of some courts – such as the Beijing Higher People's Court, which has jurisdiction over the State Intellectual Property Office (SIPO), which is responsible for administration of patents in China, and the Patent Re-examination Adjudication Board (PRAB) – usually have persuasive authority in IP-related cases.

Separate infringement and validity proceedings
Unlike the patent court systems of the United States and the United Kingdom (but in common with Germany), issues of validity and infringement are dealt with in separate proceedings and by separate bodies in the PRC. Infringement proceedings are handled by the People's Court. The court will only decide whether the patent has been infringed, and will not make any decision on the validity of the patent. Invalidation proceedings are handled by the PRAB within the SIPO.

Where the validity of a patent is challenged before the PRAB, the court hearing the infringement claim has discretion to stay the infringement proceedings. Courts are less likely to stay infringement actions for invention patents (which are substantively examined before grant) and it is usual for parallel infringement and invalidity actions to be held in the courts and SIPO, respectively.

An infringement action may last between six and 18 months in the first instance if not stayed, and between six and nine months on appeal.

Before starting proceedings

Jurisdiction

Patent infringement cases must be filed where the infringer is domiciled, or in the place of infringement. Places of infringement include not only the places where the infringing acts actually occurred but also any place affected by the consequences of infringement. To avoid litigation in a defendant's home court, and in order to benefit from the technical expertise of specialised IP courts (eg, Beijing and Shanghai), claimants may wish to join distributors of the infringing product located in these more favourable jurisdictions as additional defendants.

Limitation period

China has a two-year statute of limitations for patent infringement, which runs from the date on which the patentee knew or should have known about the infringement.

Court procedure

Most patent cases start in the Intermediate People's Courts, and appeals are heard in the Higher People's Courts.

The People's Courts have broad powers and discretion in the way cases are conducted, so there are variations in the procedures used in the different courts. In general, the proceedings start upon filing the claim at the competent court. The claim must be drafted carefully to contain all the relevant facts that substantiate the claim. In relation to an infringement action, this must include a detailed description of the infringing product, the request for relief, and the facts and reasons on which the request is based. In addition, the main evidence proving the rights claimed and infringement must also be filed at this stage.

The claimant must pay a fee, which unlike most jurisdictions is calculated as a proportion of the claim value. The defendant will be ordered to pay this fee if the claimant wins the case. The court is responsible for serving the claim on the defendant and must do so within five days.

The defendant must file its reply and defence (if any) within 15 days of receipt of the claim, which the court will send to the claimant within five days of receipt. However, failure to file a reply does not result in default judgment and the court will still proceed with the claim.

Before trial, there usually is at least one evidence exchange hearing at which the parties exchange the evidence on which they intend to rely and explain the relevance of the evidence to the disputed issues before the court. Parties may question or object to each other's evidence. In complex cases, three or four evidence hearings are common before trial. Judges will often decide the merits of the case at these preliminary evidence hearings before trial. These are very short – typically lasting less than one day.

The oral hearing

The trial takes the form of a hearing in open court before the judge(s). There is no jury. In general, the media are allowed to report on the case.

There is no provision for the parties to submit skeleton arguments or materials to the judge before the hearing. The judge will already be familiar with the case through various preliminary evidence hearings and typically will already have made a decision on the merits.

The format of a Chinese civil trial is set out in the PRC Civil Procedure Law. In summary, each party's lawyer gives an opening statement. (All PRC lawyers have higher rights of audience.)

Evidence is then presented. Oral evidence should be given first, although oral evidence is rare in practice. Documentary evidence, including expert witness statements, is read. New evidence may be presented and witnesses questioned with the court's permission. Again this is rare in practice. The evidence is not contested while it is being presented.

After this, the parties' lawyers may debate the case. Closing speeches are then made first by the claimant, then the defendant (and any third parties).

Judgment

Judgment will be issued in documentary form several months after the trial.

Appeal

China has a 'two-instance' judicial system – the decisions of the court of first instance can be appealed to the court at the higher level, which makes what essentially is a final decision. However, if the appellate court was not the Supreme Court, the parties can also seek a review by the Supreme People's Court.

Patent filing strategy for inventions mark in China

Confidentiality examination for inventions completed in China

The PRC Patent Law now requires that all applicants (ie, domestic and overseas) seeking a foreign patent filing for an invention completed in China must apply to the SIPO to undergo a 'confidentiality examination' prior to proceeding with the foreign patent filing. The confidentiality examination applies to both onshore and offshore entities.

The Implementing Rules of the PRC Patent Law clarify the procedure for any entity intending directly to file an overseas patent application for an invention completed in China. The applicant should first file a request (together with details of the invention) with the SIPO for a confidentiality examination. Upon receipt of the request, the SIPO will examine the invention and notify the applicant if the relevant invention involves a 'national security or interest' and requires confidentiality protection. If the SIPO does not notify the applicant of this finding within four months of the request, the applicant may proceed with filing an overseas patent application. However, SIPO may extend the deadline for another two months if it deems necessary.

Article 8 of the Implementing Regulations states that the meaning of "an invention completed in China as provided under Article 20 of the Patent Law" refers to the fact that the *substantive contents of the relevant technical solution are completed within China*. This phrase is not defined in the Patent Law, or in the Implementing Rules.

Non-compliance of confidentiality examination requirement

Thus, under local practice it is entirely up to the applicant to determine whether an invention is 'completed' in China and to decide whether a confidentiality examination will be required. The consequences for non-compliance of the confidentiality examination can, however, be severe, which will jeopardise any subsequent Chinese patent application in respect of the invention and result in loss of patent rights in China.

According to the latest PRC Patent Examination Guideline (published on January 21 2010), if an examiner is of the view that an invention disclosed in a PRC patent application does not comply with Article 20 of the Patent Law (ie, an applicant proceeds with foreign filing for an invention completed in China without conducting a confidentiality examination), the examiner is entitled to require the applicant to prove that the invention was not made in China (ie, the burden of proof is on the applicant). If the submission provided by the applicant is not sufficient to the satisfaction of the examiner, the examiner may reject the application.

Under Article 20 of the PRC Patent Law, non-compliance with the confidentiality examination is also a ground for invalidation in China.

1. Small molecules

1.1 Product and process claims

Product and process inventions are governed by the rules of China's Patent Law, which was implemented in 1985 and revised in 1992, 2000 and 2009. Patents are classified as inventions, utility models and designs. An 'invention' means any new technical solution relating to a product or process, or improvement thereof. A 'utility model' means any new technical solution relating to the shape, the structure or their combination for a product which is fit for practical use. A patent for design protects features of shape, configuration, pattern or ornament applied to a product that appeal to the eye. Both products and processes are potentially classified as inventions in China.

The patentability of inventions is determined by criteria similar to those in the United States and Europe. These are:

- novelty;
- inventiveness;
- practical applicability; and
- no detriment to the public interest.

Under Article 22 of the Patent Law, 'novelty' means that the invention or utility model does not form part of the prior art, nor has any entity or individual filed previously before the date of filing with the Patent Office an application relating to an identical invention or utility model disclosed in patent application documents published or patent documents announced on or after the said date. 'Inventiveness' means that, as compared with the prior art, the invention has prominent substantive features and represents notable progress; and 'practical applicability' means the invention or utility model can be made or used and can produce effective results.

Under the Examination Guidelines, 'detrimental to the public interest' refers to exploitation or use of the patent "which may harm the interests of the public or society, or may affect the normal order of the State and society".

1.2 Scope of protection of claims and Markush formulae

The Patent Law provides that "the extent of protection of a patent shall be determined by the terms of the claims. The description and appended drawings may be used to interpret the claims."

The Supreme People's Court recently issued a judicial interpretation on patent infringement trial (Judicial Interpretation) which seeks to provide guidance to local courts handling patent infringement actions in China.

In order to determine the scope of the claims, the Judicial Interpretation states that Chinese courts should rely on the understanding of the claims, specifications and drawings on the part of an ordinary person skilled in the art. If the scope of the claims is not clear and the ambiguity cannot be resolved from the information provided in the patent, Chinese courts may rely on textbook or common knowledge of an ordinary person skilled in the art in interpreting the claims.

It appears that Chinese courts are more ready to accept expert opinion regarding the meaning of the description and drawings as understood by an ordinary person skilled in the art, as opposed to relying merely on the literal meaning.

(a) Application of the doctrine of equivalents

If the scope of a claim asserted by a claimant includes a technical feature which is equivalent to the feature stated in the claim, the Judicial Interpretation states that Chinese courts may determine the scope according to the equivalent features as asserted by the claimant.

(b) Application of the 'file wrapper estoppel' principle

The 'file wrapper estoppel' principle requires the courts, in interpreting claims and assessing infringement of a patent, to exclude the technical features which have been voluntarily given up by a patentee during patent prosecution or invalidation proceedings.

The Judicial Interpretation now formally adopts the file wrapper estoppel principle and states that Chinese courts may interpret the relevant contents of a claim by reference to the patent prosecution file. The Judicial Interpretation also states that if, in the course of prosecution or invalidation proceedings, a patentee has voluntarily made amendments or statements which narrow or limit the claims, Chinese courts should not support an assertion from the patentee that the scope of the claim includes the technical features which have been abandoned.

The application of the doctrine of equivalents and the file wrapper estoppel principle would appear to suggest that the Chinese patent system follows the US system rather than systems in Europe.

(c)   **Removal of the 'redundant designation' principle**

The 'redundant designation' principle allows the courts, in interpreting claims and assessing infringement of a patent, to ignore obvious and added technical elements which do not contribute to the essential features of the claimed product or process. The Judicial Interpretation now states that Chinese courts, when interpreting a claim, should consider all technical features in the claims.

(d)   **Discrepancies between the literal meaning of the specification and the understanding of persons skilled in the art**

The Patent Law provides that "the extent of protection of a patent shall be determined by the terms of the claims. The description and appended drawings may be used to interpret the claims."

In order to determine the meaning of the description and the appended drawings, the Judicial Interpretation states that Chinese courts should rely on the understanding of an ordinary person skilled in the art. If the understanding of an ordinary person skilled in the art is different from the literal meaning of the description or drawings, the understanding of an ordinary person skilled in the art shall prevail.

It appears that, in future, Chinese courts will be more ready to accept expert opinion regarding the meaning of the description and drawings as understood by an ordinary person skilled in the art, as opposed to relying merely on the literal meaning.

(e)   **Markush formulae**

Under the Examination Guidelines, Markush claims (ie, claims to a generic chemical formula that covers potentially millions of compounds) are allowed, provided that they meet the criteria on unity of invention under Article 31 of the Patent Law. If the alternative elements in a Markush claim are of a similar nature, they shall be regarded as being technically related and having the same or corresponding special technical features, and accordingly the claim may be considered to meet the requirements of unity.

The Examination Guidelines clarify that Markush elements for alternatives of compounds shall be regarded as being of a similar nature and will possess unity if the following requirements are met:

- all alternative compounds possess a common property or activity; and
- all alternative compounds possess a common structure, which constitutes the distinguishing feature between the compounds and those in the prior art, and is essential to the common property or activity of the compounds of general formula; or, where they cannot have a common structure, all of the alternative elements belong to the same recognised class of compounds in the technical field to which the invention pertains.

The Examination Guidelines clarify that a 'recognised class of compounds' means "there is an expectation from the knowledge in the art that members of the class belong to the same class of compounds with the same performance in the context of the claimed invention, ie, each member may be substituted by another, with the expectation that the same intended result will be achieved".

1.3 Metabolites

The Examination Guidelines do not set out specific criteria regarding patentability of metabolites. The general rules under China's Patent Law will apply.

2. Second-generation inventions

2.1 Combinations[1]

Combination inventions are patentable subject (primarily) to issues of inventiveness. Under the Examination Guidelines, the following factors will usually need to be taken into account when determining the inventive step of an invention by combination:

- whether the combined technical features functionally support each other;
- the degree of difficulty or ease of combination;
- whether there is any technical motivation to make the combination in the prior art; and
- the technical effect of the combination.

As stated in the relevant Examination Guideline, a combination invention will be considered an obvious combination which does not involve an inventive step if it is:

merely an aggregation or juxtaposition of certain known products or processes, each functioning in its routine way, and the overall technical effect is just the sum of the technical effects of each part without any functional interaction between the combined technical features, that is, the claimed invention is just a mere aggregation of features.

A combination invention will be considered non-obvious and will be regarded as involving an inventive step as follows:

[i]f the combined technical features functionally support each other and produce a new technical effect, or in other words, if the technical effect after combination is greater than the sum of the technical effects of the individual features, then such combination has prominent substantive features and represents notable progress.

In assessing inventiveness it is irrelevant whether any of the technical features of the invention are completely or partially known to the public.

2.2 Enantiomers

The Examination Guidelines do not set out specific criteria regarding patentability of enantiomers. The general rules under China's Patent Law will apply. This is evident from case law determined by the PRAB and Chinese courts, as shown in the case of *CSPC Ouyi Pharmaceutical Co Ltd v PRAB* (2006).

---

1   These are governed by Examination Guidelines, Part II, Chapter 4, para 4.2.

The facts were as follows:

- The PRAB determined that a patent for a method to resolve amlodipine enantiomers using DMSO-d6 was invalid due to lack of inventiveness. The applicants filed an administrative action at the Beijing Intermediate People's Court.
- Amlodipine is a medicine for treating hypertension and angina. Pfizer developed the medicine Norvasc incorporating amlodipine. The patent for Norvasc expired in 2007.
- The court determined that the relevant prior art was the use of DMSO or reagents incorporating DMSO to resolve amlodipine enantiomers developed by Pfizer. DMSO and DMSO-d6 are chiral auxiliary reagents, have identical chemical properties and similar other properties. But DMSO-d6 is an improved version (ie, it has a higher optical purity).

The court ruled as follows:

- Under Article 22 of the Patent Law, 'inventiveness' means that "as compared with the prior art, an invention has prominent substantial features and represents notable progress". When determining whether an invention has prominent substantial features, it is generally necessary to identify the prior art, identify the features of the invention which are different as compared with the prior art, and identify the actual solution to a technical problem represented by such different features. After this, the claims setting out the invention should be analysed as to whether each would be obvious to a person skilled in the art.
- An invention will be considered not obvious, possessing prominent substantial features and representing notable progress if it involves an unexpected technical solution which, compared with the prior art, represents (i) a 'qualitative change' and possesses new function/performance, or (ii) a 'quantitative change' which exceeds expectations.
- The patent (claims 1, 2 and 3) was held to be invalid.
- Regarding claim 1, a person skilled in the art would easily think to use DMSO-d6 to resolve amlodipine enantiomers, given its similarities to DMSO. The invention did not result in unexpected technological results. Prior use of DMSO-d6 in a different field (eg, in nuclear magnetic resonance), high unit price and the fact that such substituted use was not published before the application date were not barriers to such substitution, and the invention was not 'not obvious'.
- Also, the technical solution in claim 1 essentially related to the optical purity of the resulting mixture, which was only slightly increased as compared with methods in the prior art. This technical solution was therefore not considered to represent a sufficient qualitative or quantitative change.

2.3 Selection inventions[2]

Selection inventions are patentable subject to issues of inventiveness, the main issue

---

2   These are governed by Examination Guidelines, Part II, Chapter 4, para 4.3.

being whether the selection invention can bring about 'unexpected technical effects'.

Under the Examination Guidelines, an invention will not be considered to involve an inventive step if:

- the invention consists merely in choosing among a number of known possibilities, or merely in choosing from a number of equally likely alternatives, and the selected solution does not produce any unexpected effect; or
- the invention resides in the choice of particular dimensions, temperature ranges, or other parameters from a limited range of possibilities, while such choice could be made by a person skilled in the art through normal design procedures and does not produce any unexpected technical effect; or
- the invention can be arrived at merely by a simple extrapolation in a straightforward way from the known art.

2.4   Methods of use/secondary indications[3]

Medical use is not patentable in China when being applied to diagnose or treat disease. It is only patentable when being applied to drug preparation or manufacturing.

The Examination Guidelines provide that an application relating to the medical use of a substance shall not be granted if the claim is drafted as 'use of substance X for the treatment of diseases', 'use of substance X for diagnosis of diseases' or 'use of substance X as a medicament', as such forms of wording will be considered an application for a method of treatment or diagnosis which is not patentable under Article 25(3) of the Patent Law.

However, "since a medicament and a method for the manufacture thereof are patentable according to the Patent Law, it shall not be contrary to Article 25.1(3) if an application for the medical use of a substance adopts pharmaceutical claim or use claim in the form of method for preparing a pharmaceutical, such as 'use of substance X for the manufacturing of medicament', 'use of substance X for the manufacturing of a medicament for the treatment of a disease' and so on".

In determining inventiveness of 'second medical use' type applications, the following two factors will usually need to be taken into account: proximity of the technical field of the new use to that of the prior known use; and the technical effect of the new use.

As stated in the Examination Guidelines, "[i]f the new use utilises a newly found property of a known product and can produce an unexpected technical effect, then the invention of use has prominent substantive features and represents notable progress, and thus involves an inventive step". Accordingly, in order to demonstrate inventiveness, it is necessary clearly to set out in the patent specifications the nature of the second medical use, including:

- function/performance;
- objective;
- scope of use;

---

3   These are governed by Examination Guidelines, Part II, Chapter 4, para 4.5; Part II, Chapter 10, para 4.5.

- method; and
- requirements for use.

2.5 Methods of treatment[4]

Methods of diagnosis or treatment of diseases are not patentable under Article 25(1) of the Patent Law. Under the Examination Guidelines, "'[m]ethods for diagnosis or for treatment of diseases' refer to the process of identifying, determining, or eliminating the cause or focus of diseases which are practised directly on living human or animal bodies ... Methods of treatment for diseases refer to the processes of intercepting, relieving, or eliminating the cause or focus of diseases so that the living human or animal bodies may recover or gain health or relieve pain".

Methods of treatment include methods which serve a 'treatment purpose' or which are of 'treatment nature'. Methods of treatment include treatment by pharmaceutical therapy, prophylactic methods and immunisation. An application for a method which may serve both a treatment purpose and a non-treatment purpose must clearly state that the method serves a non-treatment purpose, otherwise it will not be granted a patent.

2.6 Formulations and physical forms

Formulations are patentable as pharmaceutical preparation products. The protection scope is always limited to the application of the formulation. Novelty is closely related to the components of the formulation and the contents thereof. Inventiveness usually depends on a new active compound contained in a formulation or an unforeseen synergism between the different known components in a formulation. Sufficient examples and data regarding the pharmaceutical application or effect of a formulation are essential.

3. DNA, biologicals and personalised medicine

3.1 Discoveries[5]

Scientific discoveries are not patentable, under Article 25(1) of the Patent Law. Under the Examination Guidelines, "'[s]cientific discoveries' refer to the revelations of substances, phenomena, transformation processes and their features and laws, which objectively exist in nature. Scientific theories are the generalisation of understandings of nature, and are discoveries in a broader sense."

It should be noted that the Examination Guidelines clearly distinguish between invention and discovery but note that they are closely interrelated, stating that "such a close relationship between invention and discovery is especially prominent in 'use invention' of chemical substances. When a special property of a certain chemical substance is discovered, usually a use invention utilising this property will be made accordingly."

---

4   These are governed by Patent Law, Article 25(3); Examination Guidelines, Part II, Chapter 1, para 4.3.
5   These are governed by Patent Law, Article 25; Examination Guidelines, Part I, Chapter 1, para 7.4 and Part II, Chapter 1, para 4.1.

3.2 Gene patents and industrial application[6]

Under the Examination Guidelines, *prima facie*, genes and DNA sequences are merely types of chemical compounds and are patentable. However, genes and DNA sequences obtained from the natural environment and in a natural state comprise 'scientific discoveries' and are not patentable. But genes and DNA sequences first isolated or derived from nature are capable of being patented if they:

- do not exist in the prior art;
- can be precisely characterised; and
- have industrial benefit or value.

However, under Article 5 of the Patent Law, no patent right shall be granted for any invention/creation where acquisition or use of the genetic resources, on which the development of the invention/creation relies, is not consistent with PRC law. According to the Patent Law Implementing Rules, 'genetic resources' means material derived from the human body, animals, plants, microorganisms etc, which contains functional unity of heredity and have actual or potential value.

3.3 Antibodies[7]

Antibodies are patentable subject matter provided that they meet the requirements for novelty, inventiveness and industrial applicability. Regarding novelty, according to the Examination Guidelines if an antigen A is new, then its monoclonal antibodies will also be new. However consider where a known antigen A' has the same epitope as new antigen A, and such similarity leads to a presumption that the monoclonal antibodies of antigen A' will bind to antigen A. In this case, the monoclonal antibodies of new antigen A will not possess novelty. The exception is if the applicant can prove in the application, or compared with the prior art, that the monoclonal antibodies claimed are different from known antibodies.

Regarding inventiveness, according to the Examination Guidelines, where a known antigen has immunogenicity, an invention comprising monoclonal antibodies of such known antigen will not be considered to have inventiveness. But if the invention claims other features which generate unforeseen effects, then such monoclonal antibody will be considered to have inventiveness.

3.4 Stem cells[8]

Human embryonic stem cells and preparations thereof are not patentable under Article 5(1) of the Patent Law. Article 5(1) provides that a patent shall not be granted for any invention which violates law or social morality, or which is detrimental to the public interest.

3.5 Bioinformatics systems

In China, patents are potentially available for software or computer-implemented

---

6   These are governed by the Examination Guidelines, Part II, Chapter 10, para 9.1.2.2.
7   These are governed by the Examination Guidelines, Part II, Chapter 10, paras 9.4.1/9.4.2.1(5).
8   These are governed by the Examination Guidelines, Part II, Chapter 10, para 9.1.1.1.

inventions, such that a bioinformatics programme or system could be patented. Also sequences – DNA or protein – identified by computer programmes as having possible therapeutic value, but without a specific application having been found, can be potentially protected.

3.6   Copyright and sequence protection
Copyright laws potentially protect DNA or protein sequence information in China.

3.7   Sufficiency issues
The Examination Guidelines have a strict requirement for sufficiency. In particular for a new pharmaceutical compound or pharmaceutical composition, its specific medical use or pharmacological action, effective amount and method of application must be disclosed.

The Examination Guidelines further require: "If a person skilled in the art is unable, on the basis of the prior art, to predict that said use or action stated in the invention can be carried out, the qualitative or quantitative data of the laboratory test (including animal test) or clinical test shall be sufficiently provided for the person skilled in the art to be convinced that the technical solution of the invention can solve the technical problem or achieve the technical effect as expected."

4.   Acts of patent infringement

4.1   Infringement

(a)   Clarification of the scope of product patents
The Patent Law states that a product patent is infringed if a defendant manufactures, uses, sells, offers for sale or imports infringing products. The Judicial Interpretation clarifies that:
- 'manufacture' of an infringing product includes assembling components to form the product, except for commercial staple products commonly assembled by users or retailers; and
- 'use' of an infringing product includes using the product as a component to manufacture another product. Sale of that final product will also be held as 'sale' and constitutes an infringing act of a product patent.

(b)   Clarification of the scope of process patents
The Patent Law states that a product patent is infringed if a defendant uses a patented process or uses, sells, offers for sale or imports any products 'obtained directly from patented process'. The Judicial Interpretation clarifies that products 'obtained directly from patented process' include raw materials obtained from the patented process or any subsequent products obtained from the raw materials through further processing.

There is currently no further guidance from the PRC laws as to what constitutes a product 'directly obtained from a patented process'. Some PRC cases suggest that a product would be found to infringe a process claim only if the infringing process step is the last process step in the chain. However, the Judicial Interpretation seems to suggest that a product will be found to infringe a process claim as long as the infringing process step forms part of the process in the chain.

(c)   *Reversal of burden of proof for 'new' product*

Under Article 61 of the Patent Law, if the invention is a process for obtaining a new product, where the same product is produced by a person other than the patentee, that person is obliged to prove in a patent action that its products were obtained from a process which is different from the patented process. The provision helps patentees in proving infringement of a process patent.

The Judicial Interpretation states that the term 'new' product means a product which was not known to the public (inside or outside China) before the filing date of the patent. This contradicts the corresponding provision in the Patent Opinion issued by the Beijing Court, which states that 'new' product means "product first manufactured in China and is obviously different in composition, structure or quality, performance or function as compared to the products that existed before the filing date of the patent".

For example, in a pharmaceutical patent action, if a process patent was filed before any publication of its corresponding compound patent, the compound resulting from the patented process will be deemed as a 'new' product. The patentee, if intending to enforce the process patent, will be able to rely on Article 61 and demand that an infringer proves that its compound was obtained from a process other than the patented process. This reversal of the burden of proof is particularly useful after the expiry of the corresponding compound patent.

4.2   Contributory infringement

The Patent Law does not provide any legal basis for asserting contributory infringement. There is a wide consensus that contributory infringement can be based on a provision in the SPC Judicial Interpretation on the Civil Procedure Law, which states that a person who instigates, or assists others to perform, an act of infringement shall be co-infringer and shall be held liable jointly and severally.

The Judicial Interpretation now establishes the concept of contributory liability for patent infringement and states that parties in a conspiracy for using a patented product as a component to manufacture another product will be held jointly liable for manufacturing a patented product.

Under PRC laws, there is no direct legal basis for establishing indirect infringement of a patent. The Judicial Interpretation now formally establishes the legal basis for indirect infringement and states that a defendant will be held liable for infringement if it supplies the means which it knew can only be used to practise an invention (eg, raw materials, intermediates, parts or equipment) to a third party for conducting acts of direct infringement. If the conduct of the third party is for commercial purpose, both the supplier and the third party will be held liable for infringement. If the conduct of the third party is not for commercial purposes, only the supplier will be held liable.

For example, in a pharmaceutical patent action an API (active pharmaceutical ingredient) manufacturer or exporter may be held liable under the Judicial Interpretation even if it is not involved in the production of the final product. However, the Judicial Interpretation does not adopt an extraterritorial provision as in the Patent Opinion issued by Beijing Court (ie, an action against indirect infringement may be pursued in China if the act of direct infringement occurs outside China). The authors suggest lobbying the SPC to adopt such a provision, as this would assist in preventing export-related indirect infringement when acts of direct infringement take place overseas.

4.3   Defences

(a)   Prior-art defence
The Patent Law provides that there is no infringement if a person can prove that the technology he exploits is an existing technology. The Judicial Interpretation states that Chinese courts should rule that a defendant has successfully established a prior-art defence if the defendant can prove that the technical features of the alleged infringing technical scheme have no substantive difference when compared with the corresponding features in the prior art.

(b)   Prior-use defence
The Patent Law states that there is no infringement if a person has made an identical product, used an identical process, or made necessary preparations for the manufacture or use of the identical product or process before the filing date of the patent application and continues to make or use the product or process within the original scope.

The Judicial Interpretation provides the following guidance for the prior-use defence:
- Chinese courts will not support such a defence if a defendant relying on a prior-use defence obtains the technology illegally.
- Chinese courts will deem that a defendant 'made necessary preparations for the manufacture' in the following scenarios:
  - main technical drawings or technical documents needed for implementing the invention have been completed; or
  - main equipment or moulds needed for implementing the invention have been manufactured or purchased.
- Chinese courts, in determining whether a defendant makes or uses a product or process within the 'original scope', will consider the magnitude of manufacture before the date of patent application, or the magnitude of manufacture which can be achieved based on the existing manufacturing equipment or the existing preparation.
- Chinese courts will not support a prior-use defence if the prior-use rights holder assigns or licenses the existing technologies to others, unless the technologies were assigned or licensed together with the prior-use rights holder.

5. Patent enforcement

5.1 Obtaining information on the infringer and the infringement

(a) Investigation

Before proceeding with any enforcement action, a patent owner must ensure that it has obtained evidence of the subject's infringing activity. This step is particularly important in China as the court relies almost exclusively on documentary evidence and will examine its quality and validity in detail.

Investigation firms are often employed to collect this evidence. Investigators can help determine (i) whether the potential infringer actually manufactures and/or sells the allegedly infringing products; (ii) the volume of infringing products produced or sold; (iii) the possible location of the infringing products; and (iv) which targets to sue (and where). Evidence collected by investigators is admissible only if its authenticity is established. Proper handling of evidence is critical in PRC litigation and is discussed further below.

(b) Disclosure/discovery

There is no general discovery obligation imposed on the parties in China. The burden is solely on the claimant to collect and submit its own evidence to prove patent infringement and damages (among other things). When applying for a discovery order or evidence preservation order, the applicant must prove that the respondent has in its possession the evidence sought.

Proper and thorough gathering of evidence before and during the initial stages of litigation is critical to Chinese practice, and the importance of evidence-related planning and strategy to the overall success of any patent litigation in China cannot be overstated.

PRC courts generally accept evidence only in its original form. A notary public is often required to authenticate evidence, including evidence from the internet. If the evidence was obtained in breach of the law, it will not be admissible and, if admitted, may constitute good grounds for appeal.

While evidence obtained in foreign countries is admissible in PRC courts, it must be notarised by a local notary public in the foreign country and then legalised by the applicable Chinese embassy or consulate. Any documentary evidence in a foreign language must be translated into Chinese by a court-authorised translation company. Evidence obtained from previous administrative proceedings or preliminary injunction proceedings can sometimes be used in subsequent infringement litigation.

Evidence must be submitted to the court within a prescribed time limit. Generally, the time limit will be designated by the court and must not be less than 30 days from the day after the parties have received notice of the court's acceptance of the case and notice to respond to the suit. The deadline can be extended by the parties' agreement and with the court's approval. In most cases, new evidence may not be submitted beyond the time limit.

(c)   *Discovery order*

Articles 64 and 65 of the PRC Civil Procedure Law empower the People's Court to investigate, collect and order the production of evidence from a defendant. However, the onus is on the claimant to prove that the defendant is in possession of the evidence sought.

(d)   *Evidence preservation order*

Article 74 of the PRC Civil Procedure Law provides that if it appears that evidence may be destroyed, lost, or difficult to obtain later, a party may seek an *ex parte* court order to preserve such evidence. The court may order the requesting party to post a bond.

Evidence preservation orders can be very effective. The respondent generally will not be notified in advance and may be required to comply by providing the relevant documentation and evidence on the spot. Any evidence preservation order is typically enforced by the judges themselves. In the execution of the order, the court may question the respondent, order production of documents, take samples of the infringing product and conduct an inspection of premises. Any evidence obtained from evidence preservation efforts should be admissible in the subsequent court proceedings.

To prevent its abuse, however, most courts will require that the applicant present some preliminary evidence showing ongoing or imminent infringement before issuing such an order.

(e)   *Adverse inference*

Where there is evidence to prove that a party possesses evidence which has been ordered for production but refuses to produce it without proper reasons, the court may draw an adverse inference against that party that the evidence is unfavourable.

(f)   *Witnesses and technical experts*

The court may allow an expert to present testimony on specialised issues, particularly technical ones, as a party witness. The judges and opposing parties may question the party expert witness. More commonly, however, courts will appoint their own experts (sometimes after consultation with the parties) to assess technical issues and to produce a written report of their findings. The opinions of court-appointed experts are more likely to be adopted by the courts than those of the party experts.

5.2   Interim relief

An application for a preliminary injunction can be made independently of and in advance of any proceedings. The only requirement is that if the application is successful, patent infringement proceedings must be initiated in the court within 15 days of issuance of the injunction or the injunction will be lifted. A documentary application is made and there is usually no oral hearing.

A court is required to make a ruling within 48 hours if it finds that all procedural requirements have been met upon receiving a request for a preliminary injunction. However, in practice it is generally considered swift if a ruling is made in about one week. Once issued, the injunction is enforceable immediately.

The patentee or other interested parties – which includes licensees – may apply for an injunction. An exclusive licensee can apply independently for an injunction.

PRC courts must consider the following factors in determining whether to issue preliminary injunctions:

- whether there is patent infringement;
- whether the patent holder will be irreparably harmed to the extent that monetary damages will be inadequate compensation if the infringing act is not enjoined;
- whether the patent holder has provided an adequate bond; and
- whether issuance of a preliminary injunction would prejudice the public interest.

It is still common for goods which are manufactured in China to be exported internationally via Hong Kong, and the owners of patents which are protected in Hong Kong may consider taking parallel action in that jurisdiction. Hong Kong is a separate common-law jurisdiction largely based on the English legal system. Many PRC entities also operate trading companies or keep assets in Hong Kong, and various interim remedies are available to stop the dissipation of evidence and assets (eg, an Anton Piller order or Mareva injunction).

5.3 Unjustified threats

The Judicial Interpretation states that if a patentee issues a warning alleging infringement of its patent, but does not institute a legal action or withdraw the warning either (i) within one month upon receipt of a written notice demanding that the patentee enforce its patent or (ii) within two months from the date of the issuance of the written notice, the recipient of the warning or any interested parties may apply to the courts for a declaration of non-infringement.

5.4 Remedies

The two most common remedies for patent infringement are permanent injunctions and monetary damages. Once infringement is established, a permanent injunction is generally issued as a matter of law.

(a) Damages

Infringement damages are assessed on the basis of the following factors, in descending order of importance:

- the actual loss suffered by the patentee;
- the profits made by the infringer due to infringement;
- a multiple of reasonable royalty; and
- statutory damages.

If the infringer's profits are to be used as a basis for assessment of damages, evidence preservation becomes an essential tool to enable the patentee to obtain the necessary sales and accounting information from the defendant. In practice, the assessment of damages is often a difficult and complicated process, which explains why damages awards in China are often very low (as they tend to be in most countries with civil law systems and limited or no discovery, such as Germany).

(b)   *Calculation of damages for patent infringement based on defendant's profit*
Article 65 of the Patent Law provides that compensation for patent infringement should be calculated based on actual loss incurred by a patentee as a result of the infringement. If it is difficult to determine the actual loss, the compensation can be calculated based on the 'profit made due to the infringement by the infringer'.

The Judicial Interpretation states that profit made due to the infringement by the infringer should be limited to the profit made as a result of the infringement only. If the profit is due to a combination of other factors, the court shall discount the profit which was made due to factors other than the infringement.

If the infringing product is a component of another product, the court should take into account of such factors as the value of that component and its importance to the final product in determining compensation. If a component is key to the final product, the court may determine the compensation based on the profit of the final product.

6.   Ownership, inventors and compensation

6.1   Ownership of IP – employee inventions
The Patent Law classifies employee inventions as service invention-creations or non-service invention-creations. A service invention-creation is defined as any invention created or discovered by an employee (i) in the course of employment or (ii) created mainly using the material or technical means or facilities (including capital, parts, raw materials or undisclosed technical information) of the employer. It includes inventions made:

- in the course of an employee performing his normal employment duties;
- in execution of any task, whether or not within an employee's employment duties, which is assigned to him by the employer; or
- within one year from his retirement, change of work or termination of employment, where the invention-creation relates to his previous employment duties or tasks assigned to him by his previous employer (the 'one-year clawback').

The one-year clawback provision means that there is a potential threat that an ex-employer of an employee-inventor will own an invention.

Ownership of service invention-creations vests in the employer, who has the right to apply for patent protection. Ownership of non-service invention-creations vests in the employee-inventor and the inventor can (subject to certain conditions) apply for a patent in his own name.

It should be noted that in determining employee-inventorship decisions the Chinese courts often appear to be employee-friendly. In a 2008 case in the SPC (the highest-level court in China), the court discussed (i) the criteria for an invention-creation to qualify as a service invention-creation and (ii) ownership of clinical data. The court ruled in favour of an inventorship claim raised by an employee-inventor, even though the employer participated in and financially contributed to the clinical testing of the compound in question.

6.2 Ownership of IP – inventions made by way of consultancy or commission

The Patent Law states that an invention created or made in execution of a commission belongs to the person or entity which was commissioned to make the invention, subject to any agreement to the contrary. This is reflected in the PRC Contract Law which states that unless otherwise agreed by the parties, the right to apply for a patent in relation to an invention/innovation resulting from a commissioned development belongs to the developer.

6.3 Inventorship

Under the Implementing Rules, which offer more detailed guidelines for interpretation of the Patent Law, an inventor is defined as "any person who makes innovative contributions to the substantive features of an invention-creation". The Implementing Rules specifically exclude anyone responsible only for organisational work, or who offers facilities for using material and technical means, or who takes part only in 'auxiliary functions'.

An originator should note that inventors have the right to be named as such in patent applications for the invention. Inventorship also relates to identifying who has the right to be remunerated for the invention (see further below on employee remuneration).

6.4 Employee-inventor remuneration

Article 16 of the Patent Law provides that any entity that is granted a patent right shall award to the inventor and creator of a service invention-creation a reward and, upon exploitation of the patented invention-creation, shall pay the inventor or creator a reasonable remuneration based on the success of the invention and the economic benefits secured by the invention.

The Implementing Rules provide that the patent owner (employer) may agree with its employee-inventor or specify in a legally binding company policy the amount and payment method of the reward and remuneration as stipulated in Article 16 of the Patent Law. In the absence of agreement, the following statutory levels of reward and remuneration will apply:

- A reward of not less than Rmb3,000 (approximately US$450) for an invention patent should be provided to the employee-inventor within three months of the publication date of the patent.
- Annual remuneration of not less than 2% of the net profit derived from exploiting the invention patent during the patent term should be awarded to the employee-inventor.
- Where the patent is licensed to a third party, the employee-inventor should be awarded not less than 10% of the licence fee.

The Chinese courts can take issue with proving that there has been prior agreement with employee-inventors. When carrying out due diligence, issues to look out for include whether the contract research organisation (CRO) has set out in its employee handbook and employment agreements:

- criteria for qualifying as an inventor and the entitlement of inventor remuneration;
- the timeline for payment; and
- the calculation method of the remuneration – and whether the agreement contains an acknowledgment from the employee that the proposed remuneration is reasonable.

7.   Branding and designs

7.1   Trade marks

To be registered in the PRC, trade marks must be distinctive and not in conflict with prior rights. For composite marks with descriptive elements, disclaimers of the descriptive elements should be offered upon filing. Trade mark registration is valid for 10 years, calculated three months from the date of publication of the application in the PRC *Trade-mark Gazette* (absent opposition), and renewable for successive 10-year periods.

As the PRC adopts a first-to-file system which makes it difficult to challenge prior registered rights, undertaking an availability search prior to adopting a new mark is recommended. It should be noted that while multi-class applications are common in other jurisdictions, they are not yet available in the PRC. In addition to filing the marks in their original language, trade mark proprietors should consider adopting and filing for Chinese equivalent marks to broaden the scope of protection.

8.   Protecting valuable information

8.1   Employee pre-emptive rights

Article 326 of the PRC Contract Law purports to allow an employee an option to purchase an invention created by the employee-inventor in the course of his employment. This option is triggered on an assignment of the invention out of the employing company.

It is possible to contract out of these pre-emptive rights and typically this would be dealt with in a CRO's employment agreements with its employees. However, care should be taken that the way this is done would be enforceable – would the Chinese courts consider that the provision is fair in the circumstances, including in relation to the nature and value of the invention and any employee remuneration scheme?

Due diligence on the CRO's employment arrangements should also include whether relevant employees have signed an employment agreement incorporating provisions dealing with Article 326; and how the CRO deals with potential ex-employee inventors (eg, ex-employees who are discovered to have created an invention, after they have left that employer).

8.2   Preserving confidentiality

As with any outsourcing arrangement, an originator must ensure that adequate legal and practical steps are taken to protect confidentiality, before commencing negotiation with CROs and during (and after) the contract term. When conducting due diligence, ensure that employment agreements between the CRO and its employees contain adequate non-disclosure provisions. In addition, the outsourcing agreement should include a provision that the CRO will ensure that its employees sign a separate, detailed confidentiality agreement between originator and CRO employee. As with any outsourcing agreement, an originator should not rely on legal measures and should take practical precautions to restrict access to confidential information.

There is, however, the China-specific and very real risk of possible leakage of confidential information to government agencies – in particular:

- when undergoing confidentiality examination at SIPO for applicants seeking a foreign patent filing for an invention 'completed in China'; and
- during a technology transfer (handled by the Chinese Ministry of Commerce): the transfer of permitted technology requires a record of a technology transfer contract at MOFCOM; the transfer of restricted technology involves completing an application form for approval, which has significant disclosure requirements.

9. Counterfeiting

9.1 Patent counterfeiting

Where an entity counterfeits a patent, the entity shall, in addition to assuming civil liability, be subject to an administrative order by the patent administrative authority. In circumstances where a crime is constituted, the entity shall be held criminally liable.

The following activities are deemed to constitute patent counterfeiting:

- affixing patent markings to products or packaging for which a patent right has not been granted, continued affixing patent markings to products or packaging after the patent right has been invalidated or expired, or affixing a patent number belonging to another person to products or packaging without authorisation;
- referring to a technology as a patented technology while no patent has been granted in materials such as manuals, or referring to a patent application as a patent or using another's patent number without authorisation to lead the public into a mistaken belief that the relevant technology is a patented technology.
- marking, without permission, another's patent number on the product manufactured or sold by the defendant or on the packaging of such product; and
- using, without permission, another's patent number in advertisements or other advertising materials, thus misleading other people into believing that the technology involved is a patented one.

Patent counterfeiting may attract both civil and criminal liabilities.

9.2 Trade mark counterfeiting

Where a registered trade mark is used on crudely manufactured goods that are passed off as being of high quality, thus deceiving consumers, the administrative departments for industry and commerce shall order rectification of the situation and may circulate a notice on the matter or impose a fine, or the Trade Mark Office may revoke the registered trade mark:

- if the trade mark is falsely represented as being registered; or
- if the trade mark is used on crudely manufactured goods that are passed off as being of high quality, thus deceiving consumers.

Trade mark counterfeiting may attract both civil and criminal liabilities.

10. Collaborative models – 'open innovation'

For a long time in the past, the innovation system in China was a socialist plan-based regime. The government research institutes played a dominant role in performing innovation activities, with the government acting as the key coordinator by issuing long-term national science and technology plans; industrial players only played a trivial role. Following economic reforms, China has seen the introduction of a competition-based funding system from industry in order more efficiently to commercialise R&D results. China has begun to see industry–science partnerships – for example, spin-offs set up by PRC government research institutes and universities; contract research collaboration between industry and government research institutes and universities; and joint publication of scientific papers between universities and industry.

In addition to the changes driven by the economic reforms, the Chinese government has tightened up laws and regulations concerning IP protection of novel technologies which attract foreign direct investment in R&D of high-tech industries (including life sciences) in China. Such cooperative efforts not only provide local universities and research institutes with additional funding and advanced equipment, but also allow local parties to become more informed about the international research frontier, creating significant incentives for mutual learning opportunities among domestic and foreign enterprises.

China still faces a steep learning curve in order to understand open innovation. It requires a national strategy which focuses on the promotion of domestic innovation capabilities by the introduction of favourable policies for domestic scientists and which adopts an open mindset and an approach based on knowledge sharing through international links and partnerships.

11. Hot topics

Patent linkage, data exclusivity and new-drug monitoring-period exclusivity are three more tools for the protection of pharmaceuticals in China.

Chinese law provides for 'regulatory' protection of pharmaceuticals, which can be used in conjunction with a patenting strategy.

It also provides for patent linkage. Under the Drug Registration Measures, the SFDA must not accept any generic applications until at least two years before the patent expiry date of the reference product. However, the latest amendment of the Patent Law introduced a 'Bolar' type of exemption whereby applicants manufacturing a patented product are not infringing if this is "for the purposes of providing information necessary for administrative approval". This wording is ambiguous and gives scope for infringers to argue that their acts are in fact legitimate. In practice, a more proactive approach involving active market monitoring and engaging the SFDA may be warranted.

In China, unpublished proprietary data submitted by a company for registration for a drug containing a new chemical entity has data exclusivity for six years from the date of the approval for drug registration. Data exclusivity requires the presence of a new chemical entity and does not appear to apply to new indications or formulations. There is also no provision for an extension or new period of protection for new formulations/indications or other changes in products.

New drugs in China are also subject to monitoring-period exclusivity by the SFDA. Ostensibly this is for the purpose of checking a product's safety and efficacy, but it also provides another method of regulatory exclusivity. The definition of 'new drug' does not require the presence of an NCE, and includes new administration methods and new indications of drugs previously on the market in China. The SFDA may impose a monitoring period of three to five years from the date of drug registration approval of the product being monitored. During this period, the SFDA must not grant approval to any other company to manufacture or distribute any similar type of drug, except where before the drug registration application date of the new drug another company has obtained approval for conducting clinical trials in China for a competing drug. The criteria for determining eligibility and duration of the monitoring period are set out in the Drug Registration Measures and cover several categories, including chemical drugs, biologicals and traditional Chinese medicine.

The author would like to thank Rita Leung, Jenna Horris Ho and Petrus Chan for their assistance in the writing of this chapter.

Denmark

Kasper Frahm
Sture Rygaard
Plesner

1. Small molecules

1.1 Product and process claims

Under Sections 1 and 3(1) of the Danish Patents Act, small molecules can be protected by patent and both product and process claims are available. It was not until 1978 that product claims without the requirement to mention in the claim the intended area of use for the product were introduced. It now follows explicitly from Section 8(2) of the Danish Patents Act that the fact that an invention concerns a chemical compound does not necessitate the mention of an intended use: the first patent claim for a technical use of the small molecule can cover any use of the compound.

1.2 Scope of protection of claims and Markush formulae

The scope of Danish patent protection is determined by the claims, but guidance for interpretation can be found in the description (see Section 39 of the Danish Patents Act, cf EPC Article 69).

The starting point for the interpretation of patent claims has, in Danish case law, been established to be the natural understanding of the words in the claims as they read. However, in all cases where an infringement of claims is not obvious, the description should be consulted as a guide to interpretation.

In cases where it is not clear whether or not a product falls within certain patent claims, there is an issue as to whether such claims will be interpreted in a wide or narrow sense. Danish case law is limited, but it would appear that Danish courts try to combine fair protection for the patent holder with a reasonable degree of legal certainty for third parties by trying to establish the invention's actual contribution to technological development, all the while acknowledging that the first patent claim for a chemical compound can protect all uses of the said compound. Danish courts have therefore not simply looked at the wording of the claims, but have also in some cases looked back at the application procedure and on occasion given weight to issues that have only been mentioned in the patent holder's correspondence with the Danish patent authorities. It is generally accepted that the application file can be relevant for interpretation, in particular if the applicant has had to limit the patent claims in order to obtain the patent.

The doctrine of equivalence is accepted in Denmark. It has also been established in Danish case law that, although a certain product or action may not be covered by the wording of patent claims, it may still be within the scope of that patent's protection by being a "technically equivalent solution which is obvious for a professional".

Patent claims containing Markush formulae are accepted in Denmark and they should have a significant structural element which is shared by all of the alternatives. The general requirements must be fulfilled in order for the claims to be patentable. The description must satisfy the usual requirements for exemplification, so that a professional can carry out the invention and enablement must be substantiated over the scope of the claim.

1.3 Metabolites

Where it is a metabolite of a given compound which actually works the desired mechanism in the body, any of the compounds on the way to the actively working metabolite can probably be patented provided the general criteria are fulfilled.

The scope of a patent claim for a given chemical compound will probably also cover a metabolite of the patented compound, at least if it would be obvious to the skilled person that the metabolite would be formed when the patented compound was metabolised and it would also be obvious for the skilled person how the metabolite compound could be made. Likewise the scope of a patent for a metabolite compound could probably also cover a compound where it would be obvious to a skilled person that it would be metabolised into the patented compound and it was known or obvious to a skilled person how to make the previous compound.

2. Second-generation inventions

2.1 Combinations

Combinations of different elements or compounds can be patented if they fulfil the normal requirements of novelty and inventive step. However, where the claim is merely an aggregation of the effect of two known elements, it will not be inventive. The combined technical effect must be different from the sum of the technical effects of the individual features and produce a synergistic effect.

The right conferred by a patent is the right to keep others from using the invention – not a right to use the invention covered by the patent as such.[1]

Thus, to have a patent granted within the life sciences industry does not necessarily mean that the patented invention can be used commercially, as one might be dependent on other patent rights. If molecule A is patented by company I, molecules A and B combined may be patented by company II (if the conditions for obtaining a patent are met). However, company I can prevent company II from using its combination of A and B if company II has not obtained permission (a licence) from company I. If the AB invention constitutes an important technological advance, company II will be able to obtain a compulsory licence for the use of molecule A on "reasonable market terms".[2] If the AB invention does not constitute an important technological advance, company II will have to await the expiry of company I's patent

---

1   Section 3 of the Danish Patents Act.
2   Section 46 of the Danish Patents Act.

if a licence cannot be obtained. Compulsory licences are, however, extremely rarely applied for, or given, in Denmark.

2.2 Enantiomers

Enantiomers can be protected by product claims in Denmark. The description of a racemate does not take away the novelty of a specific enantiomer which can be patented as a selection invention, provided the other requirements including inventive step are fulfilled.

2.3 Selection inventions

The decision on whether to grant a patent to a selection invention depends on whether it is found that the invention (ie, the selected element), is already disclosed individualised in the prior art and whether the selection (eg, a smaller range of specific alternatives), has an effect which involves inventive step over the first patent.

2.4 Methods of use and secondary indications

The issue of whether to grant patents for a new use of a known substance was for some time a matter of debate in Denmark. However, in 2007 the Danish Patents Act was changed and it now follows from Section 2(5) that patents can be obtained for specific uses of a known substance. It is therefore not necessary to have process claims (Swiss-type claims) for the manufacture of the known substance for new uses.

If there is already a patent in effect for the known substance, the secondary-indication patent will be dependent on this patent (see the discussion in section 2.1 above).

The new use for the known substance has to be documented and explained in the patent application and the patent claims obviously have to meet the conditions for obtaining a patent (ie, novelty and inventive step).

2.5 Methods of treatment

There is a belief that the general knowledge and know-how of the medical community should not be monopolised, and it follows from Section 1(3) of the Danish Patents Act (like EPC Article 53(C)) that methods for surgical or therapeutic treatment or for diagnosing which can be used on people or animals cannot be patented.

In line with the Enlarged Board of Appeal's decision in G 1/04, the prohibition against patent methods for diagnosing only includes claims that relate to diagnosing in a narrow sense of the word.[3]

2.6 Formulations and physical forms

Danish law allows for patents on the formulation of a pharmaceutical (ie, claims that describe the contents and structure of a pharmaceutical besides the active pharmaceutical ingredient).

Physical forms such as salts are patentable when the claim describes a specific

---

3   See also G 2/06 (*use of embryos*) and G 1/07 (*dosage regime*).

purpose – for example, a better release or stability profile (see the discussion in section 2.4 above).

2.7 Reach-through claims

A reach-through claim – that is, a claim for substances with certain properties, which according to the patent claim can be found by using certain methods – will not be patentable as applicants are only granted patent protection for their actual contributions to their society's technological progress. Such patent claims would under Danish law be considered an attempt to obtain patent protection for a future invention and would normally not fulfil the requirements of reproducibility without undue burden – see T 1063/06 and T 1140/06.

3. DNA, biologicals and personalised medicine

3.1 Discoveries

As a general rule, discoveries cannot be patented in Denmark,[4] nor protected under any other intellectual property law. Whereas an invention is considered a contribution to society, a discovery is viewed as an acknowledgement/establishment of what already exists.

There are, however, significant exceptions to this general principle with regard to the patenting of chemical compounds. Many chemical compounds exist in nature and granting patents with product claims for these compounds can be considered as granting patent rights to mere establishment of what already exists. However, the development of Western societies and the desire for a strong pharmaceutical industry has necessitated the granting of patents for chemical compounds that also exist in nature. The inventive step of the 'invention' is that the patent holder identifies the existence of the chemical compound, explains a method for its manufacture and states a practical use for it (in other words, the patent holder places the compound at the disposal of industrialised society for the first time).

Today, the patenting of chemical compounds already existing in nature is no longer controversial, but in recent years the demarcation between discoveries and inventions has again become a live issue in more than one regard with the question of patenting of genes, stem cells and organic tissue (see below).

3.2 Gene patents and industrial application

In Denmark, plant varieties and animal varieties as well as mostly biological methods for the manufacture of plants and animals are not subject to patent protection (see Sections 1(4) and (5) of the Danish Patents Act, which implement rules in the EU Directive on the legal protection of biotechnological inventions). Plant varieties can be protected according to Council Regulation (EC) No 2100/94 of July 27 1994 on Community plant variety rights and the Danish Act on novel plants No 190/2009. However, microbiological methods and products of such methods can be patented.

The mere identification of a specific gene (be it from the human or another

---

4   See Section 1(2)(1) of the Danish Patents Act.

organism) does not warrant a patent. However, an explanation of a method for artificially producing such a gene and the provision of an industrial application of the gene can result in patent protection for the gene. The gene (or rather the industrial application of the gene) is therefore considered novel.

The so-called Biotech Directive, EU Directive 98/44 on the legal protection of biotechnological inventions, was implemented in Danish law in May 2000 and the stringent rule from Article 5(3) of the Directive regarding the industrial application of a sequence or a partial sequence of a gene having to be disclosed in the patent application also applies in Danish law. Thus, it follows from Section 18(1)(5) of the Danish Patent and Supplementary Protection Certificate Order No 93/2009 that an 'express explanation' for the industrial application of a gene is required.

Although Danish law is therefore formally in accordance with the Biotech Directive, in Danish legal literature there is some criticism of the fact that this essential requirement of an express explanation for the industrial application of a gene is not mentioned in the Danish Patents Act itself, but merely in the above-mentioned Order.

3.3   Stem cells and other organic material

In accordance with Article 3(2) of the Biotech Directive, biological material which is isolated from its natural environment or produced by means of a technical process may be the subject of an invention even if it previously occurred in nature, including in the human body.[5]

Thus, organic tissue is patentable if it is isolated from the human body or artificially produced and if it is reproducible.

Under Section 1(b)(1) of the Danish Patents Act, inventions whose industrial application would conflict with public morality or *ordre public* are not patentable. Therefore, inventions for unlawful applications (eg, a hallucinatory drug) cannot be patented if there are no legal intended uses for it.

With the huge leaps of innovation within biotechnology (ie, gene and stem cell research), the issue of *ordre public* has grown in importance. Section 1(b) of the Danish Patents Act mentions a couple of examples of unpatentable inventions – processes for cloning human beings and uses of human embryos for industrial or commercial purposes (and similar examples are found in Article 6 of the Biotech Directive). However, the question of whether less 'radical' inventions than those mentioned in Section 1(b) may be patented may depend on which European country one is in. Thus, unlike some of the Catholic countries in Europe, Denmark is quite liberal in terms of religious dogma and it is not unthinkable that, in future, patents will be granted in Denmark for products which might be considered to be against *ordre public* in more religious societies.

The patenting of stem cells has been subject to debate. The prevailing opinion is that totipotent cells (zygotes) are not patentable, as these cells have the ability to develop into human beings and this would come into conflict with Section 1(a) of the Danish Patents Act which excludes patenting of the whole or parts of the human body in all the different stages of its origin and development.

---

5   See Sections 1(6) and 1(a)(2) of the Danish Patents Act.

Pluripotent stem cells which can give rise to any fetal or adult cell type but cannot develop into a fetal or adult animal by themselves and multipotent stem cells which are determined to a specific function are, however, considered to be patentable.[6] In March 2011, the field was harmonised by the Biotech Directive.

In the case of *Brüstle v Greenpeace* at the Court of Justice of the European Union (CJEU),[7] Advocate General Bot suggested that Article 6(2)(c) of that Directive must be interpreted to the effect that the concept of a human embryo applies from the fertilisation stage to the initial totipotent cells and to the entire ensuing process of the development and formation of the human body and that unfertilised ova into which a cell nucleus from a mature human cell has been transplanted or whose division and further development have been stimulated by parthenogenesis are also included in the concept of a human embryo, in so far as the use of such techniques would result in totipotent cells being obtained. On the other hand, he found that pluripotent embryonic stem cells are not included in that concept because they do not in themselves have the capacity to develop into a human being.

In its decision of October 18 2011, the CJEU followed the Advocate General's opinion and held that an invention which uses human embryonic stem cells is not patentable if the destruction of a human embryo is required. The CJEU gave a broad interpretation of the term 'human embryo': essentially anything capable of commencing the process of development of a human being is included. The CJEU did not comment on the Advocate General's suggestions regarding totipotent cells and pluripotent embryonic stem cells.

The CJEU held that an invention will be excluded from patentability where the implementation of the claimed process requires either the prior destruction of the human embryos or their use as a base material, even if in the patent application itself the description of the process does not refer to the use of human embryos.

As regard stem cells obtained from a human embryo at the blastocyst stage – for instance those cells used in the invention covered by the Brüstle patent – the CJEU held that it was for the referring court to ascertain, in the light of scientific developments, whether the cells would be capable of commencing the process of development of a human being and, therefore, whether they were 'human embryos'.

The CJEU's decision may have a significant impact on the future of stem cell research in Europe. Patents will not be granted in Europe for any inventions that use embryonic stem cells if obtaining them results in the destruction of an embryo.

3.4   Bioinformatics systems

The requirement for industrial applicability will lead to some problems for biotech companies that make new inventions through bioinformatics systems. Thus, if a specific chemical compound or gene sequence has been discovered by computational biology and not through more traditional lab techniques, the actual characteristics of the compound or sequence may be unknown and claims to the

---

6   See Jens Schovsbo and Marten Rosenmeier, *Immaterialret, Jurist og Økonomforbunders forlag*, 2011, page 275 onwards.
7   Case C-34/10.

characteristics therefore based on speculation. Thus, although bioinformatics systems may be a strong tool in the identification of new compounds, the mere discovery of the compound does not warrant patent protection for that compound because the function and use of the compound should be identified in order to give the invention industrial applicability.

Biotech companies therefore have a difficult balancing act before them between being the first to file for a patent for a specific invention and having undertaken enough research in the invention actually to have at least one industrial application for it.

Under Section 1(2)(3) of the Danish Patents Act, computer programs that are solely that and nothing else are not to be considered inventions. However, if the computer program for example contains an algorithm that enables scientists to use the program as part of an apparatus that artificially produces a gene, it could be patentable.

3.5 Copyright and sequence information

Under Section 71 of the Danish Copyrights Act, a person who makes a database or the like in which a large amount of information is collated, or which is the result of a substantial investment, has the exclusive right to dispose of that work as a whole or a substantial part thereof by reproducing it and making it available to the public. This may be used to protect substantial sequence information (eg, in biobanks).

3.6 Sufficiency issues

According to Section 8(2) of the Danish Patents Act and Article 83 of the EPC, the patent must contain a description which must disclose the invention in a manner which is sufficiently clear (and complete) to enable a person skilled in the art to carry out the invention on the basis of it. Whereas this is clearly a requirement under Danish law, it does not seem to have led to any court decisions holding a patent invalid for insufficiency of disclosure.

The inventor does not have to (nor may he be able to) explain why an invention works, but the patent should explain how the invention is to be worked.

With regard to patents in the life sciences industry, patent holders of chemical compound patents will sometimes try to extend their inventions by stating analogue compounds even though the existence of these analogue compounds may not yet have been determined. The Danish patent authorities have established case law from which it follows that a patent application only has to render probable (ie, does not have to prove) that the stated technical effect takes place for the range of compounds covered by the patent.

4. Acts of patent infringement

4.1 Infringement

A Danish patent gives the rights holder the possibility of preventing others from exploiting the patent in different ways. The protection covers direct infringing acts, indirect infringing acts and contributory infringing acts.

The directly infringing acts and indirectly infringing acts covered by the Danish

Patents Act imply that no one, except the proprietor of the patent, may without permission exploit the invention:

- by making, offering, putting on the market or using a product which is the subject matter of the patent, or by importing or possessing the product for these purposes; or
- by using a process which is the subject matter of the patent or by offering the process for use in Denmark, if the person offering the process knows, or it is obvious in the circumstances, that the use of the process is prohibited without the consent of the proprietor of the patent; or
- by offering, putting on the market or using a product obtained by a process which is the subject matter of the patent or by importing or possessing the product for these purposes.

The third bullet point above covers so-called 'indirect' infringing acts and makes it possible for the patent holder to prevent sales of, for example, a pharmaceutical drug imported from India if a process which is patented in Denmark was used for making the drug in India.

See also section 1 above regarding small molecules, where some generally applicable principles on the scope of patent protection are discussed.

4.2   Contributory infringement

Under Section 23 of the Danish Penal Code, anybody who instigates, participates in or provides advice in respect of a breach of a given provision of the Penal Code is to be considered a contributory offender.

It is generally acknowledged in Danish literature on intellectual property that an analoguous notion of contributory infringement applies to intellectual property infringements. Thus, although a company may not in itself be infringing a patent, it may still be found to have contributed to an infringement.

Section 3(2) of the Danish Patents Act states that contributory patent infringement takes place by supplying, or offering to supply, someone who is not authorised to use the invention with the means to use the invention in Denmark. However, this only applies if these means are an essential element of the invention and the person knowingly supplies, or offers to supply, the means, or it is obvious in the circumstances that these means are suitable and intended for such use. If the means are staple commodities (ie, products occurring normally in the trade for other purposes), it will only be considered contributory infringement if the person supplying, or offering to supply, the means induces the person supplied to commit acts which exploit the patent.

This provision on contributory patent infringement was introduced in 1978, but the first court decision which interpreted it was made on March 9 2011. In that decision the Bailiff's Court (the court of first instance) in Copenhagen stated that the 'intent' requirement will normally be fulfilled if the supplier "states, suggests or in other ways recommends" the patented use in its marketing. The Danish court found that it was sufficient that 'a part' of the class of ultimate customers (not necessarily the customer of the first supplier) intended to use the products for the patented use. The preliminary injunction is under appeal and is expected to be heard in March 2012.

4.3 National treatment of 'Bolar'-type provisions

Section 3(3)(4) of the Danish Patents Act exempts from patent infringement actions that are limited to the subject of the patented invention if these actions are necessary to obtain a marketing authorisation for a pharmaceutical intended for humans or animals in the European Union, in an EU member state or in other countries.

4.4 Experimental-use exemptions

Experimental use of the subject matter of the patented invention is not considered an infringement of the patent (see Section 3(3)(3) of the Danish Patents Act) even if the experiments serve a commercial purpose. The exemption at least covers acts which are done in order to determine how the invention works and development possibilities.

4.5 Submitting authorisations and offers to supply

Prior to the sale of a pharmaceutical product in Denmark, a marketing authorisation must be obtained for that specific product. Obtaining a marketing authorisation is thus one of the first steps in the process of selling the products.

According to Section 642 of the Danish Administration of Justice Act, a plaintiff may obtain a preliminary injunction – the remedy by which a plaintiff may enjoin actions by a defendant prior to trial on the merits – if the plaintiff can establish *prima facie* or show that it is probable that the following criteria are met:

- the actions at which the injunction is directed infringe the rights of the plaintiff;
- the defendant would perform the actions against which the injunction is directed (the requirement for actuality); and
- the purpose would be lost if the plaintiff had to resort to ordinary court proceedings.

The requirement for actuality in the second bullet point above implies that there must be reason to assume that the generic company will start marketing the generic product within a relatively short timeframe (ie, that there is an actual, not just a hypothetical, threat of infringement).

It has long been acknowledged that the mere filing for a marketing authorisation, or its grant, does not in itself constitute infringement and that it can therefore not form the basis of a preliminary injunction.

Under Section 28 of the Danish Pharmaceuticals Act, a marketing authorisation expires after three years if it has not been used. However, the sunset period is suspended by a valid patent in force. Therefore, the generic company may only have applied so as to be ready to launch its product at the moment of patent expiry.

It is also possible that Denmark is only used as a reference member state under the decentralised application procedure and the product is not intended to be marketed on the Danish market. The Danish Medicines Agency is known to have a fairly quick handling of cases, and applying in Denmark may therefore simply be done for the purpose of obtaining marketing authorisations in various other relevant EU member states, where the decentralised application is also to lead to marketing authorisations.

However, the grant of a marketing authorisation may of course also be the actual first step of a planned launch prior to patent expiry and it is therefore not uncommon for patent holders to ask marketing authorisation holders to confirm that they will not launch their generic product prior to expiry of patent or supplementary protection certificate (SPC). Specific circumstances which may indicate an imminent threat of marketing of the product in Denmark may constitute grounds for a preliminary injunction.

Two recent decisions have shed light on the legal situation in Denmark.

In a case between Eli Lilly and Nomeco (a distributor of all brands of medicine) several generic producers had filed applications for marketing authorisations for generic products. Eli Lilly requested confirmation from the marketing authorisation holders that they would not commence marketing of their products prior to the expiry of Eli Lilly's patent. In their various replies the marketing authorisation holders stated that they had "no current plans to market" their products but that if they did market, then such marketing would not infringe any valid patents. On April 28 2008, the Bailiff's Court in Frederiksberg ruled that some of the statements had been conditional and that Eli Lilly had therefore shown that it was probable that one or more of the marketing authorisation holders intended to launch their products prior to patent expiry.[8] Nomeco subsequently appealed the decision of the Enforcement Court to the Eastern Division of the High Court. However, in its decision of January 21 2009 the High Court confirmed the interlocutory injunction granted by the Enforcement Court.[9]

Secondly, on November 18 2008 the High Court changed a decision of March 11 2008 from the Bailiff's Court in a case between Novartis and Teva and granted an injunction against Teva's marketing of a generic product for which a marketing authorisation had been obtained. Novartis contacted Teva, which informed Novartis that it had no current plans to market the product and that the primary purpose of the application submitted for a marketing authorisation was to use Denmark as a reference member state. Teva did not, however, want to declare that it would not start marketing its product prior to expiry and this fact, combined with the very early stage of Teva's application for a marketing authorisation (more than three years prior to the expiry of SPC), made the High Court find that infringement had been rendered probable.

In the following confirmatory action case on the merits (where it has to be substantiated that the injunction was lawfully made) before the Maritime and Commercial Court, Teva acknowledged and accepted Novartis' claims 2 and 3 (ie, that the production, offering for sale, sale or use of pharmaceuticals containing the active pharmaceutical ingredient would constitute an infringement of Novartis' Danish patent and SPC rights as long as these are in force). On that basis the Maritime and Commercial Court cancelled the injunction. It held that the timing of Teva's application for a marketing authorisation did not in itself indicate a launch prior to patent expiry. Further, the court held that by acknowledging and accepting Novartis' claims 2 and 3 during the confirmatory action, Teva had expressed that it

---

8   Case FS 1-1306/2007.
9   4, aff'd Kaeresag nr B-1294-08.

would respect Novartis' patent rights as long as these were in force, which according to the court was in all essentials the same as Teva had done during the parties' correspondence prior to the case.

Novartis appealed the judgment to the Danish Supreme Court which on June 9 2011 confirmed[10] the Maritime and Commercial Court's result. The Supreme Court noted that Teva had stated that it had no current plans to market its product and that there were in the case no specific circumstances to the contrary which could create doubt to this statement.

When a marketing authorisation has been granted, the product may be marketed when a price for the product has been notified to the Danish Medicines Agency and entered on the Danish Medicines Agency's price list of pharmaceutical products in Denmark (in Danish, *Medicinpriser*). The price does not have to be approved by the Danish Medicines Agency. When the price has been notified to the Danish Medicines Agency, the product is automatically entered on the price list. Pharmacies are obliged to sell all products entered on the price list, and the offer to supply a generic product encompassed by a patent or an SPC in force by entering it on the price list would in Denmark therefore be considered an act of infringement.

4.6 Summaries of product characteristics (SmPCs)

It has been established in Danish law that statements of intended use in SmPCs are relevant to infringement.

In a case decided by the Copenhagen Bailiff's Court on July 24 2008,[11] it was held that Aventis could obtain an injunction against several different statements in the SmPCs and package leaflets for Merck's product indicating that the product was particularly well suited for avoiding inducement of certain heart arrhythmia, which was the subject matter of the second-indication type of patent claim. This decision shows that the statements in SmPCs are considered to be marketing statements in relation to prescription drugs.

5. Patent enforcement

5.1 Obtaining information on the infringer and the infringement

The most common way in Denmark of obtaining pre-trial information on the infringer and the infringement is through the rules on preservation of evidence. The Danish rules on preservation of evidence entitle the claimant to ask the Bailiff's Court to perform a search to secure evidence of a possible infringement at the alleged infringer's business premises (see the Danish Administration of Justice Act, Chapter 57a). Under Section 653(1) of the Danish Administration of Justice Act, a claimant must demonstrate the probability that (i) the defendant has infringed the rights in question and (ii) there is reason to believe that evidence of the alleged infringement and its possible extent can be found at the premises.

The court may allow this search without notifying the defendant and, due to the

---

10   Case 266/2009, reported in the *Danish Weekly Law Journal*, 2011, p2501.
11   *Aventis Pharmaceuticals v Merck NM AB*, case FSF3-7418/2008.

risk of the evidence being lost if the defendant has been notified in advance, prior notification is usually dispensed with.

During the securing of evidence, the court will need to strike a balance between the claimant's interest in the investigation and the defendant's right of privacy, trade secrets or other considerations; the principle of proportionality applies. It is stated in the preparatory works to the Administration of Justice Act that the proprietor of a process patent has a particularly strong interest in the preservation of evidence if a shifting of the onus of proof in relation to a process patent for a new compound in the Danish Patents Act does not apply.

The claimant must bring a confirmatory action against the defendant no later than four weeks after notice given by the Bailiff's Court that the investigation has been finalised.

At the time of writing (mid-2011), the rules have been in force for almost 10 years and have shown to be an effective means for rights holders to establish the extent of possible infringements. They have mainly been used in relation to trade mark infringement matters, but they have occasionally also been used in relation to patent infringement issues.

5.2   Interim relief

The most commonly applied interim relief in matters of patent infringement is a preliminary injunction. Pursuant to Section 642 of the Administration of Justice Act, a plaintiff can obtain a preliminary injunction by substantiating or demonstrating the probability that:

- the actions at which the injunction is aimed infringe the rights of the plaintiff;
- the defendant will continue to perform the above-mentioned actions; and
- the purpose of the application for injunction would be forfeited if the plaintiff had to resort to regular court proceedings.

As with the preservation of evidence, the principle of proportionality applies so that an injunction may be refused if it would cause the defendant disproportionate harm or inconvenience. This is, however, very rarely used in IP matters, where it almost always depends on an evaluation of whether IP infringement is considered to be taking place or not (ie, the decision is not dependent on an assessment of a balance of convenience, as is known for example in the United Kingdom).

During preliminary-injunction proceedings, the defendant may claim invalidity of the patent-in-suit. However, Danish case law shows that Danish Bailiffs have only found patents to be invalid if the defendant has presented novelty-destroying prior art which has not previously been assessed by relevant patent authorities.

If a preliminary injunction is granted, the plaintiff must bring a confirmatory action on the merits against the defendant no later than two weeks after the Bailiff's Court's decision.

5.3   Springboard/post-patent expiry injunctions

There are no cases in Denmark which allow an injunction to be issued in a patent matter after the expiry of the patent based on 'springboard' principles. Further, no damages awards have been made in patent matters based on 'springboard' arguments – that is, that the infringer has obtained a head start due to the use of the patent in the patent period, for which the infringer should pay damages based on the advantage the infringer obtained on that basis. Springboard considerations have, however, been considered in relation to the exploitation of confidential information by an ex-employee and, from the reasoning in that matter, it seems that the court would accept springboard considerations, at least based on the Marketing Practices Act and probably also in relation to patent infringements, although the remedies should perhaps also be found in the Marketing Practices Act after the expiry of the patent.

5.4   Unjustified threats

There are no specific rules in Danish law regarding unjustified threats, but a threat which is consciously made even though it is known to be unjustified would usually be a violation of good marketing practice (see the Danish Marketing Practices Act), and/or in violation of Danish and EU competition law.

5.5   Remedies

Danish courts can grant the following remedies: preservation of evidence; injunctions; fine or imprisonment in cases of negligent or intentional infringement; damages and tort; the recall or removal of products from the trade; destruction or alteration of products; delivery-up of products to the plaintiff; and a publication of a court decision in national media.

The remedies normally requested in matters of infringement of a pharmaceutical patent will be claims for an injunction and a recall of the products. If an injunction is obtained, the plaintiff will usually in the following case on the merits also claim damages and request that the infringing products be destroyed. Occasionally, the plaintiff will also ask for publication of the decision in national media and to have its litigation costs covered.

The claim for damages will normally be divided between a reasonable remuneration for the use of the invention (normally calculated as a reasonable royalty) and actual damages for the further loss incurred by the plaintiff. Section 58(2) of the Danish Patents Act states that when calculating the damages for further loss, courts should among other things take into consideration the loss incurred by the plaintiff and also the unjustified earnings that the infringer has obtained by selling the infringing product.

6.   Compulsory licensing

There are several situations under Danish law where a compulsory licence can be granted. The Danish Patents Act, Chapter 6, concerns licensing and the transfer of patent rights, including, among other things, if a patented invention has not been used after three years from the grant of patent and four years after the application for patent, in the case of a combination invention (see section 2.1 above), or when general public interests necessitate it.

A compulsory licence is only granted if the licensee has not been able to obtain an agreement with the licensor about a licence on fair market terms and it is only granted to those that are considered to be able to use the invention in a proper way.[12]

The Maritime and Commercial Court decides as a first-instance court on all matters regarding compulsory licensing.

7.   Ownership, inventors and compensation

The rules regarding ownership, inventors and compensation are found in the Danish Act on Employees' Inventions.

Under Section 3 of this Act, an employee has the right to inventions he has made. However, if the invention has been made through the employee's regular work for the employer and it falls within the employee's normal area of business, or if the invention is the result of a specific assignment given by the employer to the employee, under Section 5 of the Act the employer is entitled to obtain the right to the invention. Therefore, an employee must inform his employer in detail about an invention, and the employer then has four months to decide whether or not to obtain the right to the invention.[13]

If the invention turns out to be of greater value to the employer than what the parties had anticipated the employee would perform, the employee is entitled to reasonable remuneration.[14] This remuneration should be based on the value of the invention for the employer, the conditions of employment for the employee and the amount of work the employee has put into that specific invention.

Two Danish Supreme Court decisions from 2004 concerned the issue of employee transfer of inventions. In 2004.1802H,[15] the Supreme Court found that a forest ranger's invention of a pallet system for the handling of, among other things, Christmas trees should be considered part of his regular work and the invention was transferred to his employer without extra compensation. In 2004.1018H the Supreme Court found that a sales consultant's invention of a new radiator valve lay beyond his normal work and that he was therefore entitled to remuneration from his employer for the invention. On the one hand, the court noted that his employer had found the invention so interesting that they had applied for a patent and spent more than Dkr1 million; on the other hand, the consultant's contribution to the invention was a mere idea. The remuneration was therefore set at Dkr50,000 (about €7,000).

Similar rules to the above apply for inventions made at public research institutions. Thus, according to the Act for Inventions Made at Public Research Institutions, the employee has the right to inventions he has made but the research institution has the right to obtain the inventions that are made as part of the employee's work. As opposed to the Act on Employees' Inventions, where remuneration is paid, the Act for Inventions Made at Public Research Institutions operates with payment to the employee which is dependent on the commercial success of the invention.

---

12   See Section 49 of the Danish Patents Act.
13   Act on Employees' Inventions, Sections 6 and 7.
14   *Ibid*, Section 8.
15   *Danish Weekly Law Journal* 2004, page 1802.

8.   Branding and designs

8.1   Trade marks in life sciences

The issue of trade marks in life sciences is particularly interesting in relation to the huge market for parallel import of pharmaceuticals within the European Union.

It follows from case law from the Court of Justice of the European Union that repacking of pharmaceuticals, including placing the original producers' trade marks on the repackaged products, is only allowed if the actions are necessary in order to market the parallel imported products and if the trade mark owners' interests are sufficiently protected.

The Danish Supreme Court has in a number of cases determined that repacking is not necessary – and not, therefore, a trade mark infringement if the pharmaceuticals were already in package sizes that can be sold in Denmark. However, if the package size makes it impossible or very hard for the parallel importer to sell the product in Denmark, repacking and placing the original trade mark on the new repackaged product can, under some circumstances, be allowed.

Co-branding (ie, placing the parallel importer's trade mark next to the original trade mark), has not been allowed by the Danish Supreme Court in cases where the parallel importer's trade mark appeared prominently on the packages, as it was not found necessary in order to market the product.

8.2   Extending protection using trade marks and designs

In accordance with the so-called *Philishave* decision in C-299/99, it is established in Section 2(2) of the Danish Trade-mark Act that trade mark rights cannot be obtained for signs consisting of: a form which merely follows from the type of product; a form of the product which is necessary to obtain a technical result; or a form by which the product gains a significant value. It is therefore difficult to imagine the extension of protection for a previously patented product by obtaining trade mark rights to the product.

8.3   Protection using unfair competition

Under the Danish Marketing Practices Act, some patented inventions may have protection for the specific form and design they have been given, as an identical copy of the product might be considered unfair competition except as stated in relation to springboard remedies described in section 5.3 above.

However, with pharmaceuticals the patented invention is often the active pharmaceutical ingredient or the process for manufacturing the pharmaceutical, and it is unlikely that a patent holder will, after the expiry of a patent, prevent others from entering the market by claiming unfair competition.

8.4   Comparative advertising and advertising restrictions

Under Section 5 of the Danish Marketing Practices Act, comparative advertising is allowed if the advertisement: is not misleading; relates to similar types of goods or services; is objective in the compared characteristics; does not create risk of confusion as to any connection between the compared goods; and does not discredit the competitor's goods.

Advertising of pharmaceuticals is subject to the same rules and restrictions as advertising of other goods. However, pharmaceuticals are subject to a few further restrictions – see the Danish Act on Pharmaceuticals. According to Section 66 of this Act, pharmaceuticals may not be advertised to the public if: (i) they are prescription drugs; (ii) they are unfit for use without the patient having sought medical help; or (iii) they are covered by the Danish Act on Euphoriants. The term 'the public' includes any person who is not a doctor, dentist, veterinarian, pharmacist, veterinary nurse, pharmacologist, or a student to become one of the above.

8.5 Protection of online content

Online content is generally protected under the relevant intellectual property rules, depending on what kind of content one is dealing with. The most relevant rules are the Danish Copyright Act and the Danish Designs Act.

9. Protecting valuable information

9.1 Preserving confidentiality

It is possible in Denmark to preserve confidentiality regarding a new invention. If a company is able to keep secret confidential information about, for instance, a process for manufacturing a pharmaceutical ingredient because everything happens behind closed doors and because reverse engineering is not possible, there are obvious advantages in this. First, the company does not have to spend the money on applying for patents and, second, if the company later decides to seek a patent on the invention (and obtains it) one could argue that the patent protection has been extended in time.

However, not applying for a patent on a new invention also has its disadvantages. The secret may get out, another company might develop a similar invention and it will be harder to give licences to the invention than if it was patent protected.

9.2 Know-how and ex-employees

Know-how is protected under Danish law and can be transferred by agreement.

The extent of the protection of know-how cannot be determined specifically, as it will depend on the type of know-how and the specific circumstances of a case. However, Danish case law provides support for claiming that know-how enjoys protection from unauthorised use from ex-employees.

Section 19 of the Danish Marketing Practices Act provides a general rule stating that an employee or company partner may not (i) improperly gain access to company secrets or (ii) pass on to third parties or (iii) use any knowledge of company secrets obtained legally during the employment/collaboration, for three years after the end employment/collaboration.

Section 8 of the Danish Patents Act states that the patent document should contain a clear description of the invention (see also section 3.6 above). However, it is generally accepted in Denmark that, although the description of the invention should be fair and quite substantive, there is no requirement for the inventor to publish all his know-how in connection with the description; nor is it a requirement that the description covers the best-known application of the invention.

9.3 Copyright, patient information leaflets and packaging

There are very specific rules relating to the content of patient information leaflets (PILs) and the appearance of packaging. These rules have been set by the Danish Medicines Agency and leave little room for differences in PILs and packaging.

Although the question has to our knowledge not been decided in Danish law, we find it unlikely that a company could claim copyright to PILs in relation to a generic producer later using a similar or identical leaflet in connection with the sale of its generic product, as the contents of a PIL will mainly be factual information which is closely regulated.

10. Counterfeiting

All the Danish Acts on intellectual property contain provisions for criminal penalties.[16] Further, Section 299b of the Penal Code contains a provision applicable in cases of particularly serious IP infringements under especially aggravating circumstances. It follows from these provisions that intentional or grossly negligent acts of intellectual property infringements are punishable. The sanctions vary from fines (in most cases) to up to six years' imprisonment.

In a recent decision from a Danish lower court, a person who over several years had repeatedly sold counterfeit goods on flea markets was sentenced to five months' imprisonment and – more importantly – the judge banned him from selling any type of goods at flea markets for two years. This case was appealed, and on June 15 2011 the High Court ruled to confirm the decision in its entirety whilst also increasing the term of imprisonment from five to six months.

Danish police and prosecutors have become more aware of intellectual property infringements, but the area could still use more focus from enforcement officials considering the vast number of intellectual property crimes that take place every day.

Danish customs officials are effectively enforcing Council Regulation (EC) No 1383/2003 of July 22 2003 concerning customs action against goods suspected of infringing certain intellectual property rights and the measures to be taken against goods found to have infringed such rights.

There is currently an ongoing discussion in Danish academic intellectual property circles as to whether rights holders can only act against products intended for commercial use in Denmark, or if they can also demand destruction of seized counterfeit products which are intended for personal use but which have been sent from a country outside the European Union to Denmark. In fact the Danish Maritime and Commercial Court decided the question on November 8 2011 in a judgment of very principle in the so-called *Rolex III* case.[17] The facts were that the defendant had purchased a counterfeit Rolex watch on the internet, well knowing

---

16   See Section 57 of the Danish Patents Act, Sections 76 to 80 of the Danish Copyright Act, Section 42 of the Danish Trade-mark Act, Section 36 of the Danish Design Act and Section 54 of the Danish Utility Patents Act.

17   Case V29/10.

that the watch was fake. It was shipped from Hong Kong and stopped by the Danish customs authorities, which asked Rolex whether it should be held back. The Court ruled in favour of Rolex and found that the watch could be required to be destroyed even though it was only imported for private use. The court found that it was decisive whether the manufacturing of the watch, if it had been done in Denmark, would have constituted a violation of Rolex's intellectual property rights in Denmark. This interpretation was mainly based on the 8th consideration in Customs Regulation 1383/2003. This decision has thus clarified a question of practical importance – although the judgment may still be appealed. The question will likely be decided by the CJEU at a later stage.

11.   Collaborative models – 'open innovation'

Open innovation projects are very interesting from an intellectual property perspective, but raise various issues.

If the inventor who agrees to transfer his invention to a company through a collaborative project is employed elsewhere, and the invention falls within the area of his regular work, there is an obvious conflict as to which company will obtain the rights to the invention.

Further, although an inventor might, through an open innovation agreement, have obligated himself to execute all necessary paperwork to ensure that the purchasing company obtains all rights to the invention, the company will still be reliant on the inventor actually performing these actions.

Open innovation agreements can also lead to conflict in cases where the company uses a solution presented by an inventor but does not pay the inventor compensation because the company finds the solution is already public common knowledge and therefore not a solution the inventor would be able to protect.

The area is still quite new in Denmark within the life sciences industry.

12.   Hot topics

One of the hot topics in Denmark in relation to life sciences patent infringements is the issue of the burden of proof in cases regarding possible process patent infringements.

There have, in recent years, been a substantial number of Danish cases regarding process patent infringements and the issues arising from those cases have been the following:

- the issue of whether the finding of certain chemical impurities (traces of intermediaries) in the defendant's product can constitute a fingerprint substantiating the use of the plaintiff's patented process when that specific impurity is known to be a by-product of the plaintiff's process;
- the issue of how well documented an experiment, which the parties themselves or their experts have performed to substantiate or repudiate infringement, should be;
- the issue of how much lower the yield from the defendant's claimed process can be than the plaintiff's process before there must be an economic motive for using the plaintiff's process and thereby infringement; and

- the issue of how a possible inspection of the process used at the manufacturing sites of the defendant should take place.

These issues will probably be decided upon over the coming years as a number of process patent matters will be decided.

On the procedural side, consideration is currently being given as to whether to gather all the preliminary injunction matters in the Maritime and Commercial Court where patent matters on the merits are currently heard. This would be a welcome development, as it would make for shorter oral proceedings and faster decisions if judges were to become more knowledgeable about the often quite technical aspects of life sciences cases.

France

Denis Schertenleib
Cabinet Schertenleib

1. Small molecules

1.1 Product and process claims

Product claims under French law are traditionally protected *per se*. Thus, any use, disposal, sale, offer for sale or importation will be an infringing act (Article L 613-3 of the Intellectual Property Code (IPC)).

Similarly, the protection of process claims follows the common system adopted in the rest of the European Union, according to which the direct product of a process is protected (Article L 613-3 IPC).

Acts of infringement of process claims by direct products are defined in Article L 613-3(c) of the IPC under which the "Offering, putting on the market or using the product obtained directly by a process which is the subject matter of the patent or importing or storing for such purposes" are prohibited. The definition of a direct product has not been settled or even appreciably discussed by French case law.

It should be noted that, for biotechnology products, specific legislation derived from the Biotechnology Directive 98/44 has been put in place.[1] This states that:

the protection conferred by a patent on a product containing or consisting of genetic information shall extend to any material in which the product is incorporated and in which the genetic information is contained and performs its stated function.

And

The protection conferred by a patent on a process that enables a biological material to be produced possessing specific characteristics as a result of the invention shall extend to biological material directly obtained through that process and to any other biological material, derived from the latter, by reproduction or multiplication and possessing those same characteristics.

As a result, it is clear that in a biotechnology context the protection afforded to the direct product of a process is extended to various acts of propagation.

It should also be noted that the protection afforded to a biotechnological product (especially a DNA sequence) could be limited to the specific use disclosed in the description, under Article L 613-2-1 of the IPC which provides:

The scope of a claim concerning a gene sequence shall be confined to the part of such sequence that is directly related to the specific function disclosed concretely in the description.

---

1   Articles L 613-2-2 and L 613-2-3 of the IPC.

The rights created by the grant of a patent including a gene sequence may not be called upon against a later claim on the same sequence if this claim satisfies the requirements of Article L. 611-18 and if it discloses any other particular application of this sequence.

This provision clearly departs from the normal rules on the protection of products *per se*.

While there is still some debate as to whether these provisions constitute a proper implementation of the provisions of the Biotechnology Directive, there is currently no case precedent that has tested these provisions.

1.2 Scope of protection of claims and Markush formulae

Claims have to be interpreted as understood by the skilled worker on the basis of the description and his common general knowledge, pursuant to both Article L 613-2 of the IPC and Article 69 of the European Patent Convention (EPC).

While this proposition does not depart much from the position in other member states, it should be noted that this commonly results in the patent being interpreted by French courts as limited to specific examples disclosed in the description.[2]

Similarly, when faced with a claim that potentially encompasses wide subject matter, the approach of the French courts is not to deal with this issue under the assessment of validity, but rather of infringement.

Thus, a claim that may be infringed solely because it is drafted so widely that it also encompasses embodiments that do not solve the technical problem forming the subject matter of the patent will not lead to an issue with the validity of the claim but, rather, to limitation in the scope of the claim.[3]

The French courts will not deal with such issues under sufficiency or inventive step, but rather as a matter of claim interpretation and restriction of scope. As a result a court could limit a Markush claim that is deemed too wide to describe embodiments.

1.3 Metabolites

While there is no specific case law on metabolites or precursors, the issue of whether there could be an infringement of a patent over a drug, merely by providing a precursor of a drug, would be dealt with either through the issue of claim interpretation (as described in the previous section), or the issue of contributory infringement.

Indeed, under the provisions of Article L 613-4, there can be contributory infringement if one provides the essential means to work an invention. Thus, under certain circumstances it is conceivable that the provision of a precursor of a drug could be seen as the provision of an essential means to work the invention.

However, such interpretation could be defeated by considerations involving the proper interpretation of the claims based on the description. Indeed if the

---

2   *Sofresud v Bertin* Court of Appeal of Paris, September 8 2006.
3   See *HILITI* case, Court of cassation, Commercial chamber, June 4 2002, Number of appeal: 00-11857. See also *SDP* case, Court of Appeal of Douai, Chamber 1, section 2, December 15 2010, Docket number 09/05106, dismissing the appeal lodged against High Court of Lille, June 18 2009, Chamber 1, Docket number 08/04156.

description of the relevant patent does not provide any description of the use of a precursor of a drug, then conceivably the French courts may reject an interpretation of the patent that would cover an undisclosed metabolite.

The issue of active metabolite is largely untried in France. However, any interpretation of the claims that substantially departs from what is actually described by the patent is unlikely to be met favourably by the French courts.

2. Second-generation inventions

2.1 Combinations

Combination patents are afforded protection and thus there is no impediment *per se* to patenting a combination and enforcing such a patent, provided that the combination satisfies all the requirements of patentability.

However, it should be noted that the French courts often apply an 'obvious to try' doctrine, which frequently results in combination or selection patents being revoked. For example, in the *Solvay* case it was held that an invention which resulted from the selection of a molecule on which to operate a substitution from a list of more than 30 molecules, and then the selection of a substitution position, was not inventive on the ground that the combination was obvious to try.[4] Similarly, in the *Teva v Merck* (Alendronate 10 mg) case, the court held that there was no inventive step involved in the selection and the testing of several combinations of potential modifications to an existing molecule.[5]

In these cases, even the existence of a surprising effect in the combination was not sufficient to overturn such a finding of obviousness to try. Thus, under such case precedents, a patent protecting a combination of products will have to overcome the hurdle of such combination being not only novel, but also not obvious to try in light of the prior art.

2.2 Enantiomers

Several cases have dealt with enantiomers such as *Teva v Sepracor* and *Hexal AG and Sandoz v Boehringer Ingelheim Pharma*.[6] These cases have ruled that it was necessary for a patent claiming specific enantiomers, to contain some experimental results supporting the alleged effects, in order to avoid the nullity of the patent either for lack of inventive step or for lack of sufficient description.

Nevertheless the *Lundbeck* case confirmed that enantiomers can be patented under French law, provided they satisfy the requirements of patentability.[7]

---

4   High Court of Paris, 3rd chamber, 1st section, July 1 2008, *SAS Solvay-Fluores France and Company Solvay Fluor GmbH v Company El Dupont de Nemours and Company*, Docket number 05/09022.
5   High Court of Paris, 3rd chamber, 2nd section, *Company MSD Somerset Ltd v Company Teva Classics and Company Teva Sante*, Docket number 05/07130, p 9 of the ruling.
6   *Company Hexal AG and SAS Sandoz v Company Boehringer Ingelheim Pharma GmbH & CoKG*, Docket number 08/12537; High Court of Paris, 3rd chamber, 1st section, October 6 2009, *Companies Teva v Sepracor*, Docket number 07/16446.
7   See, for instance, High Court of Paris, 3rd chamber, 4th section, September 30 2010, *Company Ratiopharm GmbH v Company H Lundbeck A/S*, Docket number 10/08089, judging valid litigious claims of patent EP 0 347 066 B1.

2.3   Selection inventions

The case law discussed above for combinations applies equally to selection inventions.

In addition, the older Amoxicillin case[8] held, in 1980, that to select from six different combinations, one combination that conferred an improved effect was obvious. Thus, selection inventions will have to face the high hurdle that the selection should not have been obvious to try.

2.4   Methods of use and secondary indications

While there may have been some controversy in the past about whether second-medical use claims could be valid,[9] this issue is regarded as settled following the adoption of the European Patent Convention 2000. Nevertheless, this issue may still be arguable for patents that were issued before the new patent convention.

The recent case of *Actavis v Merck*[10] has, however, indicated that the French courts may not be ready to follow the enlarged Board of Appeal decision Enlarged Board G2/08[11] and may disregard dosage regimen from secondary indications. Thus, under this case precedent, features belonging to a dosage regimen cannot be taken into consideration when assessing patentability.

2.5   Methods of treatment

Methods of treatment of the human or animal body by surgery or therapy and diagnostic methods practised on the human or animal body are excluded from patentability under both the application of the European Patent Convention (Article 53(c) of the EPC) and national law (Article L 611-16 of the IPC).

This exclusion is largely based on public interest, public health and ethical considerations.[12] It aims notably at preventing an eventual hindrance in the practice of such methods by physicians or veterinaries.[13] Thus, it is likely that the French courts will continue to uphold these exclusions vigorously.

2.6   Formulations and physical forms

Under the recent case law of *Laboratoires Negma v Biogaran* concerning the purity of chemical compounds,[14] it was held that a product is not deemed novel only because it is prepared in a purer form. This was confirmed by the case of *Hexal AG & Sandoz v Boehringer Ingelheim Pharma*,[15] which held that "a known compound does not become novel only because of the fact that it is prepared in a form more pure".

---

8   *Beecham v Allard* Court of Appeal of Paris, 4th chamber, October 17 1980, PIBD, III, 267.
9   Frédéric Pollaud-Dulian, "La Propriété Industrielle" ("Industrial Property"), *Economica*, 2011, n° 239 and following, p 168 and following.
10   High Court of Paris, 3rd chamber, 1st section, *Company Actavis Group and Company Alfred E Tiefenbacher GMBH v Company Merck Sharp & Dohme Corp*, Docket number 07/16296; see pp 7 to 9.
11   *Abbott Respiratory, Official Journal* EPO, 10/2010, p 456.
12   *Ibid.*, p 139.
13   Jacques Azema Jean-Christophe Galloux, *Droit de la propriété industrielle*, Dalloz, 6th edition, 2006, n° 211, p 132.
14   High Court of Paris, November 12 2010.
15   High Court of Paris, 3rd chamber, 3rd section, May 7 2010, *Company Hexal AG and SAS Sandoz v Company Boehringer Ingelheim Pharma GmbH & CoKG*, Docket number 08/12537, p 15 of the ruling.

It is arguable that such doctrine could apply to biotechnological products, such as proteins for which purity is both a relevant and complex issue.

2.7 Reach-through claims

There is no specific case law on reach-through claims in France. However, in the field of biotechnology, the issues discussed above for product and process claims are relevant.

3. DNA, biologicals and personalised medicine

3.1 Discoveries

Under French law, discoveries are not patentable as such. This principle is solidly stated in Article L 611-10 of the IPC.

This was reiterated by the High Court of Paris in the case *Institut Pasteur v Chiron Blood Testing*, where it was held that a virus was not patentable in itself but that all or part of its genome was patentable inasmuch as it enables the manufacture of a product used for diagnosis or treatment.[16]

In addition, following the Biotechnology Directive 98/44, Article L 611-18 of the IPC was incorporated expressly to exclude from patentability the discovery of elements of human body.

No ruling has interpreted this provision or its relationship to patents covering cells or genes.

3.2 Gene patents and industrial application

The provisions governing biotechnology patents and industrial applicability are derived from the Biotechnology Directive and are discussed in the Intellectual Property Code (Article L 611-18 IPC).

Although there is no case law on the industrial applicability of gene patents, it should be noted that France may have diverged in its implementation of the Biotechnology Directive, in that it has indeed excluded from patentability discoveries of components of the human body, but it has not expressly provided for the patentability of isolated gene sequences as provided for under the Biotechnology Directive. As a result, the proper scope of patents covering genetic material is still an untried legal issue.

3.3 Stem cells and other organic material

The leading case at the European Patent Office is decision G 2/06.[17] This decision has interpreted the provisions of the Biotechnology Directive on public order and morality.

Under this decision, the Enlarged Board held that, under the Biotechnology Directive, claims directed to products which at the filing date could be prepared

---

16   HCP, February 7 2007, *Institut Pasteur* 05/11023, p 6.
17   *Wisconsin Alumni Research Foundation*, November 25 2008, G 2/06, *Official Journal EPO*, 5/2009, p 306 and following.

exclusively by a method which necessarily involved the destruction of human embryos would not be patentable.

This prohibition is in fact wide ranging, as the destructive step need not even be part of the claims. Thus, inventions such as cell lines that may historically have arisen from the destruction of embryos would be unpatentable.

In view of France's implementation of the Biotechnology Directive, and as the IPC expressly refers to the concept of human dignity (Article L 611-17 IPC), which is in effect a wider concept, it is expected that French courts and the French Patent Office will apply a regimen at least as stringent as that used at the European Patent Office.

3.4   Bioinformatics systems

It should be noted as a preliminary point that computer programs may be patented under certain conditions. In order to be patentable, they need to possess a technical character.[18] In addition, in a ruling of the Court of Appeal of Paris it was held that a valid claim should not cover a program as such.[19]

Thus, while the decisions of the French courts are not as developed as those of the EPO, they are nevertheless broadly consistent with them. As such, there is no absolute impediment *per se* to patenting a program in France.

The issue relating to the patenting of products obtained from a computer system, such as a DNA sequence with a postulated function, presents specific difficulties.

First, sufficiency would be an issue in as much as it could be argued that no real invention had been made at the priority date. If, as in the *Sepracor* case,[20] it is held that the patentee was not in possession of the invention at the time of filing, or if use that is put forward in the description was speculative or mere conjecture, a finding of insufficiency could be made by the courts.

The sequence information, however, could itself conceivably be protected under copyright. An essential requirement of copyright is originality. Originality can be defined as the mark of the author's personality or, in the area of computer programs, as the mark of intellectual contribution.[21]

However, in the recent case of *Dassault Systems*, it was held that there could be no copyright lying in the mere product of a computer program.[22] Thus the patentability of the products of such computer programs may be doubtful.

3.5   Databases and search engines

Databases may be protected by a *sui generis* right under Article L 341-1 of the IPC, or by general copyright. However, it is doubtful that copyright could protect a genetic sequence in its actual physical form or in its biochemical function.

---

18   See EPO, Case Law of the Boards of Appeal of the European Patent Office, 6th edition, 2010, p 23.
19   Court of Appeal of Paris, 4th chamber, section B, June 5 2009, *Kone v Azpitarte*, Docket number 2007/20589, PIBD n° 903, III, p 1331 and following, see p 1336.
20   High Court of Paris, 3rd chamber, 1st section, October 6 2009, *Companies Teva v Sepracor Inc*, Docket number 07/16446.
21   Court of cassation, Full Court, March 7 1986 (*Pachot* case).
22   Court of Appeal of Paris, Pole 5, chamber 1, February 24 2010, Docket number 08/11742.

4.   Acts of patent infringement

4.1   Infringement

Please see section 1.1 on product and process claims for an overview of direct infringement in France.

4.2   Contributory infringement

Contributory infringement is provided for in Article L 613-4 IPC, which prevents the supply or offer to supply, in France, of the means of implementing, in France, an invention with respect to an essential element of it. There are no specific provisions concerning contributory infringement for biotechnological products.

Thus, Article L 613-4 of the IPC could apply to such subject matter. Various cases in the field of biotechnology have applied the provisions of this article.

The case *Institut Pasteur v Chiron Blood Testing and Chiron Healthcare Ireland Ltd,*[23] ruled that contributory infringement constitutes an act of infringement, provided that the supplied means relate to an essential element of the invention. According to this decision, an essential element of the invention should be understood as an element contributing to its result.

The similar case of *Institut Pasteur v Siemens Healthcare Diagnostics*[24] confirmed this position and further held that the means supplied did not necessarily have to be claimed in themselves.

4.3   Experimental-use exemptions

Under Article L 613-5 of the IPC, there is an experimental-use exemption which has recently been amended to include an express exemption for acts performed for the purpose of obtaining a marketing authorisation.

Article L 613-5 further states that there is no infringement of a patent right where the invention is used for experiments on the subject matter of the invention. Thus, there are two elements for this exception to apply: there must be an experiment (namely to gain knowledge); and the experiment must be on the subject matter of the invention. Thus, using the invention for experimentation unrelated to the invention is not exempted.

Experiments with patented drugs for the purpose of finding new modes of administration were held not to constitute infringements.[25]

In a later *Wellcome Foundation v Perexel International and Flamel* case,[26] it was held that the experimental-use exemption did not apply to commercial or industrial acts, but that bioequivalence testing for the purpose of obtaining a marketing authorisation were not industrial or commercial acts and thus were exempted.

---

23   High Court of Paris, 3rd chamber, 3rd section, February 7, 2007, *Institut Pasteur v Chiron Blood Testing and Chiron Healthcare Ireland Ltd*, Docket number 05/11023, p 11 of the judgment; Court of Appeal of Paris, 4th chamber, section A, March 4 2009, *Institut Pasteur v SAS Chiron Healthcare*.

24   High Court of Paris, 3rd chamber, 2nd section, May 28 2010, *Institut Pasteur v Company Siemens Healthcare Diagnostics*, Docket number 08/08679, p 16 of the judgment.

25   *Wellcome Foundation v Perexel International and Flamel*, Court of Appeal January 27 1999.

26   Paris High Court, February 20 2001.

In *Science Union & Laboratoires Servier v AJC Pharma*,[27] it was held that experiments involving the semi-industrial manufacture of batches for the grant of a marketing authorisation (MA) were exempted under experimental use. More recently, in the other *Union & Laboratoires Servier v AJC Pharma* case,[28] the above position was confirmed.

In addition to these exemptions, Directive 2004/27/EC added a specific exemption related to marketing authorisations. Thus, the position of French courts is necessarily predominantly to construe all experiments performed for the purpose of obtaining an MA as covered by the experimental-use exemption. Once the MA is obtained, any further production is unlikely to constitute research and would thus infringe.

4.4   Offers to supply

'Offering' within the meaning of Article L 613-3 of IPC is widely interpreted by case law.

Under the ruling of the High Court of Paris, *Bayer Healthcare AG v Zhejiang Jingxin Pharmaceutical Co*,[29] an offer includes all acts and especially promotional acts aiming to present the allegedly infringing product to clients. It does not matter that the product has not been physically presented to clients, nor that it cannot yet be marketed because it lacks the relevant authorisations.

Marketing can result from any material operation intended for bringing a product into contact with the relevant public and thus covers all acts of presentation of the product, even if marketing cannot immediately occur.[30]

Thus, under French law, marketing is a very general act defined as any material operation tending to place a product in circulation.[31]

4.5   Summaries of product characteristics (SmPCs)

In the case *Abbott v Wyeth*,[32] the Court of Appeal of Paris held that the fact that a product could inherently be used for a specified use could not constitute an infringement-of-use claim. Thus there was a requirement of a material act on the part of the infringer for the use claim to be infringed.

It is likely that a statement contained in a leaflet or a summary of product characteristics would constitute a sufficient material act that could result in the infringement of a use patent.

---

27   Paris High Court October 12 2001.
28   Paris High Court January 25 2002.
29   High Court of Paris, 3rd chamber, 3rd section, *Bayer Healthcare AG v Company Zhejiang Jingxin Pharmaceutical Co, Ltd*, Docket number 06/16242, p 4 of the judgment.
30   High Court of Paris, 3rd chamber, 2nd section, *Evysio Medical Devices ULC v Guidant France, Guidant Europe NV/SA, Abbott France and Abbott Vascular Devices*, Docket number 06/08499.
31   Court of Appeal of Paris, 4th chamber, section A, April 4 2004; High Court of Paris, 3rd chamber, 2nd section, September 19 1997, *Aktiebolaget Hassle and Laboratoires Astra France v Chong Kun Dang*, Docket number 26297/94, p 7 of the judgment.
32   Court of Appeal of Paris, October 29 2004, 4th chamber, section B, *Abbott Laboratories v Wyeth Nutrition and SA Candia*, Docket number 2003/01748; High Court of Paris, 3rd chamber, 3rd section, March 14 2000, Docket number 98/9979.

5. Patent enforcement

5.1 Obtaining information on the infringer and the infringement

The predominant method of obtaining evidence on infringement in France is by way of seizure proceedings (see Article L 615-5 of the IPC). Under this procedure, it is possible to obtain a court order to effect a search of premises and the seizure of goods for the purpose of evidencing infringement.

Following the implementation of EC Directive 2004/48 on the enforcement of intellectual property rights, further evidential means were provided, such as the right of information, which enables a court to compel the disclosure of documents required to determine the origin and the distribution networks of infringing goods.

This provision has led to varying uses by the courts. A court may reject such a request on the ground of the preservation of commercial secrets and competition.[33] Conversely, a court may rule that the 'commercial secret' and the 'considerable harm' relied on by the defendant does not constitute a legitimate impediment.[34]

In addition, the existence and the conduct of prior seizure proceedings can influence a court's willingness to grant such remedy.[35]

5.2 Interim relief

Under French law, it is possible for a patentee to request the court to grant an interim injunction if there is an infringement or an immediate threat of an infringement. The criterion for granting an interim injunction is that an infringement must appear likely. It should be noted that such an assessment involves a full review of validity and infringement.

Thus, proceedings for an interim injunction appear similar to a full trial on the merits, except that they are held in a much shorter timeframe. Typically, interim injunctions will last between four and 12 weeks.

5.3 Other remedies such as *ex parte* injunctions and customs seizures

It should be noted also that *ex parte* proceedings exist under French law, thus enabling a patentee to obtain an injunction against an infringer without the infringer being aware of the proceedings. However, while these proceedings have already been used in trade mark matters, they are still extremely rare for patents.

In addition, the existence of customs seizures should be noted. Such seizures are carried out by French Customs and are effective against goods entering the European Community through France.[36]

5.4 Post-patent expiry injunctions

It is generally accepted that injunctions can only be granted and take effect while the

---

33   *Ibid*, p 5.
34   High Court of Paris, 3rd chamber, 2nd section, Order of the pre-trial judge, September 25 2009, *Company Promiles and Company Decathlon v Company Trading Innovations*, Docket number 07/13634, p 4.
35   For an example concerning the production of accounting elements, see High Court of Paris, 3rd chamber, 1st section, Order of the pre-trial judge, April 29 2009, aforementioned, p 5 (last paragraph) and 6 (first paragraph) of the Order.
36   See in particular Council Regulation (EC) No 1383/2003.

infringed intellectual property right is still in force.

This doctrine is indirectly confirmed by a recent Supreme Court decision, which held that even seizures cannot be carried out after the patent has expired, notwithstanding the fact that the patentee still had a legitimate interest in carrying out the seizure to evidence past infringement.[37]

5.5　Unjustified threats

Liability for threats of infringement arises from the general law of unfair competition under Article 1382 of the French Civil Code. There are no specific provisions governing such threats and so any guidance can only be found in specific case precedents. In certain circumstances, such threats of infringement can constitute acts of unfair competition. As an example, the intimidation of the clients of a competitor through threats of infringement proceedings was deemed to be an act of unfair competition.[38]

Unfair competition can arise even in the absence of an express threat. Thus, the act of informing distributors of the defendants of the existence of an infringement action and presenting the defendants' products as being infringing products, while the proceedings were still ongoing, can constitute an act of disparagement and thus of unfair competition.[39]

Any form of communication with customers of a suspected infringer should, therefore, be handled with caution.

5.6　Remedies

The main remedies available under French law include damages and injunctions.

(a)　Damages

Damages are computed on the basis of a loss-of-profits assessment, or an indemnifying royalty. Loss of profits equate to the profits that the patentee could have made by working the invention in the place of the infringer.

The indemnifying royalty is available upon request by the patentee, or if the patentee could not have exploited the patent. This royalty is often set above a commercial royalty negotiated at arm's length.

Article 13 of EC Directive 2004/48 (the Enforcement Directive) has been transposed into Article L 615-7 of the IPC and has *inter alia* provided that the courts may take into account the profits made by the infringer when awarding damages. While there is no possibility for a patentee directly to recover the profits of the infringer, this provision nevertheless enables the courts to take such evidence into account.

The provision has recently been applied by the Court of Appeal of Paris in a ruling dated March 19 2010, in which profits made by a manufacturer were expressly taken into consideration for setting the damages awarded.[40]

37　Court of Cassation, Commercial chamber, December 14 2010, Number of appeal: 09-72946.
38　Court of Appeal of Paris, 4th chamber, section B, January 12 2007, SARL *Cabac Logistique and Mr Jean-Michel Falieres v Company Sevim*, Docket number 05/08799, p 5 of the ruling.
39　High Court of Paris, 3rd chamber, 4th section, May 14 2009, *Company Quest Technologies Inc and Company Distrisud v SARL AHT Sud and Mr Sylvain C*, Docket number 09/03665.
40　Court of Appeal of Paris, Pole 5, 2nd chamber, *SAS Sempa and SA of Spanish Law Zumex Maquinas y Elementos*, Docket number 06/16476.

(b)   *Injunctions*

Injunctions are available as of right when the court finds in favour of the patentee. If the court so provides, such injunctions can be enforceable pending appeal.

It should be noted that breaching an injunction in France does not result in criminal liability but in the award of high damages designed to deter non-compliance.

(c)   *Other remedies*

Further remedies are provided for by Article L 615-7-1 of the IPC, which partially transposes Articles 10 and 15 of the EC Directive 2004/48. These provide for the delivery of infringing products and the recall of such products from commercial channels. In addition, the court may order publication of the judgment (or extracts from it) at the infringer's expense.

6.   Compulsory licensing

Compulsory licences are provided for in Articles L 613-11 to L 613-19 of the IPC. There are two types of compulsory licence:

- judicial licences granted upon application by third parties; and
- *ex officio* licences granted at the initiative of the French state.

There are three sub-types of compulsory judicial licence. These can be granted by courts on the grounds of a lack of exploitation of the patented invention, or because a patent cannot be worked without a licence of a dominant patent.

Ex officio licences are provided for by Articles L 613-16 to L 613-19-1 of the IPC and in Article L 5141-13 of the Public Health Code. These provide for licences being ordered by the state on the ground of public health, export of drugs to countries with public-health problems, for the good of the economy, for national defence or specifically in the field of semiconductors.

However, in spite of the existence of several distinct legal provisions allowing for various compulsory licences, in practice these provisions are almost never used. Therefore, such licences are exceedingly rare.

7.   Ownership, inventors and compensation

In practice, inventors are afforded extensive protection and rights to share in the benefits of their inventions. Article L 611-7 of the IPC, which is the relevant provision concerning inventions made by employees, defines two broad categories of invention:

- mission inventions, defined as inventions created in the normal course of the employee's duties, or expressly requested by the employer – such inventions belong to the employer; and
- all other inventions – on a first analysis, these belong to the employee.

As regards mission inventions, the employee shall enjoy 'additional remuneration' (Article L 611-7(1) of the IPC).

However, included with those inventions which are not regarded as mission inventions, the law provides two further categories: assignable inventions and non-assignable inventions.

Assignable inventions are defined by Article L 611-7(2) IPC as inventions which were made either in the course of the employee's work, or within the field of activity of the employer, or by using the resources of the employer. In such circumstances, the employer will be entitled to have the invention assigned to it and the employee will be entitled to obtain a 'fair price' in exchange for the assignment of his rights.

All other inventions, which are non-assignable inventions, belong to the employee.

In practice, there has been a convergence between the fair price and the additional remuneration provided for either mission inventions or assignable inventions. Both can now be substantial.

8. Branding and designs

The name of a drug may be protected by trade mark law provided that such name meets traditional requirements of trade mark law and complies with specific legal requirements under the Public Health Code.

For instance, under Article R 5121-2 of the Public Health Code the invented trade-mark name of a drug should not be such that it may be confused with its common name. Likewise, the shape of a drug may also be protected by trade mark law.

However, a shape which has a technical effect cannot be protected as a trade mark. The cases *Lego*[41] and *Lexomil*[42] in the field of pharmaceuticals have expressly denied the possibility of protection by trade marks of a shape having a technical effect.

9. Protecting valuable information

9.1 Preserving confidentiality

While some case precedents have ruled that under certain circumstances an obligation of confidentiality can arise in the course of business, obligations of confidentiality are not readily implied into contracts. It is thus necessary for contracting parties to make express provision for confidentiality.

9.2 Know-how and ex-employees

Know-how can be protected by rules of civil liability. It may also be protected by criminal law, if it can be held to constitute a 'manufacturing secret' (Article L 1227-1 of the Labour Code). Know-how which is not expressly protected by any intellectual property right may be protected by Article 1382 of the French Civil Code.

To obtain compensation, the holder of the know-how must prove a tortious act, a loss and a causal relation between the wrongful act and the loss.

A tortious act may, for instance, take the form of the misappropriation of another

---

41   Court of Cassation, Commercial chamber, October 7 2007, Number of appeal: 95-15859.
42   Court of cassation, Commercial chamber, January 21 2004, Number of appeal: 02-12335.
43   Court of cassation, Commercial chamber, February 25 2003, Number of appeal: 00-21542, dismissing the appeal lodged against a ruling handed down by the Court of Appeal of Paris on September 27 2000. See also: Court of Cassation, Criminal chamber, June 20 1973, Number of appeal: 72-92270.

company's know-how. Under the *Chantelle* case,[43] it should further be noted that the misappropriation is sufficient to trigger liability and damages, even if no use is made of that know-how.

9.3 The disclosure of manufacturing secrets

Even after employees leave the company, they are subject to a general but limited obligation of confidentiality under employment law. Thus, it is necessary to include explicit confidentiality provisions in employment contracts. Separately, under Article L 621-1 of the IPC, it is a criminal offence for an employee or a director of a company to reveal trade secrets.

In practice, however, the use of such criminal remedies is extremely rare.

9.4 Copyright patient information leaflets and packaging

Patient information leaflets and packaging can be protected under all intellectual property rights, including copyright, trade marks and designs.

10. Counterfeiting

Patent infringement potentially carries a criminal penalty (Articles L 615-12 to L 615-16 of the IPC). Thus, under Article L 615-14 of the IPC, "any person who has knowingly infringed the rights of the owner of a patent as defined in Articles L 613-3 to L 613-6 shall be liable to a three-year imprisonment and a fine of €300.000".

These penalties are increased, if the infringement was committed by an organised criminal group or if the counterfeit goods are dangerous to health, or to human or animal safety.

However, cases involving criminal penalties are extremely rare. The reason for such limited use of penal provisions may arise from the fact that it is necessary to prove that the alleged infringer knowingly infringed the patent and also from the complexity of the relevant facts.[44]

It should also be noted that it is possible for customs staff to effect seizures under their general powers or under Regulation EC 1383/2003.

11. Collaborative models

Any collaborative model should include clear and unambiguous contracts dealing with intellectual property. An obvious risk is that a contributor could claim co-ownership of any inventions or other intellectual properties rights.

It should be noted that in French law, even remuneration for services rendered often does not imply the assignment of any intellectual property rights. Similarly, a general assignment of copyright is not permitted in French law.[45]

It should also be noted that the validity of such contracts can be called into question if proper consideration is not given. Thus, contracts should be drafted in such a way as to deal with each intellectual property right that might be created through the collaboration.

---

44   Frédéric Pollaud-Dulian, "La Propriété Industrielle" ("Industrial Property"), *Economica*, 2011, n° 837, p 435.
45   See Article L 131-3 of the IPC; see also Article L 131-1.

12.   Hot topics

The field of supplementary protection certificates (SPCs) constitutes an area of law 'in turmoil'. Recent cases have helped, or will help, to clarify both the requirements for the grant of such additional intellectual property rights and their scope.

12.1   The requirements for the grant of an SPC over a combination of active ingredients

Under Article 3(a) of Regulation 469/2009, an SPC shall be granted if the product is protected by a basic patent in force. The issue of whether a product was indeed 'covered' by a basic patent has caused many difficulties.

Under several rulings of the Court of Appeal, including the recent *Daiichi Sankyo* case, it was held that a combination of active ingredients would meet this requirement only if this combination was claimed as such (and thus disclosed by the patent).[46]

Thus, so far the Court of Appeal has always refused to apply the 'infringement test'. Under such test, a combination of active ingredients may be deemed as 'protected by a basic patent' in force if using such a combination would infringe the basic patent. This test diverges from that applied by the Court of Appeal, notably when a product could be protected by a basic patent through the infringement of a process claim or when a combination could be protected, even if it is not described in the basic patent.

The *Daiichi Sankyo* case has been appealed to the Supreme Court. In addition, there are several pending references for preliminary rulings from national courts to the Court of Justice of the European Union concerning this issue, in cases involving medicinal products comprising more than one active ingredient.[47]

Finally, in a recent case where the infringement test was relied on in order to obtain the nullity of a decision of the French PTO refusing to issue an SPC on a vaccine comprising multiple antigens, the aforementioned appeals and references led the Court of Appeal[48] to order a stay of proceedings pending the decision of the French Supreme Court concerning such appeal and the ruling of the CJEU in the *Medeva* case.

12.2   The requirements for the grant of a paediatric extension

In the so-called 'Losartan' case,[49] the Court of Appeal ruled on the question of the validity of a pediatric extension granted by the French Patent Office on the ground that the company involved (Dupont) did not demonstrate that all pediatric marketing authorisations had been issued in the 27 states of the European Union at the time of filing the application for the pediatric extension.

The Court of Appeal ruled that Article 36 of Regulation EC No 1901/2006 on

---

46   Court of Appeal, November 6 2009, *Daiichi Sankyo*, 09/06530; Court of Appeal, January 19 2005, *Abbott Laboratories*, 04/14435; Court of Appeal, February 8 2006, *EI du Pont de Nemours*, 05/20525.

47   Case C-6/11, January 5 2011, *Daiichi Sankyo Company*, Case C-630/10, December 24 2010 *University of Queensland, CSL Ltd*, Case C-322/10, July 5 2010, *Medeva BV* (joined with Case C-422/10, *Georgetown University*).

48   *Medeva BV*, 10/16069, 10/16063, 10/16059 and 10/1605511, May 11 2011.

49   High Court, February 12 2010, *EI Dupont de Nemours and Company*, 10/51453.

pediatric extensions does not provide that, when submitting an application, all the marketing authorisations must have been obtained. According to the Court of Appeal, to require the production of all 27 pediatric marketing authorisations on the day of filing of the application would be impracticable in the absence of a centralised procedure for registration.

Thus, the Court ruled that Article 36 can only be concerned with the requirement to provide the initial marketing authorisations. Consequently, pediatric marketing authorisations can be provided in the course of the procedure for granting an extension.

12.3 Scope of protection of an SPC protecting a single active ingredient

The so-called 'Losartan'[50] and 'Valsartan'[51] cases have held that the holder of an SPC concerning a single active ingredient may prohibit the exploitation of a drug consisting of a combination of active ingredients containing the active ingredient.

Thus, these rulings demonstrate that, under ordinary patent law, it is necessary to distinguish what is protected by the patent (the subject matter of the patent – in these cases the single active ingredient) from the scope of the patent (that is to say the effects of the patent or the rights conferred by the patent).

It should be noted also that in the *Losartan* case the Court of Appeal has further held that only the marketing of the generic drugs before the expiration of the protection could be prevented.[52] This suggests that the production and storing of a generic drug may not infringe the SPC, even if the basic patent could have been infringed by such actions.

---

51 High Court, January 28 2011, *Novartis* AG 11/50892.
52 Court of Appeal March 15 2011, *SAS Mylan* 10/03075.

Germany

Joachim Feldges
Birgit Kramer
Field Fisher Waterhouse

---

1. Small molecules

Most patent applications for Germany in the biotech field are filed with the European Patent Office (EPO), and therefore the case law of the EPO largely determines the scope and requirements of patent protection for small molecules.

1.1 Scope of protection of claims

In accordance with Section 14 of the German Patent Act (GPA), which corresponds with Article 69 of the European Patent Convention (EPC), the scope of protection conferred on a patent is determined by the terms of a patent claim. The description and drawings are considered when construing the claims. The prosecution file is not an admissible source for claim construction in German patent law. To a very limited extent, the principle of good faith may prevent a patentee from claiming a specific scope of protection if this is inconsistent with the pleadings of the patentee in a previous nullity case between the same parties.[1] See also section 12, 'Hot topics'.

1.2 Markush formulae

Patent claims worded in the form of Markush formulae are allowable. It is not required that the application supports all possible combinations covered by the Markush formulae, as long as there are examples for compounds covered by the formulae in the patent fulfilling all requirements of the claims.

1.3 Metabolites

Metabolites may be patentable where they have not previously been available to the public. However, an invention claiming a metabolite may be obvious if the properties of the metabolite are pre-described in the context of a known substance being transformed into the metabolite in the human body. In the Terfenadine case[2] the Higher Regional Court Munich decided that a patent for the metabolite MDL 16,455 is not infringed by a drug using its prodrug Terfenadine. Terfenadine had been subject matter of an earlier patent that had expired in the meantime. At the time the patent for MDL 16,455 had been applied for, it was – in contrast to the time of the litigation – unknown that it is the active metabolite of Terfenadine. The court reasoned that the new information that MDL 16,455 is the active metabolite of

---

1 German Federal Supreme Court, judgment of June 5 1997 – X ZR 73/95 – *Weichvorrichtung II*.
2 Higher Regional Court Munich, judgment of June 3 1993 – 6 U 5155/92 – *Terfenadine*.

Terfenadine does not change the fact that the use of Terfenadine protected by the senior patent has become patent-free and accordingly free for the public use.

2. Second-generation inventions

2.1 Combinations

Combinations of known substances may be patentable if it was not obvious for the skilled person to combine the two substances. The bonus effect as such will not be sufficient to support an inventive step.[3]

2.2 Enantiomers

Enantiomers are novel over racemates. In addition, the grant of a supplementary protection certificate (SPC) on a racemate does not prevent the grant of an SPC on an enantiomer. An inventive activity to isolate an enantiomer may be acknowledged where specific difficulties had to be overcome.[4]

2.3 Selection inventions

As a general rule, the disclosure of a group of compounds also anticipates the individual compounds falling under that definition of the group.[5] However, in its more recent judgment for Olanzapine,[6] the Federal Supreme Court tightened the requirements for an anticipating disclosure. Hence, although – unlike the EPO case law – the selection invention as such is not patentable in Germany, there may be space for argument as regards the anticipating disclosure. Moreover, even where the substance as such may not be novel, an invention may be based on specific uses of properties of a selected member of a group of chemical compounds.

2.4 Formulations and physical forms

Where the formulation of a known compound is novel and not obvious, it can be patented. The same applies to specific physical forms such as polymers and crystals.

2.5 Dosage claims

In its judgment in *Carvedilol II*,[7] the German Federal Supreme Court decided that a specific dosage of a medicament is not patentable in view of the exemption of therapeutic methods from patent protection. Moreover, neither novelty nor inventive activity can be based on a specific dosage. However, this judgment was rendered before the Enlarged Board of the EPO. The Enlarged Board of Appeal handed down its decision G2/08 – *Dosage regime/Abbott Respiratory* on February 19 2010, allowing that inventions are based on a specific dosage.

It is possible that the German Federal Supreme Court will harmonise its case law with this recent EPO decision.

---

3   German Federal Supreme Court, judgment of December 10 2002 – X ZR 68/99 – *cosmetic sun protection*.
4   German Federal Supreme Court, judgment of September 10 2009 – Xa ZR 130/07 – *Escitalopram*.
5   German Federal Supreme Court, January 16 1988 – X ZB 18/86 – *Fluoran*.
6   German Federal Supreme Court, December 16 2008 – X ZR 98/07 – *Olanzapine*.
7   German Federal Supreme Court, December 19 2006 – X ZR 236/01 – *Carvedilol II*.

Joachim Feldges, Birgit Kramer

2.6 Methods of use and secondary indications

Provided that a specific indication is novel and inventive, it is patentable. The same applies to specific methods of use. However, as explained above, the case law of the German courts is more strict as regards the exemption of medical treatments from patent protection.

2.7 Reach-through claims

Reach-through claims are not allowable in accordance with the position taken by the EPO (T 1063/06 of February 3 2009). This rejection is based on the fact that reach-through claims only define medical compounds which they want to cover as a problem to be solved, instead of determining the solution of such problem.

3. DNA, biologicals and personalised medicine

3.1 DNA sequences

Section 1a(1) of the GPA exempts the human body in its phase of generation and development, including germ cells, from patent protection. Moreover, the sole discovery of parts of the human body including the sequence or partial sequence of a gene cannot be patented. However, according to Section 1a(2) of the GPA, an isolated part of the human body including the sequence or partial sequence of a gene can be patented, even if its structure is identical to the structure of a natural part of the human body. According to Section 1a(3) of the GPA, the industrial application for a gene sequence has to be disclosed in the application, including the function fulfilled by such sequence.

If the subject matter of an invention is a sequence or partial sequence of a gene which is concurrent with the structure of a natural sequence or partial sequence of the human gene, the use disclosed as an industrial application must be included in the patent claim, according to Section 1a(4) of the GPA. This restriction applies to national German patents granted under the German Patent Act, but not to German parts of European patents. Hence, the scope of protection of the German part of the European patent is – unlike the scope of a German national patent – not limited to the function of the sequence disclosed in the application.

3.2 Stem cells

The use of stem cells is subject to severe restrictions under German law. As regards patentability of inventions on stem cells, the German Federal Supreme Court, in a decision of November 12 2009, referred a case to the ECJ and submitted several questions on the ethical limits of patentability of stem cells. On March 10 2011, the Advocate General issued a negative opinion according to which stem cells carrying within them the capacity to evolve into a complete human being must legally be classified as human embryos and must therefore be excluded from patentability.[8] The ECJ (now the CJEU) has not yet rendered its decision.

---

8   ECJ case C-34/10,

195

3.3 Inventions based on non-wet laboratories

According to the case law of the EPO, the application of non-wet laboratory techniques, and in particular computer programs for the identification of a chemical compound, does not exclude its patentability. However, the sole identification of a compound, or the sole allegation of specific properties based on non-wet laboratory techniques, will often not be sufficient to demonstrate that there is a completed invention. Increasingly, the EPO and the German Patent and Trademark Office require supporting data for asserted properties of compounds. Consequently, bioinformatics may help to identify specific compounds but do not replace the need to produce laboratory data to support an alleged invention.

A lack of supporting data may also be regarded as a lack of sufficient disclosure for the alleged invention and consequently result in non-patentability.

4. Acts of patent infringement

4.1 Direct infringement

Product claims confer protection with regard to the manufacture, commercialisation, use, possession and import of the protected compound. Even if the process applied for by the infringer to manufacture the compound was unknown at the priority date of the patent, or if the use of the patented compound applied for by the infringer was unknown at that date, a product claim will cover such manufacture and use.

Process claims confer protection as regards the use of such process. In addition, the scope of protection extends to any product obtained directly through the patented process.

Use claims confer patent protection as regards the patented use. If a medicament is purposely prepared (*sinnfällig hergerichtet*) for the patented use (eg, through a package leaflet), a court may even grant an injunction covering such manufacture of a medicament for a patented use.

If the attacked acts fulfil all features of the patent claim, this will constitute a direct infringement of such patent claim. A direct infringement entitles the patentee or exclusive licensee to an injunction, even if the patentee or exclusive licensee does not have its own product covered by the patent on the market. There is no review of balance of interests by the court to that extent.

4.2 Contributory infringement

The offer or supply of means which are suitable and determined by the recipient to be used in accordance with the patent may constitute a contributory infringement in accordance with Section 10 of the GPA. In addition to the objective requirement that the offered or supplied means must be suitable for the patented use or suitable to be used for building a patented device, the supplier must know, or it must be obvious to the supplier that the supplied means will be used in accordance with the patent. The sole fact that a high percentage of patients to whom a drug is administered will fulfil features specified in a patent claim will not suffice for a contributory infringement, as long as the supplier does not include such features in its package leaflet.[9]

In general, an injunction based on contributory infringement will not prevent a defendant from offering or supplying the means being suitable for patented use generally. However, the defendant will have to include a warning to the recipient that the offered or supplied means may not be used without the consent of the patentee for the patented use.

4.3 National treatment of 'Bolar'-type provisions

In accordance with Section 11(2)(b) of the GPA, studies and trials which are necessary for obtaining regulatory approval for the commercialisation of a medicament in the European Union or in any third country are exempted from patent protection. The same applies to all 'practical means' in the context of such studies and trials. The German legislator implemented the 'Bolar' exemption without any restriction to generic compounds. It also applies to any studies and trials on innovative drugs. Moreover, the exemption is not limited to studies and trials which are needed for regulatory approval in the European Union, but also extends to studies and trials for such approval in third countries.

The exemption is limited to what is needed for carrying out such studies and trials. Consequently, the manufacture of clinical material is exempted, whereas the manufacture of commercial material for the first launch during the patent term constitutes a patent infringement. Also, the announcement of the launch of a generic compound after the expiry of the patent constitutes a patent infringement if such announcement is made during the patent term.[10]

4.4 Experimental-use exemptions

In accordance with Section 11(2) of the GPA, experimental acts which relate to the subject matter of the patent are exempted from patent protection. The German Federal Supreme Court has defined the scope of this exemption in two judgments: *Clinical Trials I*,[11] which was also confirmed by the German Federal Constitutional Court in a decision of May 10 2000,[12] as well as *Clinical Trials II* of April 17 1997.[13]

In its judgment in *Clinical Trials I*, the German Federal Supreme Court held that clinical trials on medical uses contribute to learning more about the subject matter of an invention. Consequently, they are exempted from patent protection. In its judgment in *Clinical Trial II*, the German Federal Supreme Court held that clinical trials on a protein obtained from different host cells from those described in the patent are also covered by the exemption of experimental use.

After the introduction of the 'Bolar' exemption into the German Patent Act, the exemption for experimental use will still be of relevance for all acts not falling under such 'Bolar' exemption. This is particularly the case for the development of medical devices, as well as for any other experimental acts which do not relate to studies or trials needed for a regulatory approval.

---

9   Regional Court Düsseldorf, judgment of February 24 2004 – 4a O 12/03 – InstGE 4, 93 – *Ribavirin*.
10   German Federal Supreme Court, judgment of December 5 2006 – X ZR 76/05 – *Simvastatin*.
11   Federal Supreme Court, judgment of July 11 1995 – X ZR 99/92 – *Clinical Trials*.
12   1 BvR 1864/95.
13   X ZR 68/94.

4.5 Submitting authorisations and offers to supply

The submission of an application for regulatory approval does not constitute a patent infringing act.[14] In accordance with the 'Bolar' exemption, this also applies if the submission for regulatory approval includes the need to submit a sample with the regulatory authority.

Any offer to supply medicaments will constitute a patent infringing act if covered by a patent claim. This also applies to an offer during the patent term to supply a medicament after the expiry of the patent.[15]

5. Patent enforcement

5.1 Obtaining information on the infringer and the infringement

Three options for the preservation and seizure of evidence exist under German patent law, namely the independent procedure for the taking of evidence under Section 485 *et seq* of the German Code of Civil Procedure, preliminary injunction proceedings under Section 935 *et seq* of the German Code of Civil Procedure and, ultimately, seizure pursuant to Section 94 of the German Code of Criminal Procedure.

In practice, the independent procedure for the taking of evidence under Section 485 *et seq* is of particular importance. These proceedings are 'independent', since they do not require the pending of infringement proceedings on the merits and can therefore be commenced prior to commencing such proceedings. In practice, this is advisable if the requirements for the independent procedure for the taking of evidence are met, since the outcome of these proceedings will often improve the chances of an amicable solution. To institute these proceedings, a motion must be filed with the court having personal and subject-matter jurisdiction for infringement proceedings on the merits. The motion is founded if the applicant can show a recognisable legal interest, which presupposes that the independent procedure for the taking of evidence under Section 485 *et seq* of the German Code of Civil Procedure serves the purpose of avoiding a lawsuit or of preserving evidence. Under these circumstances, the applicant can request the court to order the inspection of the accused device by a court-appointed expert, who renders an opinion on the properties of the accused device as far as relevant for the patent dispute. This expert testimony will be eligible evidence in any subsequent infringement proceedings on the merits.

Independent thereof, or simultaneously thereto, preliminary injunction proceedings under Section 935 *et seq* of the German Code of Civil Procedure can be commenced if the specific circumstances of the particular case under consideration warrant the assumption that the potential infringer for example envisages disposing of, or modifying, the accused device in order to obstruct an inspection or otherwise to frustrate the proof of a given patent infringement.

---

14   German Federal Supreme Court, judgment of March 24 1987 – X ZR 20/86 – *Rundfunkübertragungssystem*.
15   German Federal Supreme Court, judgment of December 5 2006 – X ZR 76/05 – *Simvastatin*.

5.2   Interim relief

Preliminary injunction proceedings are regulated by Sections 916 to 945 of the German Civil Procedure Act. An injunction will be granted if there is first an injunction claim (*Verfügungsanspruch*) and secondly a particular reason for preliminary measures (*Verfügungsgrund*). The applicant must provide *prima facie* evidence (*Glaubhaftmachung*; Section 920(2) in connection with Section 294 of the German Civil Procedure Act) for the injunction claim and must state the reason for requesting an injunction.

Claims that may be asserted in preliminary injunction proceedings are claims for an injunction under Section 139(1) of the GPA and claims for disclosure of the origin and the distribution channels of the infringing goods under Sections 140b(1) and (2) of the GPA (the latter only in cases of obvious infringement, however). Claims for destruction can be secured under certain circumstances by way of a preliminary injunction requesting that the infringing goods be handed over to a bailiff for custody. Claims for damages and for rendering accounts cannot be asserted in preliminary injunction proceedings, because this would be in breach of the rule that preliminary proceedings may not anticipate the proceedings on the merits.

Preliminary injunction proceedings in the context of patents are characterised by particularities which arise from the fact that the assessment of patent infringement is based on technical facts which regularly require comprehensive clarification of technical questions in order to enable the court to have a reasonable basis for reaching a decision. Such comprehensive clarification of technical questions is often difficult if not impossible in summary proceedings. In particular, a defendant may often be unable to search comprehensively the prior art and attack the validity of the patent at short notice. In the meantime, a preliminary injunction has severe consequences for the business of the attacked party and means a definite fulfilment of an asserted claim for an injunction. In order to avoid the risk of an erroneous decision with far-reaching consequences, the courts grant preliminary injunctions only in cases where the question of validity of the patent and of infringement look so likely to be answered in favour of the applicant that an erroneous decision is not seriously to be expected. Nevertheless, there is a growing tendency for courts to grant preliminary injunctions in patent disputes.

The applicant must establish that it needs a preliminary injunction, rather than being referred to proceedings on the merits. This legal requirement is known as 'the urgency of the matter' (*Dringlichkeit*). Most courts do not define such urgency by referring to a fixed period of time, but assess it according to the circumstances of each particular case.

In considering whether to issue preliminary injunctions, the courts will generally consider the circumstances on a case-by-case basis. Indications weighing in favour of issuing a preliminary injunction are: clear technical facts; a clear infringement; the likelihood that the patent is valid; previously won infringement processes; attempts by the defendant to circumvent an earlier infringement judgment; a short remaining patent term; the threat of essential disadvantage to the market position of the patentee; and the attacked product being offered within a short timeframe.

Indications likely to militate against issuing a preliminary injunction may comprise: complex technical facts; lack of certainty as to whether there has been an infringement (in particular if no literal infringement is asserted and the issue of equivalence cannot easily be assessed by the court); doubts as to the validity of the patent; the threat of an important disadvantage to the market position of the defendant; the applicant's failure to use the patent; and hesitant behaviour on the part of the applicant.

5.3 Post-patent expiry injunctions

An injunction cannot be based on an expired patent, or used as compensation for an infringement which occurred during the patent term as a *post mortem moratorium*.[16]

5.4 Unjustified threats

Before starting infringement proceedings, the patentee is entitled (but not obliged) to send a warning letter to the alleged infringer. If such warning letter is not sent prior to filing a complaint with a court and the defendant acknowledges the claims immediately, the plaintiff may be held liable for the costs of the proceedings.

A warning letter will contain a request to the other party to give a declaration to cease and desist from certain allegedly infringing behaviour, along with a deadline for the provision of the declaration to cease and desist. The letter will also threaten court action if the declaration is not issued. It is a means of asserting patent rights out of court without being a necessary prerequisite for court action.

In the context of patent law, unjustified warning letters concerning an alleged patent infringement are of particular relevance. Such warning letters may be unjustified – for example if the attacked embodiment is not infringing, the patent right is not valid, or the warning letter is addressed to the wrong party.

An unjustified warning letter has two main consequences for the recipient. An unjustified letter entitles the addressee to file an action for a declaratory judgment of non-infringement. If the addressee ceases to deal with the allegedly infringing products on a commercial basis and it is later held by a court that the patent is invalid and/or not infringed, then the addressee may be entitled to claim damages. Irrespective of whether the addressee refrains from the allegedly infringing acts, it will also be entitled to a reimbursement of the costs of defending the challenges. As a point of principle, such unjustified warning letters are considered to be an interference with the established business of the addressed party, which enjoys protection as an absolute right under Section 823(1) of the German Civil Code. This case law is a feature of protective rights such as patents, utility models, trade marks and design patents. For unjustified warning letters that are based on alleged violations of the Act against Unfair Competition (ie, the majority of warning letters in practice), a claim for damages will be declined. It is considered that the rejection of unjustified threats is part of general business risk and as such costs should be borne by the relevant company.

---

16   Federal Supreme Court, judgment of February 21 1989 – X ZR 53/87.

5.5   Remedies

In the event of patent infringements, the patentee has various remedies under German law. Pursuant to Section 139(1) of the German Pact Act, a party using a patented invention in an infringing way is subject to a claim for injunctive relief if there is a risk of reoccurrence (*Wiederholungsgefahr*) or a threat of a first-time infringement (*Begehungsgefahr*). An entitlement to an injunction can arise regardless of whether or not there is wilfulness or negligence on the part of the infringer.

Pursuant to Section 139(2) of the GPA, in the event of wilful or negligent patent infringement, the infringer is obliged to compensate the infringed party for the damages suffered. While wilful patent infringement can rarely be proven, it is almost impossible for an infringer to show that he did not act with negligence, as anybody intending to manufacture or deal with goods on a commercial basis is expected to enquire whether such acts may infringe any third party's patents. German courts expect diligent companies to be aware of the grant of a patent and to examine potential infringements within one month following publication of the grant.[17]

Damages can be calculated in accordance with three methods which are well established in German case law.[18] The infringed party may claim lost profits, the profit obtained by the infringer through the patent infringement, or a reasonable royalty. From a procedural perspective, in the infringement proceedings a court will generally not decide on the amount of damages awarded, but only rule that the infringed party is entitled to claim damages from the infringer. The actual amount of damages to be paid will then be subject to secondary proceedings.

Further, pursuant to Section 140a(1) of the GPA, the infringed party may claim from any infringing party the destruction of all patented products or products that are manufactured by means of a patented process, but only to the extent that such products are the property or in the possession of the infringer. Under Section 140a(2) of the GPA, the claim for destruction will also apply to material and devices which are in the possession of the infringer and which serve the purpose of manufacturing the infringing products.

Section 140a(3) of the GPA provides that the injured party may claim from any infringing party the recall of products which are the subject matter of the patents, or to remove those products permanently from the distribution channels. Claims for recall and removal also apply to products directly manufactured by means of a process which is the subject matter of the patent. The claims for destruction, recall and removal are excluded if they are considered unreasonable in a given case, and the legitimate interests of third parties will be taken into account.

Pursuant to the first sentence of Section 140e of the GPA, if a complaint was filed based on the German Patent Act, and if requested by the winning party (who must show a legitimate interest), the court may order publication of the decision at the expense of the losing party.

---

17   Federal Supreme Court – X ZR 28/85 – *Formstein*.
18   Federal Supreme Court – I ZR 18/61 – *Kreuzbodenventilsäcke III*.

6. Compulsory licensing

In Germany, compulsory licences may be granted for four reasons:

- an outstanding public interest (Section 24(2) of the GPA);
- in certain cases of dependent inventions (Section 24(2) of the GPA);
- in certain cases requiring export to countries having insufficient healthcare (Regulation (EC) No 816/2006 of May 17 2006); and
- reasons based on antitrust laws.

The practical importance of Section 24 of the GPA is limited. Since the formation in Germany of the Federal Patent Court in 1961, only about 20 compulsory licence proceedings have been started and only in one case (in 1991) did the Federal Patent Court grant a compulsory licence; moreover, this was later lifted by the Federal Supreme Court.[19] The importance of the regulation lies in the numerous licence agreements which have arisen from it, particularly where the public interest is related to public health and the patentee has wished to avoid being exposed to court proceedings which might demonstrate that it was not concerned about public health.

From the time a compulsory licence is granted, the licensee can raise the usual objections against infringement based on the granted licence. However, the right to a compulsory licence as such (ie, when no such licence is yet in effect) is no genuine defence against, or exception to, an infringement. This is different from the right to be granted a licence under antitrust laws (the 'FRAND' objection, ie Fair, Reasonable and Non-discriminatory), which has played a more important role in the field of technical industry standards such as GSM or MPEG. This objection may be invoked directly against infringement claims.[20]

7. Ownership, inventors and compensation

According to Section 6 of the GPA, the right to a patent shall belong to the inventor or his successor in title. It is the inventor (not the applicant) who, under German patent law doctrine, shall be rewarded for his intellectual achievements and creativity and shall be incentivised to foster technology by further inventions. It follows that any rights to the patent must be derived from the inventor. His position comprises the entitlement to the patent under material law, as well as his so-called moral rights (*Erfinderpersönlichkeitsrecht*), which include the right to be named as the inventor.[21]

However, in order to avoid the substantive examination of the patent application being delayed due to the need to determine the identity of the inventor in the proceedings before it, the Patent Office does not examine the entitlement of the applicant. Rather, pursuant to Section 7 of the GPA, the applicant will be deemed to be entitled to request the grant of a patent (and the same applies to proceedings before the EPO under Article 60(3) of the EPC). This formal presumption could be abused by a party who is not, in fact, entitled and therefore the German Patent Act grants to the true inventor the means to counter-attack such unlawful behaviour by

---

19   Federal Supreme Court, judgment of December 5 1995 – *Polyferon*.
20   Federal Supreme Court, judgment of July 13 2004 – K ZR 40/02 – *Standard-Spundfass*; May 6 2009 – K ZR 39/06 – *Orange-Book-Standard*.
21   Sections 37(1) and 63 of the GPA.

way of entitlement claim. A claim for entitlement to a patent is subject to a statute of limitation of two years after grant of the patent, except in cases where the applicant acted with bad faith.

Employers are entitled to their employees' inventions. Since a recent change of the German Act on Employees' Inventions in 2009, the employer is deemed to have claimed the invention if the employer does not release it to the employee within four months after having received notice of the invention. The employee is entitled to reasonable compensation in accordance with the standards set out under the German Act on Employees' Inventions.

8.   Branding and designs

8.1   Extending protection using trade marks and designs

Trade marks are important tools in the life sciences as a means of providing comprehensive protection for a product. The name of a chemical compound, in particular its INN, may not be registered as a trade mark. However, trade marks are often registered which are rather close to the official name of a chemical compound. Designs may also be used for obtaining additional protection for a medicament (eg, the special colour or form of a pill). However, designs are more frequent for medical devices than for medicaments.

8.2   Protection using unfair competition

Companies may assert their rights under the German Act against Unfair Competition, when competitors act against the rules of fair competition and thereby affect a company's business interests. The law against unfair competition is particularly relevant in the field of advertising.

8.3   Comparative advertising and advertising restrictions

The German Act against Unfair Competition provides detailed regulations on what is allowed in advertising and what is not. Comparative advertising is regulated restrictively in Section 6 of the German Act against Unfair Competition.

In the field of pharmaceuticals, the German Act on Drug Advertising provides specific rules for advertising on drugs. In this context the interdiction to advertise certain drugs in a certain manner to non-physicians is of particular importance. Internet content, which may not be advertised to the general public, must be secured using blocking techniques.

In this context, a submission of the Federal Supreme Court of July 16 2009 – *Arzneimittelpräsentation im Internet* to the Court of Justice led to a decision that the interdiction to advertise ethics publicly does not include the reproduction of the packaging and the patient information leaflet via the internet.[22] Although advertising and accordingly the interdiction includes factual information, it must be reviewed in every single case whether the given information aims at advertising. This was not assumed where the packaging and the patient information leaflet were reflected

---

22   ECJ, Decision of May 5 2011, C-316/09.

identically on the internet. It was relevant in that context that the internet user actively had to obtain the information and was not confronted with it unwillingly.

8.4 Protection of online content
Online content on the internet can be protected by copyright protection under German law.

9. Protection of valuable information

9.1 Preserving confidentiality
Valuable information is protected in the first place by confidentiality agreements. Employment agreements regularly contain a confidentiality clause. Further, alongside the general rules under criminal and civil law, Sections 17 to 19 of the German Act against Unfair Competition provide for specific regulations penalising the disclosure of trade secrets and other confidential information.

9.2 Know-how and ex-employees
Section 17 of the German Act against Unfair Competition requires the existence of an employment relationship. Therefore, following the termination of an employment relationship (apart from the possibility of prosecution under criminal law), contractual claims for damages will be of greatest assistance to the employer.

In such proceedings, the provision of evidence of quantifiable damage resulting from the disclosure of confidential information will often be difficult. However, even if the employer is granted a claim for damages, this may not solve the employer's problems. Often there will be a question as to whether the employee will be able to pay a damages award. Above all, the claim for damages will not reverse the disclosure of confidential information.

9.3 Copyright, patient information leaflets and packaging
Patient information leaflets and packaging can be the subject matter of copyright protection under German law.

10. Counterfeiting
In German law, the Patent Act, the Trade Mark Act and the Copyright Act all provide for criminal penalties against counterfeiting. Section 142 of the GPA, Section 143 of the German Trade Mark Act and Section 106 *et seq* of the German Copyright Act provide for imprisonment of up to five years or a monetary penalty if the counterfeiting is carried out on a commercial basis. Where this is the case, a criminal prosecution can be started. Under civil law, the most important rights of the right holder are claims for an injunction, destruction and damages.

In cases of importation of counterfeiting goods, German customs officials regularly act on the basis of EU regulation 1383/2003. On application by the right holder in the case of an obvious infringement, customs officers can seize infringing goods at the border. The right holder can also apply for a destruction of the infringing goods. If a seizure turns out to be unlawful, the owner of the seized goods has a claim for damages.

11. Collaborative models – 'open innovation'

Collaborative models have become more popular recently. In particular, the possibilities provided for by the internet foster the collaboration of parties who would not have found each other otherwise. This implies the advantage that deadlocked structures that may hinder the flow of ideas are avoided. On the other hand, such collaborative models only allow a limited control of the flow of information. It is difficult to control where certain information came from and where it goes in a virtual network of contributors. Accordingly, one of the big challenges of such networks is the regulation of knowledge and confidentiality. In this respect, existing tools such as patent pooling or co-ownership will have to be adapted to the special needs of a given situation.

12. Hot topics

In Germany the appropriate determination of the scope of patent protection has been the subject matter of recent case law. In the decision in *Okklusionsvorrichtung*,[23] the Federal Supreme Court set a limit to an overbroad construction of patent claims based on functional features, as it has been practised in particular by the infringement courts in Düsseldorf. The Federal Supreme Court held that a construction of a patent claim that is orientated at a functional feature of a claim may not undermine the prevalence of the claims over the description. In particular, embodiments which are described in the specification but not covered by the terms of the claims may not fall within the scope of protection, either literally or under the doctrine of equivalence.

---

23   Federal Supreme Court, judgment of May 10 2011 – X ZR 16/09 – *Okklusionsvorrichtung*.

India

Pravin Anand
Neeti Wilson
Anand and Anand

---

1. Small molecules

1.1 Product and process claims

The definition of 'invention' under the current Indian patent law of 2005 covers both product and processes. Section 2(1)(j) defines an invention to mean "a new product or process involving an inventive step and capable of industrial application".

The Patents Act 1970 had a very limited scope of protection, wherein the scope for patentability of an invention included the features termed 'new' and 'useful' and the manner of manufacture. Even though manufacture was not defined in the old Act, the Patent Office had established the practice of interpreting manufacture as a process resulting in a non-living tangible product. The landmark decision of the Calcutta High Court on the process of production of the Bursitis virus containing vaccine changed the practice and today the definition of 'invention' is much wider in scope than that of the old Act. This is also reflected in the change of wording from 'industrial application' to 'capable of industrial application' in relation to an invention as capable of being made or used in an industry.

Small molecules in the area of the pharmaceuticals sector have particularly been a subject of debate in the Indian patent field, with varying definitions of 'pharmaceutical substance'.

The Patents Act 1970 defined 'medicines or drugs' as including the following categories of substances:

- all medicines for internal or external use of human beings or animals;
- all substances intended to be used for or in the diagnosis, treatment, mitigation, or prevention of diseases in human beings or animals;
- all substances intended to be used for or in the manufacture of public health, or the prevention or control of any epidemic disease among human beings or animals;
- insecticides, germicides, fungicides, weedicides, and all other substances intended to be used for the protection or preservation of plants; and
- all chemical substances which are ordinarily used as intermediates in the preparation or manufacture of any of the medicines or substances above referred to.

The definition was vague and was changed in the current legislation to read: "Pharmaceutical substance means any new entity involving one or more inventive steps" thereby drawing even further debate in the matter.

The scope of protection available to small molecules is largely dependent on its overcoming the test in Section 3(d) of the Indian Patents Act. This provision does not consider the following as inventions within the meaning of the Act:

the mere discovery of a new form of a known substance which does not result in the enhancement of the known efficacy of that substance or the mere discovery of any new property or new use for a known substance or of the mere use of a known process, machine or apparatus unless such known process results in a new product or employs at least one new reactant.

The legislation further provides for an explanation for Section 3(d) that salts, esters, ethers, polymorphs, metabolites, pure form, particle size, isomers, mixtures of isomers, complexes, combinations and other derivatives of known substance shall be considered to be the same substance, unless they differ significantly in properties with regard to efficacy.

This peculiar provision of the Indian Patents Law affects almost all the ingredients of any life sciences invention, as will be shown in the topics to follow.

1.2 Scope of protection of claims and Markush formulae

A Markush claim is recognised to allow important innovations to be patented. For example, when a new organic compound that has a novel structure never created before is invented and can have many possible substituents that could be used, one can effectively group these possible substituents in Markush-type claims. Therefore, one can claim the basic structure, along with substituents such as halogens, alcohols, hydrocarbons etc.

Patent Office practice has developed towards allowing such groups of compounds, when supported by a single and definitive process. It is recognised that there are often changes in the substituent groups that do not change the original use of the compound and, thus, can be thought of as part of the original invention.

A Markush claim is usually allowed, provided there are sufficient supporting examples presented in the specification for the various substituent ranges.

1.3 Metabolites

Metabolites are recognised as the compounds that are formed inside a living body during metabolic reaction. The types of metabolites listed in the Patent Office manual for India are:

- active metabolites formed from inactive precursors (eg, DOPA and Cyclophosphamide);
- active metabolites formed from precursors that show a mechanism of action that is different from that of the parent compound (eg, Buspirone and 1-pyrimidyl piperzine Fenflouromine and norfenfleuromine);
- active metabolites which contribute to the duration of action of the parent compound (eg, Hexamethylmelamine and Clobazam); and
- active metabolites that show antagonistic effect on the activity of the parent compound (eg, Trezodone and m-chlorophenyl pierzine, Aspirin and salicylate).

A metabolite for a known drug may not be considered patentable under Section 3(d) (discussed later), since giving the drug to a patient may naturally result in formation of that metabolite. A metabolite may therefore be considered to be known for a known drug. However, there is scope for interpretation with respect to enhancements in efficacy with respect to its properties.

2.   Second-generation inventions

2.1   Combinations

The Indian Patents Act does not allow patents for a substance obtained by a mere admixture resulting only in the aggregation of the properties of the components thereof, or a process for producing such substance under the provisions of Section 3(e). However, an admixture resulting in synergistic properties of a mixture is not considered as mere admixture (eg, soap, detergent, lubricants and polymer composition) and such combinations are allowed patents.

API-carrier combinations also have to pass the test of being outside the scope of Section 3(e) to be allowed a patent in India. An explicit disclosure or experimental data to indicate that the presence of the carrier in any way influences the activity of the active ingredients or is a compatibility selection would hold the invention allowable under Section 3(e).

With respect to the assessment of inventive step involved in an invention based on a combination of features, consideration should be given as to whether or not the state of the art was such as to suggest to a skilled person precisely the combination of features claimed. The fact that an individual feature or a number of features were known may not conclusively show the obviousness of a combination.

2.2   Selection inventions

Selection inventions are recognised as patentable by the Indian Patent Office if they overcome the objection of obviousness. For example, in *Olin Mathieson Chemical Corporation v Biorex Laboratories Ltd*,[1] it was held not to be obvious that a useful drug would be obtained by substituting -CF3 for -Cl in a known drug, given the large amount of prior material which was before the skilled person at the date of the invention and which led in a number of different directions.

Selection patents are usually allowed based on the following criteria:

- the selection is based on some substantial advantage gained or some substantial disadvantage avoided;
- substantially all the selected members necessarily possess the advantage in question; and
- the selection is in respect of a quality of special character which can be peculiar to the selected group.

The advantage of the selection invention should be clearly disclosed and it should also be shown that it would not otherwise be apparent to a person skilled in

---

1   [1970] RPC 157.

the art. Unexpected bonus effects not described in the specification may not form the basis of a valid claim to a selection invention. Although the size of the class from which a member or members have been chosen is not relevant to the question of novelty of a selection invention, it may be relevant to the question of obviousness.

The technical significance of the parameters by which the product or process is selected should be considered. Where unusual parameters are used in a claim, it may be difficult to prove whether or not the prior art would inevitably have exhibited those parameters.

2.3   Methods of use and secondary indications

Second-medical-use claims are not allowed in India.

New uses for a known substance, or the mere use of a known process, are not considered patentable under Indian patent law. A discovery of a new property of a known substance is not considered patentable. For instance, it is known that paracetamol has an antipyretic property; further discovery of a new property of paracetamol as analgesic would not be patentable.

2.4   Enantiomers, formulations and physical forms

Enantiomers may fall within the scope of Section 3(d) of the Indian Patents Act. In fact a 'new form' of any known substance, if discovered, may be considered non-patentable (eg, isomers – structural isomers or positional isomers and stereo isomers, polymorphs, hydrates and homologues). However, the subject matter will be decided on a case-by-case basis. For example, a polymerisation process using a sterically hindered amine was held non-obvious over a similar prior-art process at the Indian Patent Office, as the prior art disclosed a large number of unhindered amines. Also pure forms, or a purification process or a purified compound which had never existed before due to inherent long-standing problems, can be considered patentable.

If the new form has enhanced efficacy over the previously known substance, it will be outside the scope of Section 3(d). Moreover, if the new form is invented rather than discovered, the invention qualifies for a patent.

If it is novel, inventive and has industrial application, a drug or pharmaceutical product can be claimed in a single application in light of the common inventive concept, for the following:

- a drug or pharmaceutical product;
- a modified drug or pharmaceutical of a known compound, if proved to be more efficacious than the known compound;
- a process of making the product as defined in the first and second bullets above; and
- a formulation containing the drug in the first and second bullets above.

Formulations are therefore considered patentable under Indian patent law if they satisfy the universal patentability criteria and also overcome the exclusions of Section 3(d) of the Indian Patents Act.

Some examples of formulation patents granted by the Indian Patent Office include: pharmaceutical formulations comprising an anti-obesity agent and acidulant (IN227344); nanoparticulate active substance formulations (IN228002); an aqueous solution formulation of alpha-interferon (IN228225); compounds and formulations for oral delivery of peptides (IN231358); formulation for oral administration for the treatment of type-2 diabetes (IN230111); a pharmaceutical liquid formulation of oxaliplatin for parenteral administration (IN230152); stabilised protein crystals, formulations containing them and methods of making them (IN228403); a pharmaceutical aerosol formulation and a method for the manufacture of a polypeptide aerosol formulation (IN227545); a formulation for vitamin D compounds (IN227572); oral pharmaceutical formulation in the form of aqueous suspension of microcapsules for modified release of amoxicillin (IN228060); an antiallergic formulation comprising lactic acid bacteria as an active ingredient (IN 218097); a novel anti-fungal formulation active against a broad spectrum dermatophytoses (IN226628); and a sustained-release formulation (IN225164).

2.5 Methods of treatment

Under Section 3(i) of the Indian Patents Act, processes for the medicinal, surgical, curative, prophylactic, diagnostic, therapeutic or other treatment of human beings, or any process for a similar treatment of animals to render them free of a disease or to increase their economic value or that of their products, are not patentable. For example, methods of treating malignant tumour cells are not patentable as they are regarded as treatment of human beings. The treatment of sheep for increasing wool yield (1958 RPC 85) has also been held not to be patentable at the Indian Patent Office.

An invention of a method of treatment of the human or animal body by surgery or therapy, or of diagnosis practised on the human or animal body, will be considered not patentable under Section 3(i) and may also not be considered as not being capable of industrial application.

(a) Therapy

The Indian Patent Office manual defines the term 'therapy' to include prevention as well as treatment or cure of disease. Although some medical dictionaries point towards a narrow interpretation of the term, other works of reference, including non-specialist dictionaries, indicate a more general meaning. This was preferred following the principle that words in statutes dealing with matters relating to the general public are presumed to be used in their popular, rather than their narrowly legal or technical, sense. However, for a treatment to constitute therapy, there must be a direct link between the treatment and disease being cured, prevented or alleviated. It appears that any medical treatment of a disease, ailment, injury or disability (ie, anything that is wrong with a patient and for which he would consult a doctor), as well as prophylactic treatments such as vaccination and inoculation, may be regarded as therapy.

The same considerations apply for animals as for human patients; therefore, prophylaxis and immunotherapy in animals are regarded as therapy.

It should be noted that an application of a substance to a human body purely for cosmetic purposes is not a treatment or therapy. For example, claims to a process for improving the strength and elasticity of human hair and fingernails would be allowable.

On the other hand, the application to the skin of an ointment designed to be effective to remove keratosis from the skin would be an instance of medical treatment. Here, 'treatment' in relevant senses means that the purpose of the application of a process or substance to the body must be to arrest or cure a disease or diseased condition or to correct some malfunction or amelioration of some incapacity or disability.[2]

With regard to dental treatments, there is a thin line between cosmetic and therapeutic methods. It was held in *Lee Pharmaceuticals Application* (1978) RPC 51 that since one of the reasons of grinding pits and fissures in teeth was to prevent the onset of dental decay, the purpose of the treatment was therapeutic rather than cosmetic. Claims to a method of removing dental plaque and/or caries would not be allowed,[3] nor would claims to a method of cleaning teeth which removed both plaque and stains, embracing both curative and cosmetic effects.[4]

(b)   *Surgery*

Methods of treatment of the human or animal body by surgery are excluded from patentability. 'Surgery' is defined as the treatment of disease or injury by operation or manipulation. It is not limited to cutting the body, but includes manipulation such as the setting of broken bones or relocating dislocated joints (sometimes called 'closed surgery'), and also dental surgery.

In general, any operation on the body which requires the skill and knowledge of a surgeon may be regarded as surgery and includes non-curative treatments such as cosmetic treatment, the termination of pregnancy, castration, sterilisation, artificial insemination, embryo transplants, and treatments for experimental and research purposes. If carried out by surgery, the removal of organs, skin or bone marrow from a living donor is also regarded as a surgical treatment.

As per Patent Office practice, methods of abortion, induction of labour, control of oestrus or menstrual regulation are regarded as therapy, irrespective of the reason for the treatment.

(c)   *Diagnosis*

Methods of diagnosis practised on the human or animal body are also excluded from patentability. Methods of diagnosis performed on tissues or fluids which have been permanently removed from the body are patentable, however. 'Diagnosis' is considered as the identification of the nature of a medical illness, usually by investigating its history and symptoms and by applying tests. Determination of the general physical state of an individual (eg, a fitness test) is not considered to be diagnostic if it is not intended to identify or uncover a pathology.

The provision of Section 3(i) of the Indian Patents Act relates to methods of

---

2   *Joos v Commissioner of Patent* (1973) RPC 59.
3   *Oral Health Products Inc (Halstead's) Application* [1977] RPC 612.
4   *ICI Ltd's Application No 7827383* (BL O/73/82).

diagnosis practised on the human or animal body; diagnosis in itself is a method of performing a mental act and is also excluded from patentability. Typically, the process of diagnosis involves a number of steps leading to the identification of a condition. For a claim to fall under this prohibition, it must include both the deductive step of making the diagnosis and the preceding steps for making that diagnosis involving specific interactions of a technical nature with the human or animal body. The exclusion is therefore a narrow one, and requires all the method steps of a technical nature to be practised on the body.

A patent may be obtained for surgical, therapeutic or diagnostic instruments or apparatus. The manufacture of prostheses or artificial limbs and the taking of measurements for the purposes of their manufacture are, therefore, patentable.

2.6 Reach-through claims

Reach-through claims are permissible under Indian patent law, subject to the same standards as for other patent claims. The test of sufficient disclosure is the main criterion for making successful claims. An example of a reach-through claim allowed by the Indian Patent Office is for an invention relating to a dual-specificity antibody that was described by reference to its binding affinity with interleukin. The granted claim read as:

> A dual-specificity antibody, or antigen-binding portion thereof, that specifically binds interleukin-1α and interleukin-1β, wherein said dual-specificity antibody is not a fully mouse antibody.

There are no regulations or training materials or manual points issued to guide the examination process for reach-through claims; nor is there any case law relating to the patentability of reach-through claims

3. DNA, biologicals and personalised medicine

3.1 Discoveries

The Indian Patents Act establishes in Section 3(c) that the mere discovery of a scientific principle, or the formulation of an abstract theory, or the discovery of any living thing or non-living substances occurring in nature are not patentable.

The distinction between discovery and invention is important when applying this statute to patent applications. A claim for discovery of scientific principle is not considered patentable, but such a principle when used with process of manufacture resulting in a substance or an article may be patentable.

3.2 Gene patents and industrial application

Gene sequences may be claimed and will be allowed if they are not mere discoveries (outside the scope of Section 3(c)). Gene sequences or any DNA sequences without disclosure of their functions are not considered patentable, for lack of inventive step and industrial application.

Gene therapy would fall within the scope of Section 3(i) (preventing method-of-treatment patents) and hence not be granted a patent. Personalised medicines as such may not be allowed patents, as they would not fulfil the criterion of industrial application – though the method of arriving at personalised medicines may be patentable in India.

3.3 Stem cells and other organic material

The Indian Patents Act does not allow patents for an invention the primary or intended use or commercial exploitation of which could be contrary to public order or morality or which may cause serious prejudice to human, animal or plant life or health, or to the environment. There are no guidelines as to what might be considered as contrary to public order or morality, except the examples provided in the Patent Office manual of terminator technology (eg, the GURT technique for plant varieties) and method of adulteration of food. Processes for cloning human beings or animals, processes for modifying the germ line, the genetic identity of human beings or animals, or uses of human or animal embryos for any purpose are also not patentable as they are considered against public order and morality by the Patent Office.

The Intellectual Property Appellate Board (*Novartis Glivec* case)[5] and the court (*Roche v Cipla*)[6] have taken a step further by linking pricing with public interest for patented medicines, though no conclusion can be drawn from the fact that both cases are *sub judice* at the Supreme Court of India.

Section 3(j) of the Indian Patents Act forbids the patenting of plants and animals in whole or any part thereof, other than micro-organisms but including seeds, varieties and species and essentially biological processes for the production or propagation of plants and animals. Under Indian Patent Office practice, stem cells/animal tissues have been regarded as part of an animal/plant, though there are to date no court cases to this effect. Furthermore, the growth and use of such cells/tissues are patentable only if they are outside the purview of Section 3(c) (not a mere discovery), Section 3(i) (not relating to method of treatment) and also of Section 3(j) (not within the scope of being essentially biological processes). Micro-organisms are allowed patents; therefore, a bacterial cell or microbial cell mass would be patentable in India.

The landmark Indian case relating to live forms was decided by the Kolkata High Court in *Dimminaco – AG v Controller of Patents and Designs and Others*.[7] The issue was the patenting of a process for preparation of a vaccine, which was invented for protecting poultry against the infectious bursitis virus. The Controller held that the process of separation of the vaccine which had a living entity cannot be considered a manufacture and hence not patentable under Section 2(1)(j) of the Patents Act. He also held that since the vaccine contained a living organism, it cannot be patented. The court held that the matter involved was a new process of preparing a vaccine under specific scientific conditions. The vaccine was useful for protecting poultry against contagious bursitis infection and there was no statuary bar to accepting a manner of manufacture as patentable, even if the end products contain living organisms.

Thus, the Indian Patent Office allows processes relating to life forms, as long as

---

5   2009, TA/1-5/2007PT/CH.
6   2008 (37)PTC 71.
7   See AID No 1 of 2001.

they are not essentially biological processes. The term 'essentially biological process' is not defined and is subject to interpretation.

3.4 Bioinformatics systems

The scope of bioinformatics today is vast, going beyond just the use of computer science for biological/medical applications. Bioinformatics includes the creation and advancement of databases, algorithms, computational and statistical techniques, composing and executing a series of computational or data manipulation steps and theories to solve formal and practical problems arising from the management and analysis of biological data.

A mathematical or business method, a computer program *per se* or algorithms are not patentable under Section 3(k) of the Indian Patents Act. Therefore, claims relating to software programs are considered as for computer programs *per se* simply expressed on a computer-readable storage medium, and as such are considered not allowable. An invention consisting of hardware, along with software or a computer program in order to perform the function of the hardware, may be considered patentable (eg, embedded systems).

Also, under Section 3(n) of the Indian Patents Act, the presentation of information is not patentable. Any manner, means or method of expressing information – whether visual, audible or tangible by words, codes, signals, symbols, diagrams or any other mode of representation – is not patentable. For example, a means of speech instruction, in the form of printed text where horizontal underlining indicated stress and vertical separating lines divided the words into rhythmic groups, was held not patentable.

Bioinformatics systems are patentable if they are found to be outside the scope of Section 3(k) and fulfil the patentability criteria of novelty, inventive step and industrial application.

3.5 Industrial applicability

In order to be granted a patent in India, an invention should be 'capable of industrial application'. This term has been defined in relation to an invention to mean that the invention is capable of being made or used in an industry.

The term 'industrial application' was introduced in the Patents Act through an amendment in 2002. Under the definition of 'invention' prior to the amendment, an invention had to be new and useful for grant of patent. Section 64(1)(g) of the Patents Act states that lack of utility is a ground for revoking a patent. In *Lakhapati Rai and Ors v Srikissen Dass and Ors* (1917), it was held that 'utility' does not mean improvement; it means practicability. The test of utility is whether the invention will work and whether it will do what is claimed for it. Industrial application is therefore determined on a case-by-case basis.

3.6 Copyright and sequence information

DNA/RNA/gene sequences are patentable if they are not considered a discovery under Section 3(c), and are found not to be within the scope of Section 3(j) of the Indian Patents Act. When a genetically modified gene/amino acid sequence is novel, involves an inventive step and has industrial application, the following can be claimed in a single application:

- gene/amino acid sequence;
- a method of expressing that sequence;
- an antibody against that protein/sequence; and
- a kit made from the antibody/sequence.

All of these claim sets are linked by the common inventive concept of the sequence.

Section 3(l) of the Patents Act does not allow the patenting of a literary, dramatic, musical or artistic work or any other aesthetic creation, as these fall within the ambit of copyright. Molecular sequences (eg, DNA/RNA/amino acid) can be considered as literary works as computer software which is granted copyright and, by analogy, such sequences may also be considered as copyrightable. This argument was used in *Emergent Genetics v Shivam Sundaram and Ors*,[8] which is still *sub judice* at the Delhi High Court. However, the court had indicated in the interim application to vacate the *ex parte* stay that copyright on the DNA of a plant variety may be too wide a protection. Only after the trials and final judgment in this case will it be clear whether or not the courts consider copyright to subsist in a DNA sequence.

Copyright to databases has been well recognised and the sequence databases obtained from genome sequencing or genome analysis using bioinformatics clearly qualify for copyright protection.

3.7   Sufficiency issues

Sufficient disclosure of the invention is the consideration on which a patent monopoly is granted. Therefore, it is incumbent upon the applicant to make a clear, precise, honest and open disclosure of the invention. The patent specification should describe the invention clearly. It is the duty of a patentee to state clearly and distinctly the nature and limits of what he claims. If the language used by the patentee is obscure and ambiguous, no patent can be granted, and it is immaterial whether the obscurity in the language is due to design, carelessness or want of skill. Lack of clarity is a ground for opposition. An ambiguously drafted description or one that lacks clarity, either by negligence or lack of skill, will result in failure to obtain a patent.

The Patent Office expects unnecessary use of technical jargon to be avoided. The use of technical terms well known in the art is preferred. Little-known technical terms are allowed, provided they are adequately defined in the specification.

If the applicant mentions biological material in the invention and it is not possible clearly to describe the same in the complete specification, the requirement of sufficiency of disclosure can be completed by depositing such material in an international depository authority under the Budapest Treaty. The deposit should be no later than the date of filing; however, the reference number to the deposit can be made in the specification within three months from the date of filing the application, giving all the available characteristics of the material required for it to

---

8   CS(OS) 50/2004 Delhi HC.

be correctly identified. Access to the material should be available in the depository institution only after the date of the patent application in India, or after the date of the priority. The complete specification should also contain the details of such deposition and the source and geographical origin of the biological material specified in the specification.

The international depository authorities in India are:

- Microbial Type Culture Collection and Gene Bank (MTCC) – IMTECH Chandigarh; and
- Microbial Culture Collection (MCC) National Centre for Cell Science (NCCS) University of Pune Campus, Ganeshkhind Pune.

Sequence listing, when present in the application, may be numbered in the specification if necessary. Sequence listings are required to be submitted in electronic form along with the paper copy.

4.   Acts of patent infringement

4.1   Infringement

The offence of infringement is not defined in the Indian Patents Act; therefore, the meaning of infringement is derived from the rights conferred on the patentee. It is the violation of these rights which constitutes infringement.

Section 48 of the Patents Act 1970 provides that a patent shall confer upon the patentee:

- where the subject matter of the patent is a product, the exclusive right to prevent third parties, which do not have the patentee's consent, from the act of making, using, offering for sale, selling or importing for those purposes that product in India;
- where the subject matter of the patent is a process, the exclusive right to prevent third parties, which do not have the patentee's consent, from the act of using that process, and from the act of using, offering for sale, selling or importing for those purposes the product obtained directly by that process in India.

The grant of patent is subject to certain conditions specified in Section 47 of the Patents Act which are not considered as infringements:

- use of any patent by the government solely for its own use;
- use of any patent merely for the purpose of experiment or research, including imparting instructions to students;
- use of a patent between the date on which the patent ceases to have effect and the date of the advertisement of the application for restoration of the patent;
- use on foreign vessels/aircraft etc which are temporarily or accidentally in India, including its territorial waters;
- Bolar-type provision (ie, the act of making, constructing, using, selling or importing a patented invention solely for uses reasonably relating to the

development and submission of information for regulatory approval);

- parallel importation (ie, the importation of patented products by any person from a person who is duly authorised under the law to produce and sell or distribute the product); and
- continued manufacture of a product, even after the grant of a patent, by enterprises which have made significant investment and were producing and marketing the relevant product covered by a patent application under Section 5(2) before January 1 2005. Patent holders are entitled to receive a reasonable royalty from such enterprises. This provision is applicable until 'black box' applications are disposed of.

Patent litigation has not developed to its fullest extent in India. There have been relatively few suits filed for patent infringement in the past, and most never went beyond the stage of an interim injunction. However, after the 2005 amendments to the Patents Act, patent litigation has picked up substantially in Indian courts, with patentees from various technological fields instituting suits for patent infringement.

Courts have recently interpreted patent claims literally and not purposively. There is, however, a Supreme Court judgment (*Bishwanath Praad Radhey Shyam v Hindustan Metal Industries AIR* 1982 SC 1444) which states that claims should be given an 'effective meaning'. The Supreme Court held that the proper way to construe a specification is not to read the claims first and then see what the full description of the invention is, but instead to read the description of the invention first, in order that the mind may be prepared in respect of the invention that is to be claimed, as the patentee cannot claim more than he desires for the patent. The claims must be construed so as to give an effective meaning to each of them, but the specification and the claims must be looked at and construed together.

4.2   Contributory infringement

The Patents Act does not define contributory infringement, but the concept is recognised and it is accepted that apart from the person who actually manufactures the infringing article or uses the patented process, others who are indirectly connected with the manufacture or sale of the article may also be involved in the infringement. Manufacturers and agents, repairers of patented articles, inducers and procurers of infringement, distributors *et al* may all fall within the ambit of contributory infringement.

Manufacturing parts of a claimed invention with the objective of creating that invention would be infringement, whereas making parts of a combination may not be considered infringement. There is as yet no case law in the area of life sciences relating to contributory infringement.

4.3   Submitting authorisations and offers to supply

The Indian Patents Act is silent on the definition of infringement and hence requests for authorisations to supply will be interpreted by the courts as to whether or not such requests would constitute infringement.

It is likely that the courts would take into consideration the intent of the person submitting marketing approval requests and might consider such requests as infringement. In fact, the Delhi High Court granted interim injunctions in two instances,[9] where a generic manufacturer had applied for marketing approvals at the office of the Drug Controller General of India (DCGI) for patented products. The interim injunctions were later vacated in both the matters. These cases have not yet been decided on the merits so as to determine how the courts will finally interpret the linkage between the Patents Act and the regulatory authorities.

The act of making offers for sale which fall within the ambit of the patent holder's rights will amount to infringement.

4.4 Summaries of product characteristics (SmPCs)
Marketing approval for a drug will be granted by the office of the DCGI. The information to be submitted is as per the data relating to the application form and the common technical document (CTD) dossier is prepared for submission along with the said forms. There are currently no exclusive rights on information in relation to intellectual property.

5. Patent enforcement

5.1 Obtaining information on the infringer and the infringement
Discovery is possible only after legal proceedings have commenced. Under the Code of Civil Procedure 1908 (CPC), discovery can be by interrogatories, delivered with the leave of the court, for the examination of the opposing parties. Interrogatories are allowed, provided they are relevant to the proceedings.

Discovery will not be ordered against anyone who can give information as to the identity of an infringer or wrongdoer. Discovery to find the identity of an infringer is available against any entity against which the plaintiff has a cause of action in relation to the wrong. It is not available against a person who has no connection with the wrongdoing other than the fact that he was a spectator, or has some document relating to it in his possession. If through no fault of his own, a person becomes involved in the tortious acts of others so as to facilitate their wrongdoing, he may incur no personal liability but he comes under a duty to assist the person who has been wronged by giving him full information and disclosing the identity of the wrongdoers. It does not matter whether he became so involved by voluntary action on his part, or because it was his duty to do what he did. It may be that this duty to provide information or disclose the identity of wrongdoers causes him expense, in which case the person seeking the information should reimburse him. Justice requires, however, that he should cooperate in righting the wrong if he unwittingly facilitated its perpetration. Discovery will not be ordered against a mere witness.

The form of interrogatories is prescribed. A party can object to answering interrogatories on the ground that they are scandalous or irrelevant or are not *bona fide*, or that the matters inquired into are not sufficiently material at that stage, or on

---

9   *Bayer and Ors v Cipla* 2009 (41)PTC 634 (Del) and *Bristol-Myers Squibb v Hetero* IA 15772/2008 in CS(OS) No. 2680/2008.

the ground of privilege. If allowed, the interrogatories must be answered on affidavits. Where a party who is interrogated omits to answer or answers insufficiently, he may be required to answer either by affidavit or by *viva voce* examination.

At any time during the pendency of the suit, the court may order the production, upon oath, of such documents as are in the possession or power of a party. Parties to a suit are also entitled to give notice to each other to produce a particular document for inspection.

Where a party from whom discovery of any kind is sought objects, the court may, if satisfied that the right to discovery or inspection depends upon the determination of any issues in the suit, order that such issues be decided first. Where the claimant fails to comply with the order of interrogatories, discovery or inspection of documents, the suit is liable to be dismissed for want of prosecution and if it is the defendant who fails to comply, his defence is liable to be struck out.

In *M Sivasamy v Vestergaard Frandsen A/S and Ors* (FAO(OS) 205, 206 and 211 of 2009 Delhi High Court) , the court held that in order to balance the rights of both the parties, such a procedure should be adopted as to maintain the confidentiality of the documents produced and not prejudice either of the parties.

5.2 'Springboard'/post-patent expiry injunctions

The rights of a patentee on a claimed invention subsist only during the life of the patent. After this time, there will be no patent infringement.

However, the 'springboard' doctrine may be applied for the trade secrets or confidential information available to an ex-employee under common or contract law.

5.3 Unjustified threats

Section 106 of the Patents Act provides a remedy against unjustified threats.

Where any person threatens any other person, by circulars or advertisements or by oral or written communication addressed to that or any other person, with proceedings for infringement of a patent, any person aggrieved thereby may bring a suit against him for the following remedies:

- a declaration to the effect that the threats are unjustifiable;
- an injunction against the continuance of the threats; and
- damages, if any harm has been sustained.

The court may grant to the plaintiff all or any of the requested remedies unless in such suit the defendant proves that the acts in respect of which the proceedings were threatened would constitute or, if done, constitute an infringement of a patent or of rights arising from the publication of a complete specification in respect of a claim of the specification now shown by the plaintiff to be invalid.

An explanation to the statute is provided which states that a mere notification of the existence of a patent does not constitute a threat of proceeding within the meaning of this section.

5.4 Remedies

The owner of a patent has only civil remedies available to him; no criminal remedies are available. The civil remedies include:

- injunction;
- damages or an account of profits;
- delivery-up of infringing materials; and
- costs.

Additionally, under Section 108(2) of the Patents Act the court may order the destruction of the infringing goods, along with any tools or implements used in the manufacture of the infringing goods.

Under Section 113 of the Patents Act, a certificate of validity may be granted when the validity of the claims of an existing patent are challenged and upheld, either by the Intellectual Property Appellate Board or the High Court.

The granting of these reliefs is discretionary. The injunction may be subject to such terms as the court may think fit to impose. In all cases, it will be limited to the term of the patent. Provisions relating to reliefs in the Patents (Amendment) Act 2005 are not exhaustive. Thus, it would appear that the court is not debarred from ordering delivery-up or destruction of the infringing articles.

5.5   Interim relief

Interim injunctions, or temporary injunctions, are sought in every suit for a permanent injunction in an intellectual property case, because of the length of time a case takes to get to trial. While *ex parte* interim injunctions are granted by courts in the normal course in some jurisdictions in India, other courts tend to be wary of granting this relief. The greatest number of *ex parte* interim injunctions are known to have been granted by the Delhi High Court, while the Bombay High Court rarely, if ever, grants an *ex parte* interim injunction.

At the *ex parte* stage, the court would want to be convinced of the claimant's title to the right which is sought to be enforced against the defendant, whether the suit is properly filed and whether the court has territorial jurisdiction to entertain it. However, these issues are not examined in great depth at this stage and the hearing of an interim injunction application *ex parte* takes a relatively short time before orders are passed. There are rare occasions when a claimant lacking the requisite documentary evidence succeeds in obtaining the grant of an *ex parte* interim injunction against an innocent defendant. The *ex parte* interim relief is of immense value in intellectual property cases in India where courts, burdened with a very heavy backlog of work, may not be able to decide an application for an interim injunction until a year after the institution of the suit (although the CPC provides that the court should endeavour to make a final determination of the application within 30 days from the date on which the injunction was granted).

6.   Compulsory licensing

6.1   Section 84 compulsory licences

Under Section 84 of the Patents Act, at any time after the expiration of three years from the date of the grant of a patent, any party may make an application to the

Controller for the grant of a compulsory licence on a patent on any of the following grounds:

- that the reasonable requirements of the public with respect to the patented invention have not been satisfied;
- that the patented invention is not available to the public at a reasonably affordable price; or
- that the patented invention is not worked in the territory of India.

An application under this section may be made by any party notwithstanding that it is already the holder of a licence under the patent, and no party shall be estopped from alleging that the reasonable requirements of the public with respect to the patented invention are not satisfied, or that the patented invention is not worked in the territory of India, or that the patented invention is not available to the public at a reasonably affordable price by reason of any admission made by it, whether in such a licence or otherwise or by reason of having accepted such a licence.

Every such application must contain a statement setting out the nature of the applicant's interest together with such particulars as may be prescribed and the facts upon which the application is based. If satisfied, the Controller may grant a licence upon such terms as he may deem fit.

The compulsory licence provisions are attracted if the patented invention is not being worked in the territory of India on a commercial scale to an adequate extent, or is not being so worked to the fullest extent that is reasonably practicable; or if the working of the patented invention in any part of India on a commercial scale is being prevented or hindered by the importation from abroad of the patented article by:

- the patentee or persons claiming;
- persons directly or indirectly purchasing from him; or
- other parties against whom the patentee is not taking or has not taken proceedings for infringement.

The general purposes for granting compulsory licences are:

- that patented inventions are worked on a commercial scale in the territory of India without undue delay and to the fullest extent that is reasonably practicable; and
- that the interests of any person for the time being working or developing an invention in the territory of India under the protection of a patent are not unfairly prejudiced.

The licence granted by the Controller does not authorise the licensee to import the patented article or an article or substance made by a patented process from abroad where such importation would, but for such authorisation, constitute an infringement of the rights of the patentee.

Notwithstanding anything contained above, the Indian central government may, if in its opinion it is necessary so to do in the public interest, direct the Controller at any time to authorise any licensee in respect of a patent to import the patented article or an article or substance made by a patented process from abroad. This authorisation may be subject to such conditions as are necessary, relating among other matters to royalties and other remuneration, if any, payable to the patentee, the quantum of import, the sale price of the imported article and the period of importation.

Any person who has the right to work any other patented invention either as patentee or as licensee thereof, exclusive or otherwise, may apply to the Controller for the grant of a licence of a related patent on the ground that he is prevented or hindered without such licence from working the other invention efficiently or to the best advantage possible.

6.2 Special provision for compulsory licences on notification by central government

If the central government of India is satisfied, in respect of any patent in force, in circumstances of national emergency or of extreme urgency, or in case of public non-commercial use, that it is necessary that compulsory licences should be granted at any time after the grant of the fee payment to work the invention, it may make a declaration to that effect, by notification in the *Official Gazette* under Section 92 of the Patents Act. Thereupon the Controller shall on application made at any time after the notification by any person interested grant to the applicant a licence under the patent on such terms and conditions as he thinks fit.

In settling the terms and conditions of a licence granted under this section, the Controller must attempt to ensure that the articles manufactured under the patent shall be available to the public at the lowest prices consistent with the patentee deriving a reasonable advantage from their patent rights.

The Controller must, as soon as is practicable, inform the patentee of the patent relating to the application.

6.3 Compulsory licence for export of patented pharmaceutical products in certain exceptional circumstances

Compulsory licences may be granted for the manufacture and export of patented pharmaceutical products to any country having insufficient or no manufacturing capacity in the pharmaceutical sector so as to address public health problems. This is on the condition that the compulsory licence has been granted by such country or such country has, by notification or otherwise, allowed importation of the patented pharmaceutical products from India. The arrangement is in line with the Doha Declaration and can be authorised on the basis of Section 92A of the Patents Act.

The Controller shall, on receipt of an application in the prescribed manner, grant a compulsory licence solely for manufacture and export of the concerned pharmaceutical product to such country under such terms and conditions as may be specified and published by him.

The provisions of this section are without prejudice to the extent to which pharmaceutical products produced under a compulsory licence can be exported under any other provision of the Patents Act.

For the purposes of this provision, 'pharmaceutical products' means any patented product, or product manufactured through a patented process, of the pharmaceutical sector needed to address public health problems and shall be inclusive of ingredients necessary for their manufacture and diagnostic kits required for their use.

The provision saw Pfizer and Roche on one side and NATCO on the other exploring the possibilities of this particular section of the Indian Patents Act. NATCO applied for a compulsory licence at the Patent Office to export the patented drugs Sunitinib or Sutent (Pfizer) and Erlotinib or Tarceva (Roche) to Nepal.

The two compulsory licence cases did not go beyond the Patent Office due to withdrawal of the application by NATCO; however, it provided a precedent with regard to hearing both the parties in cases of compulsory licensing matters based on the principle *audi alterum partem* of natural justice.

6.4 Termination of compulsory licence (Section 94)

On an application made by the patentee or any other person with title or interest in a patent, a compulsory licence granted under Section 84 may be terminated by the Controller under Section 94 of the Patents Act, if and when the circumstances that gave rise to its grant no longer exist and such circumstances are unlikely to recur. The holder of the compulsory licence will have a right to object to such termination.

7. Ownership, inventors and compensation

The Indian Patent Act defines 'true and first inventor' to mean the inventor who has first filed the patent application and is not a mere importer of an invention into the country (Section 2(1)(y) of the Patents Act). The true and first inventor or his legal representative is entitled to apply for a patent. The invention may also be assigned by the inventor and the assignee will have rights arising from the patented invention. A patent application may be filed jointly by different parties.

The Patent Act does not provide any guidance as to the manner and format of an assignment of an invention, or the compensation to be given to the inventor in exchange for the assignment.

The assignment may be by way of service contract, where the employer may own the rights of the invention made by an employee during the course of employment. The employer–employee relation implies a transfer of ownership of the invention from the employee to the employer (unless it is specified otherwise in the employment contract). However, in the absence of specific mention of the ownership of IP rights in the employment contract, the ownership of the invention will be subject to the interpretation of the existing employment contract and the choice of the inventor.

The concept of inventor compensation is not explicitly dealt with in Indian patent law or in practice. The assignment terms determine the benefits which the inventor would receive in the event of a transfer of his rights over the invention to an employer or assignee. If the assignment follows the grant of the patent, it must be recorded in writing in the Patent Register.

It should be noted that if an invention in respect of which a patent has been applied for is wrongfully obtained, this will be a ground for opposition and revocation of the patent (Sections 25(1) and (2) and 64 of the Patents Act). Disputes between co-owners are also clearly dealt with in Sections 50 and 51 of the same Act.

8.   Branding and designs

8.1   Trade marks in life sciences

A trade mark may be registered under any of 45 classes as per the fourth schedule of the Trade Marks Rules 2002. Parts of an article or apparatus are, in general, classified with the actual article or apparatus, except where such parts constitute articles included in other classes. The classes under which marks relating to the life sciences fall are mainly:

- Class 1. Chemicals used in industry, science, photography, agriculture, horticulture and forestry; unprocessed artificial resins, unprocessed plastics; manures; fire-extinguishing compositions; tempering and soldering preparations; chemical substances for preserving foodstuffs; tanning substances; adhesives used in industry.
- Class 5. Pharmaceutical, veterinary and sanitary preparations; dietetic substances adapted for medical use; food for babies; plasters or materials for dressings; materials for stopping teeth, dental wax; disinfectants; preparations for destroying vermin; fungicides, herbicides.
- Class 9. Scientific, nautical, surveying, electric, photographic, cinematographic, optical, weighing, measuring, signalling, checking (supervision), life-saving and teaching apparatus and instruments; apparatus for recording, transmission or reproduction of sound or images; magnetic data carriers, recording discs; automatic vending machines and mechanisms for coin-operated apparatus; cash registers, calculating machines, data processing equipment and computers; fire-extinguishing apparatus.
- Class 10. Surgical, medical, dental and veterinary apparatus and instruments, artificial limbs, eyes and teeth; orthopaedic articles; suture materials.
- Class 29. Meat, fish, poultry and game; meat extracts; preserved, dried and cooked fruits and vegetables; jellies, jams, fruit sauces; eggs, milk and milk products; edible oils and fats.
- Class 30. Coffee, tea, cocoa, sugar, rice, tapioca, sago, artificial coffee; flour and preparations made from cereals, bread, pastry and confectionery; ices; honey, treacle; yeast, baking powder; salt, mustard; vinegar, sauces (condiments); spices; ice.
- Class 31. Agricultural, horticultural and forestry products and grains not included in other classes; live animals; fresh fruits and vegetables; seeds, natural plants and flowers; foodstuffs for animals, malt.
- Class 32. Beers, mineral and aerated waters, and other non-alcoholic drinks; fruit drinks and fruit juices; syrups and other preparations for making beverages.
- Class 33. Alcoholic beverages (except beers).

Services
- Class 42. Scientific and technological services and research and design relating thereto; industrial analysis and research services; design and development of computer hardware and software.

- Class 44. Medical services, veterinary services, hygienic and beauty care for human beings or animals; agriculture, horticulture and forestry services.

8.2 Extending protection using trade marks and designs

Patent protection provides limited-time monopoly as opposed to the unlimited-time protection available to trade marks. A patent owner may therefore be tempted to use trade mark law to extend the protection of an embodiment of the invention illustrated in an expired patent for an unlimited duration.

The duty of disclosure imposed by the Patents Act requires the inventor to give to the public a sufficiently detailed description of the invention so that a person skilled in the art, using the instructions contained in the description of the patent, should be able to use the invention. As a result, following the expiration of the patent, the patentee cannot assert any trade mark rights on the embodiments as described by the inventor in the patent. Trade mark law also prohibits the protection and registration of functional marks. Therefore, in order to claim trade mark protection for any subject matter relating to an expired patent, the mark/design relied upon should not be essential to the working of the patent and must not be functional.

As for patents and design rights on functional features for patents and aesthetic features for designs, there is a clear dichotomy. As the design monopoly is for 15 years and that for patents is 20 years, the design protection is the means by which protection can be extended. If a design cannot be used without infringing the patent, then the design monopoly may effectively extend for the 20-year lifetime of the patent. Interestingly, the trade mark right to the exclusive use of the distinguishing guise can be used to 'extend' the design monopoly of an article in respect of which the design is registered.

8.3 Protection using unfair competition

In 2002, India enacted a Competition Act to regulate unfair trade practices and ensure free and fair competition. While the Act does not have any specific provisions regarding intellectual property, it does have provisions to prevent the misuse of the dominant position by any one market participant. It also provides for a forced demerger of any company found to be abusing its dominant position. The Act is administered by the Competition Commission of India, which was set up under the same statute. The Competition Act was enacted to replace the only other previous unfair-competition legislation, which was the Monopolies and Restrictive Trade Practices Act 1969 (MRTP Act).

The Competition Act 2002, as amended by the Competition (Amendment) Act 2007, deals with the applicability of Section 3, which sets out prohibitions against anti-competitive agreements relating to IP rights with Section 3(5) specifically setting out the 'reasonable conditions' that may be necessary to protect IP rights not constituting anti-competitive behaviour. Licensing arrangements that are likely adversely to affect the price, quantity, quality or variety of goods and services will fall within the ambit of competition law. For instance, a licensee may be required to grant back to the licensor any know-how or IP rights acquired and not to grant licences to anyone else. This may be construed as augmenting the market power of the licensor in an unjustified and anti-competitive manner.

While explaining Section 3(5) of the Competition Act, in *FICCI – Multiplex Association of India v United Producers/ Distributors Forum and Ors*[10] the ruling mentioned that IP laws do not have any absolute overriding effect on competition law. The extent of the *non obstante* clause in Section 3(5) is not absolute, as is clear from the language used therein; it exempts the rights holder from the rigours of competition law only to protect its rights from infringement. It further enables rights holders to impose reasonable conditions that may be necessary to protect their rights.

There are provisions of certain other Acts which also deal with unfair competitive trade practices. These include:

- provisions of the Consumer Protection Act 1986 which protect consumers from false claims made by traders; and
- provisions of the Trade Marks Act 1999, and in particular those which prohibit a false trade description.

Common-law principles have also taken the tort of passing-off well beyond its conventional limits, either by protecting strong trade marks in fields unrelated to their own or by recognising a very small prior use and treating the mark almost like a copyright. The tort of trade libel also provides a common-law remedy against disparagement of both goods and services.

The proprietor must establish that the infringer has knowingly applied a false trade description to the goods to establish infringement.

Acts amounting to false trade description constitute a criminal offence, which may entail punishment by way of imprisonment for a term of up to two years (three years in the case of food- or drug-related offences), or a fine or both.

8.4   Comparative advertising and advertising restrictions

Comparative advertising is a form of marketing where a party advertises its goods or services by comparing them with the goods or services of another party. This concept is allowed for honest comparison between one trader's products and those of another.

Indian statutes do not provide a definition of comparative advertising, but the UK Regulation which is usually accepted by the Indian Courts defines comparative advertising as meaning any advertisement which "explicitly or by implication identifies a competitor or goods or services offered by a competitor".

The MRTP Act lists certain actions to be 'unfair trade practices' as any unfair method or unfair or deceptive practice which gives false or misleading facts disparaging the goods, services or trade of another person.

The Indian Trade Marks Act provides that a registered trade mark is infringed by any advertising of that trade mark if such advertising takes unfair advantage and is contrary to honest practices in industrial or commercial matters, is detrimental to its distinctive character, or is against the reputation of the trade mark. However, if the impugned use of the mark is in accordance with 'honest practices' in industrial or

---

10   Case CCI 01 of 2009 Del.

commercial matters, then there would be no infringement under the Act. There is no definition or explanation, however, as to what constitutes 'honest practices'.

In *Reckitt & Colman v Kiwi TTK*,[11] the Delhi High Court held that a tradesman is entitled to declare his goods to be the best in the world, even though the declaration is untrue, and that his goods are better than his competitor's even though such statement is untrue. He can even compare the advantages of his goods over the goods of others. However, the tradesman cannot, whilst claiming that his goods are better than his competitors', say that his competitors' goods are bad. If he says so, he really slanders the goods of his competitors; in other words he defames his competitors and their goods, which is not permissible.

Any complaints against an advertisement can be addressed to the Advertising Standards Council of India (ASCI). Though the code of advertising practice of the ASCI does not have legal status, it is binding on all advertisements that are displayed in India through a notification in *The Gazette of India: Extraordinary*. It applies to advertisers, advertisement agencies, and advertisements that are made abroad but displayed in India

The Consumer Protection Act 1986 allows a consumer association, central government or the state government to take up a case of unfair trade practice before a consumer forum. An action for the same can also be taken before the ASCI from the consumers. Therefore, advertisers must be careful to ensure that the advertisement does not dilute the competitor's brand value and that they have cogent facts to back up their claims.

8.5   Protection of online content

The protection of online content can be sought through two statutes, namely the Copyright Act 1957 and the Information Technology Act 2000 (the IT Act).

Since it was enacted in 1957, the Copyright Act has been amended five times (in 1983, 1984, 1992, 1994 and 1999, with the amendment of 1994 being the most substantial). Following these amendments, the Act is now compliant with most international conventions and treaties in the field of copyrights. India is a member of the Berne Convention and the Universal Copyright Convention besides being a member of TRIPS. The Copyright Bill, which will carry further amendments, is currently being debated at the houses of the Indian parliament.

Literary works under copyright protection include computer programs. Section 2(ffc) of the Copyright Act defines a computer program as a set of instructions expressed in words, codes, schemes or in any other form, including a machine-readable medium, capable of causing a computer to perform a particular task or achieve a particular result. The term 'computer' includes any electronic or similar device having information-processing capabilities.

Under the Copyright Act, 'copyright' means

the exclusive right subject to the provisions of this Act, to do or authorise the doing of any of the following acts in respect of a work or any substantial part thereof in the case of a literary work –

---

11   1996 (37)DRJ 648.

(i) to reproduce the work in any material form including the storing of it in any medium by electronic means;

(ii) to issue copies of the work to the public not being copies already in circulation;

(iii) to perform the work in public, or communicate it to the public;

(iv) to make any cinematograph film or sound recording in respect of the work;

(v) to make any translation of the work;

(vi) to make any adaptation of the work;

(vii) to do, in relation to a translation or an adaptation of the work, any of the acts specified in relation to the work in sub-clauses (i) to (vi);

(viii) to sell or give on commercial rental or offer for sale or for commercial rental any copy of the computer program, provided that such commercial rental does not apply in respect of computer programs where the program itself is not the essential object of the rental.

Online content is, therefore, protected mainly by copyrights.

The second statute in place relating to the cyber laws in India is the IT Act, which was amended in 2008. The IT Act provides for legal recognition for transactions carried out by means of electronic data interchange and other means of electronic communication. This Act plays an important role in relation to areas of interplay between information technology and intellectual property rights.

The recent decision of the Delhi High Court in *Super Cassettes Industries Limited v My Space Inc and Anr* (CS(OS) No 2682/2008) has far-reaching implications and a significant bearing on the immunity available to intermediaries such as social networking websites for hosting infringing content. In this case Super Cassettes, being the owner of a large repertoire of copyrighted works, sued My Space, a social networking website which allowed users to upload music, images etc onto the site, for copyright infringement. The court's *prima facie* view was that the acts were infringing and thus granted an injunction restraining My Space from uploading the contents of the plaintiff.

The court further observed that Sections 79 and 81 of the IT Act 2000 must be read conjointly and thus rights of owners under the Copyright Act 1957 and/or Patents Act 1970 remain unfettered by any provision of the IT Act; and thus copyright infringers cannot escape liability. The court further found that the acts committed, including modification/amendment of content and inserting advertisements, excluded networking website My Space from the purview of the definition of 'intermediaries' under Section 79 of the IT Act. Also, it was further observed that the safe-harbour provision under the Digital Millennium Copyright Act (DMCA) or any such provision of the Act would not be applicable in the Indian scenario.

As for protection relating to domain names, India has the .IN Domain Name Dispute Resolution Policy (Policy) setting out terms and conditions for resolving disputes between registrants and complainants, arising out of the registration and use of the .in internet domain name. The .IN Registry must appoint an arbitrator from the list of arbitrators maintained by the Registry and this arbitrator will conduct arbitration proceedings in accordance with the Arbitration and Conciliation Act 1996 and also in accordance with the Policy and the rules provided thereunder.

9.   Protecting valuable information

9.1   Preserving confidentiality, know-how and ex-employees

There is no specific legislation for the protection of trade secrets or valuable confidential information in India. The law relating to confidential information and trade secrecy exists through common-law principles. The broad elements of the tort are:

- that there is information which is secret;
- this information has been disclosed to another under conditions of confidentiality; and
- the confidant has misused the information.

Confidential information and trade secrets are protected by common law, unless the secret is a government secret, in which case the Official Secrets Act 1923 may apply.

India is party to the GATT, and Section 7 on protection of undisclosed information of TRIPS applies where, in Article 39, members are required to protect undisclosed information.

Information lawfully within control may be prevented from being disclosed to, acquired by, or used by others without their consent in a manner contrary to honest commercial practices, provided such information has commercial value because is secret. For the purpose of this provision, 'a manner contrary to honest commercial practices' means at least practices such as breach of contract, breach of confidence and inducement to breach, and includes the acquisition of undisclosed information by third parties who knew, or were grossly negligent in failing to know, that such practices were involved in the acquisition.

Further, Article 39 states when requiring, as a condition of approving the marketing of pharmaceutical or of agricultural chemical products which utilise new chemical entities, the submission of undisclosed test or other data, the origination of which involves a considerable effort, that members shall protect such data against unfair commercial use. The Indian Government stance has, however, contradicted this approach. The government stance is that India should not provide data exclusivity for pharmaceuticals and agro-chemicals in favour of the pharmaceuticals industry, as such exclusivity would have considerable impact in delaying the entry into the market of cheaper generic drugs.

In view of the above, confidentiality clauses will be inserted in technology transfers or other licence agreements to maintain the confidential nature of the subject matter. Employment contracts and other agreements are very important in terms of protecting confidential information and the acquired skills, knowledge and experience of a particular profession. Trade secrets are protected in India either through contract law or through the doctrine of breach of confidentiality.

10.   Counterfeiting

The best way to combat piracy and counterfeiting, both in terms of time and cost, is to use criminal proceedings. There are, however, circumstances in which civil actions might be preferred:

- When an injunction is desirable, civil action is necessary, since it is only civil courts which can issue injunctions.
- In the case of a foreign proprietor, the presence of the client or the complainant at hearings may not be possible. In a criminal action, the normal rule is that the complainant must be present at all hearing dates. It is possible to seek an exemption from appearance for a day or two, but absence must be the exception. Hence, foreign claimants either prefer civil proceedings or appoint a local attorney to attend their criminal actions.
- The police, being an additional agency in criminal actions, may be a hindrance and in some cases due to negligence or lack of knowledge may even spoil the case. Thus, in one software case, the police tied the floppy disks to the case file by passing an ice pick through them, thereby destroying the evidence.

Criminal actions are likely to be preferred in the following circumstances:
- Where there is exact copying, a civil action may not be possible as the defendant may simply deny that he manufactured the goods in question. The use of an Anton Piller order to recover the goods from the defendant can minimise this discrepancy. However, the defendant could still argue that the seized goods are originals. It then becomes a matter of complex evidence to establish the spurious nature of the goods. The defendant could also claim that he bought the goods from a named source. In these cases a criminal action is effective, since the police have the means to investigate such issues thoroughly.
- Where there is hard-core piracy, there may even be evidence of a link between piracy and more serious offences such as terrorism or drug smuggling. In such cases, a civil action would not create the right deterrent. Ordinary economic offenders fear the entry of a uniformed policeman to their premises, mainly because of the social stigma associated with being raided by the police. If there is an arrest, it is even worse and in an appropriate case if the application for bail is rejected, the pressure can be so great that it invariably leads to settlement of the entire dispute.

In a criminal action, confidentiality can be preserved by filing the action against unknown persons. In such cases, even the police, as part of the raiding team, are not informed of where the raid is to take place until a few minutes before it occurs. In a civil action, by contrast, the defendant must be named, and his full address given and the case passes through many hands, so that confidentiality is always at risk. Theoretically, the possibility of proceedings being taken exists but, in practice, judges do not favour such proceedings.

10.1 Border measures

The Indian Customs Office has formulated rules for the enforcement of IP rights and, as per Customs Notification 49/2007, there is a prohibition against import of patented products produced or made using a patented process in contravention of the patented process. Recently a patent of Ram Kumar on dual SIM cards had telecom companies such as Samsung and Spice coming under the purview of the Customs Office with their consignments containing dual-SIM technology mobiles being seized at the ports of entry in India by customs staff. The patentee Ram Kumar filed an application at the Customs Office to block any dual-SIM cell phone imports, citing his patent on dual-SIM. This was duly carried out by customs officials, although legal analysts opined that the patent did not cover all dual-SIM phones as dual-SIM technology itself is part of the prior art. One of the dual-SIM cell phone importers, Micromax, filed a declaratory suit for non-infringement before a court in Gurgaon, asking the court to declare that Micromax's imports of dual-SIM phones did not infringe Ram Kumar's patent. The court restrained Ram Kumar from preventing the export or import of dual-SIM phones by Micromax. However, the issue of the ability of the customs authorities to assess issues of patent infringement has been brought sharply into focus by this matter.

11.   Collaborative models – 'open innovation'

In 2007, Computer Sciences Corporation introduced a unique Collaborative Open Innovation Network (COIN) to drive rapid innovation in information technology. Traditionally, a company or organisation will use a COIN, which is made up of a community of people, to drive rapid results for a specific research project. The project will go beyond industry project-based collaboration, creating an ecosystem that extends beyond such innovation initiatives to include mentoring of students, collecting and aggregating ideas for the introduction of technology-enabled services, and working on government-sponsored programmes to bridge the gap between industry and academia.

An open-source model to speed up drug discovery was initiated in 2008, with the launch of an open-source initiative for developing drugs to treat diseases such as tuberculosis, malaria and HIV.

The demand for free, open systems is growing and India possesses the innovation and expertise to deliver on this demand and to benefit from it. An even stronger commitment to open-source technologies from India's government and its leading businesses can help the country to capitalise on this opportunity and move its economic development forward.

Patent pooling and 'pre-competitive' data pooling are areas which are in the early stages of evolution in the Indian patent field.

In 2011, Gilead Sciences Inc entered into agreements with four Indian generic drug makers (Hetero Drugs, Matrix Laboratories, Ranbaxy Laboratories and Strides Arcolab) for manufacturing and marketing three of its new HIV drugs, marking a new model in out-licensing of intellectual property.

Gilead did not take advantage of patent protection to sell the drug exclusively for a period and then open it up to generic companies for expanding the market presence. Instead Gilead, the IP owner, is encouraging generic companies to take the products to the market in partnership with Gilead, so that the branded launch and the generic launch will happen almost at the same time.

Along with the generic drug companies, Gilead is also entering into a licensing agreement with the Medicines Patent Pool Foundation for marketing the new drugs.

The licensing terms grant these players the rights to: elvitegravir, an investigational intergrase inhibitor; cobicistat, an investigational antiretroviral boosting agent; and the Quad, which combines four Gilead HIV medicines in a once-daily, single-tablet regimen.

12. Hot topics

12.1 Exclusive marketing rights (EMRs)

GlaxoSmithKline (GSK) filed two mailbox applications under the provisions of Section 5(2) of the Indian Patents Act 1970 for its anti-diabetes drug Rosiglitazone on August 28 1998 and filed an application for grant of exclusive marketing rights on June 30 2000. The Examiner issued an examination report in relation to GSK's EMR claim and subsequently, by an order the Controller of Patents, refused GSK's applications for EMR.

Dissatisfied with the order, GSK moved to the Calcutta High Court, which set aside the rejection by the Controller (order dated May 3 2002) and directed the Controller to reconsider the application for EMR afresh, keeping all points open.

The Controller again rejected the EMR application and GSK filed another petition before the High Court challenging this second rejection by the Controller. By this time, the Patents (Amendment) Act 2005 had already come into force and the provisions relating to EMR applications were repealed. Through an order dated February 10 2006, a single judge of the High Court ruled in favour of GSK.

On February 10 2006, the Controller of Patents and Union of India filed two appeals against the correctness of the judge's order. Two other appeals were filed by a third party to the proceedings who wished to be added as party-respondent in the writ application.

The preliminary objection of the appellants was that after January 1 2005 there was no scope for considering pending applications for EMR and, moreover, there was no scope to revive for further consideration any such EMR applications on which a decision had already been taken. On the other hand, the petitioners took the stand that on January 1 2005 there was no pending application made by them for grant of EMR.

The High Court accepted the preliminary objection regarding the maintainability of the petition, thereby allowing the appeal. The court was of the opinion that EMRs were given for a temporary period as there was a "prohibition created by law", and EMR cannot be granted afresh after the embargo was lifted. The merits of the appeal of the third parties concerned were not dealt with.

GSK moved to the Supreme Court, challenging the order passed the High Court. The appeal was allowed by the Supreme Court.

12.2 Disclosure requirements

Disclosure requirements are governed by Section 8 of the Patents Act. A patent applicant is required to provide a statement disclosing the detailed particulars of any application filed outside India corresponding to its Indian patent application. The applicant is also required to keep the controller updated about the progress of any applications filed elsewhere in the world relating to the same or substantially the same invention. This obligation on the applicant is a continuing one. In addition, the applicant must furnish the controller with details of the processing of applications filed elsewhere in the world if and when the controller requests such updates. The court's ruling in *Chemtura Corporation v UOI and Ors*[12] makes the disclosure requirements more stringent than the obligations under Section 8 itself. It has been held in said case that the Section 8 requirements are not considered fulfilled if the applicant simply provides the Patent Office with the filing details and a one-off status report on its corresponding foreign applications. The disclosures regarding the processing of corresponding applications have become mandatory and are to be made voluntarily including full particulars.

12.3 Working of patents in India

Section 146(2) of the Indian Patents Act requires every patentee and licensee to furnish a periodical statement of the working of patents on a prescribed form to the Controller of Patents. This statement should include the details of the working of the patented invention on a commercial basis in India. Failure to comply with the requirement of Section 146(2) will attract a penalty of Rs1 million (approximately US$21,000).

Following the inception of the Indian Patents Act, this provision was not enforced in its true spirit in India until December 2010 when the Controller General of Patents issued a public notice making it mandatory for all patentees and licensees to file the working statements in the prescribed format, failing which penalty provisions would be incurred.

The details of working must be presented in the prescribed format, providing information regarding the quantity and value of the patented product:
- manufactured in India;
- imported from other countries (with details of each country); and
- any licences and sub-licences granted in India.

If the patented invention is not worked, then a statement must be provided to this effect, and the statement should explain the reasons for not working. As stated previously, if a patent is not worked, it may attract compulsory licensing provisions.

---

12   Case CS (OS) No. 930 of 2009.

Israel

Liad Whatstein
Dr Shlomo Cohen & Co

Introduction

The key form of intellectual property protection relevant to the life sciences in Israel is patents.

Israel is a strict examining country. The Israeli Patent Office conducts a prior-art search and examines the application on various other grounds – for example, unity of invention and sufficiency of support for the claimed invention. In addition, the Patent Office relies on the examination and search results of parallel applications in the leading examination jurisdictions (Europe and the United States). Before examination, the applicant is required to file with the Patent Office an information disclosure statement. Examination is automatic and no formal request for examination is required. In special circumstances (eg, pending infringement or substantial investment in connection with the invention in Israel), it is possible to petition for an early initiation of the examination and an expedited examination of the patent application.

The applicant can base the Israeli application on a corresponding granted patent or allowed application in one of the 'recognised countries' (ie, Australia, Austria, Canada, Denmark, the European Patent Office (EPO), Germany, Japan, Norway, the Russian Federation, Sweden, the United Kingdom and the United States). When invoking this 'modified' examination, the claims of the Israeli application must be rendered identical to the granted claims of the foreign patent on which it is based. In such case, the application shall be deemed to have complied with the requirements of novelty, inventive step, enablement and support. The period of such 'modified examination' is shorter than regular examination and it primarily concerns formal issues specific to Israeli patent practice. Nevertheless, caution should be exercised before invoking this procedural route as it could potentially present some serious pitfalls to the patentee. Among others, examination in Israel is stayed if the corresponding patent is opposed, and the presumption of validity (in infringement proceedings) can also be jeopardised.

An Israeli patent application is confidential until the end of its examination. According to a Bill currently in the pipeline, patent applications filed in Israel will be published after 18 months from their date of filing and the prosecution file will then become open to public inspection. Acceptance of the application is published on the Patents and Trademark Office website. Following publication, there is a three-month non-extendable opposition period. If an opposition is filed, the patent will not be granted until the opposition proceedings are resolved. This pre-grant opposition system has been severely criticised as it can be abused to block the granting of a patent. Although opposition proceedings are conducted at the Patent Office, they are equivalent in nature and procedure to litigation in court proceedings. Agents also have a right of audience at the PTO but, in view of the contentious nature of opposition proceedings, it is important that they are conducted by patent litigation counsel with litigation skills and experience.

1.   Small molecules

Protection under an Israeli patent can cover the entire life cycle of pharmaceutical research and development, provided that it satisfies the substantive requirements of patentability.

1.1   Substantive law of patentability

The substantive law of validity under Israeli law is similar to the principles recognised under UK and US law. Accordingly, a patentable invention is "an invention, whether a product or a process in any field of technology, which is new and useful, can be used industrially and involves an inventive step" (Section 3 of the Israeli Patents Act).

An invention is deemed new if it was not published, in Israel or abroad, before the priority date. Publication of the invention in violation of undertakings of confidentiality will not affect the rights of the patentee and will not be considered when assessing novelty or inventive step. An express agreement as to confidentiality is not necessary. An obligation of confidence can be inferred from the circumstances or from the relationship between the parties.[1]

When assessing the novelty of the invention, it is not permitted to present a 'mosaic' of prior-art documents.[2] A prior publication will anticipate the invention only if it describes the invention in clear and exact terms so as to enable any person skilled in the art to practice the invention. A patent is valid and the invention is considered novel, even if it covers a product that has been put on the market before the priority date, provided that the product does not provide an enabling disclosure of the invention (ie, there is no 'on sale' doctrine under Israeli law).

When evaluating inventive step, all professional knowledge should be reviewed and a 'mosaic' of different publications can be presented. The addressee of the prior art is an 'ordinary skilled person' who does not exercise inventive faculties. An inventive step must not necessarily be significant and the simplicity of the invention does not prevent protection.[3] Generally, any invention requiring some research and development, and which does not directly occur to the skilled artisan, is worthy of patent protection. Israeli courts have also adopted various subtests for assessing inventiveness which focus on the economic and motivational (rather than the technical) issues. Among others, the 'long-felt need' test, the commercial success of the patented product and reaction of the scientific community to the invention are relevant in evaluating inventive step.

---

1   MLA 2015/95 *Krone AG v Inbar Reinforced Plastic* ILR 51(3) 548.
2   CA 345/87 *Hughes Aircraft Company v The State of Israel* ILR 44(4) 45 103.
3   *Hughes v The State of Israel* (see n2); CA 793/86 *Porat v ZML Ziud Moderny LeRefua* ILR 44(4) 578.

1.2 Product and process claims

Product claims of Israeli patents can be to a specific new chemical entity of a proven utility or to a broad class of compounds likely to have a similar utility. A single claim defining chemical alternatives (ie, a Markush claim) will be permitted if all alternatives are likely to have a common property or activity.

Process claims, including processes for the making of the active ingredient or its intermediates and for the making of the formulation, as well as product by-process claims, are also eligible subject matter.

1.3 Scope of protection of claims

In determining the scope of protection, Israeli courts interpret the claims on the basis of the specification and the drawings. The patent specification is given a purposive construction rather than a purely literal interpretation and the Supreme Court cautioned that 'meticulous verbal analysis' is inappropriate.[4]

As stated above, the protection afforded to the patentee also applies to the essence of the invention or its 'pith and marrow'. As consistently held, the objective of patent law is to protect 'true inventions' without being constrained by the linguistics of the drafting of the patent.[5] Therefore, guiding principles should remain flexible to address the particular facts and circumstances of each matter.

Accordingly, in order to address non-literal infringements, the Supreme Court adopted the doctrine of variants developed by the UK courts in *Rodi & Wienenberger v Showell* and in *Catnic Components v Hill & Smith Ltd*,[6] and the doctrine of equivalents as developed in the well-known judgment of the Supreme Court of the United States in *Graver Tank & Mfg Co v Linde Air Product Co*.[7] Thus, for instance, replacing one of the components of the invention is insufficient to avoid infringement if the new component is merely an insignificant variant, or if it achieves the same result using the same means.[8] In addition, the claims will be entitled to a broader interpretation if the patent represents a significant contribution to the relevant art.[9]

The doctrines of equivalents and of the infringement of the 'essence' of the invention have been applied in numerous cases, including with regard to pharmaceutical patents. For example, in *Hassle v Dexel* (unpublished), the patent claimed a bi-layer formulation. The defendant added to the formulation an internal layer composed of a sugar core. The court held that the addition of an additional internal layer does not avoid infringement as "it does not have the effect of changing the functioning of the invention which is done in the same way and also achieves the same result" and because the defendants' formulation "applies the inventive solution of the patent".[10]

---

4   *Hughes v The State of Israel* (above, note 2, p 64).
5   *Ibid*, p 69.
6   *Rodi & Wienenberger v Showell* [1969] RPC 367, 381; *Catnic Components Ltd v Hill & Smith Ltd* [1981] FSR 60, 66.
7   *Graver Tank & Mfg Co v Linde Air Product Co* 339 US 605 (1950).
8   *Hughes v The State of Israel* (above, note 2, p 73).
9   *Ibid*.
10 CF (Tel Aviv) 1240/99 *Aktiebolaget Hassle v Dexel Ltd* (not published), pp 13 to 14.

2.   Second-generation inventions

The same substantive laws of validity apply to 'secondary' patents as simple product and process claims. For instance, claims to particular salts or physical forms of a compound (different polymorphs or hydrates) or its optical isomers are available. The use of a known compound for subsequent medical indications can be claimed (in the format of a Swiss-type claim). Pharmaceutical formulations, the particle size of the active pharmaceutical ingredient, combinations of previously known active pharmaceutical ingredients, screening methods and the like are all eligible subject matter for patent protection. In fact, characterising these patents as 'secondary' may be inappropriate, as they confer protection on important aspects of the pharmaceutical R&D process.

There is no Israeli case law with regard to reach-through claims as yet. The patentability issues that reach-through claims faced in the US Patent and Trademark Office and the EPO, such as support and enablement, are likely to be raised in Israel with the main issue being whether the applicant has in fact 'over claimed' the invention.

2.1   Methods of treatment

'A method of therapeutic treatment of the human body' cannot be the subject matter of patent protection. The 'method of treatment' exclusion was narrowly interpreted. Only a method as such is excluded. Products or compositions used for the treatment of the human body are not excluded from patentability.[11] In addition, 'use' or 'Swiss-type' claims are acceptable. Such claims provide adequate protection and are also recognised for primary, secondary and any subsequent medical indications of an active ingredient.

2.2   Selection inventions

Selection inventions in Israel have been governed by a 1970 Supreme Court judgment which endorsed the well-known UK House of Lords judgment dating back to the 1930s in the matter of *IG Farbenidustrie*.[12] According to the *IG Farbenindustrie* rules, a selection patent will be valid if it satisfies the following rules: first, the selection must be based on some substantial advantage to be secured to the use of the selected members; second, the whole of the selected members must possess the advantage in question; third, the selection must be in respect of a quality of a special character which is unique to the selected group; and in addition the patentee must define in the specification in clear terms the nature of the advantage possessed by the selected sub-class or members.

Without elaborating in detail on the historical background of the *IG Farbeindustrie* rules, it is clear that the court's overall approach to selection inventions was rather hostile and it characterised them as a "special sort of patent, properly governed by special rules". The attitude towards selection inventions was substantially transformed since *IG Farbeindustrie*, and this hostile approach is

---

11   CA 244/72 *Plantex Ltd v The Wellcome Foundation Ltd* ILR 27(2) 29.
12   CA 273/70 *Pelimport Ltd v Ciba Ltd* ILR 24(2) 341; *IG Farbenindustrie AG's Patent* (1930) 47 RPC 289.

nowadays obsolete. Under current UK and EPO practice and case law, a selection patent does not differ in nature from any other patent and its validity is not evaluated under a set of more restrictive rules aimed at curtailing patent protection. Hence, it appears that the 1970 Supreme Court judgment adopting the historically restrictive approach to 'selection inventions' has become obsolete. Indeed, in a judgment given just prior to the time of writing this chapter (mid 2011), the Supreme Court questioned whether the doctrine of selection inventions as expounded in the *IG Farbeindustrie* rules remains relevant and held that the validity of selection inventions must be evaluated on the basis of the 'ordinary' novelty/inventive step rules.[13]

3. DNA, biologicals and personalised medicine

3.1 Discoveries

The decisive feature as to the patentability of biotechnological inventions is whether the material is naturally occurring. Thus, naturally occurring organisms, such as plants, animals and micro-organisms, are not eligible for patent protection. Mutants of naturally occurring organisms are afforded patent protection. Non-mutant micro-organisms which are derived from non-natural environments – for example, new adaptable strains isolated from man-made waste or pollution areas – are patentable. It is also acceptable to claim the micro-organism (natural or mutant) for use as an active component in a composition for producing a novel product (such as antibiotics, sugars, proteins, etc) by the micro-organism. Natural products (newly discovered genes, proteins, polysaccharides etc) are patentable, for use in treating a medical condition or in a particular process.

3.2 Gene patents and industrial application

In order to obtain gene patents, the applicant must identify a specific utility. According to PTO practice, *in vitro* studies showing increased biological activity constitute sufficient support. It is expected that this practice will survive judicial scrutiny in the future and that Israeli courts will not impose unrealistic requirements on patentees for the proof of *in vivo* therapeutic application at the date of filing.

3.3 Stem cells and other organic material

Local practice with respect to biotechnological inventions closely follows the European Patent Convention. The following are examples of patentable biotechnological inventions under Israeli practice:

- novel microbiological organisms which are not naturally obtainable or isolatable;
- a (recombinant) DNA/RNA (a gene or part of a gene) for encoding a particular (chimeric) protein, or a biologically active fragment thereof, and the related unmodified or mutated (modified) nucleotide sequence of such DNA/RNA;
- a vector carrying the DNA/RNA of the above bullet point;

---

13   CA 8802/06 *Unipharm v Smith Kline Beecham* (as yet unpublished).

- a micro-organism transformed by the vector of the above bullet point;
- a (chimeric) protein or a biological active fragment thereof, encoded by the DNA/RNA of the second bullet point above, and the unmodified or mutated (modified) amino acid sequence of such protein or a biologically active fragment thereof;
- a process for preparing the (chimeric) protein of the above bullet point or a fragment thereof and the use thereof;
- transgenic plants and animals, provided that their genome has not passed another genetic change other than the transformation with the (recombinant) DNA/RNA of the second bullet point above for encoding the (chimeric) protein of the fifth bullet point above or a biologically active fragment thereof;
- an antibody and the binding fragment thereof;
- a monoclonal antibody, a binding fragment thereof and the hybridoma cell producing it;
- a DNA/RNA sequence that hybridises to the (recombinant) DNA/RNA of the second bullet point above;
- a DNA/RNA fragment or a sequence that plays a role in the process of encoding the (chimeric) protein of the fifth bullet point above or a biological active fragment thereof;
- a ligand/protein/material that induces a biological or an immune response in an animal or human body;
- a (biological) material for selective enhancement or differentiation of germ cells, including germ cells of human origin, and such enhanced or differentiated germ cells, provided that these enhanced or differentiated germ cells keep their genetic line identity stable and unmodified; and
- an *in vitro* (or *ex vivo*) process for selective enhancement or differentiation of germ cells, including germ cells of human origin, and such processed germ cells, provided that the processed germ cells keep their genetic line identity stable and unmodified.

According to local practice, a claim set for an invention in the biotechnological field is usually confined to one or more original proteins, sharing a certain common activity, their use and processes for their production. Among the claims that can be included are: claims for nucleic acids; vectors encoding and expressing the active proteins; and plasmids in transformed micro-organism or other transgenic organism, such as plant or animal which carries such DNA/RNA (provided that the genome of the transgenic organism has not been changed besides inserting and carrying the DNA/RNA).

Only components that directly relate to the original protein (its structure, use, application, function, activity and process for its production) as well as the DNA capable of encoding the protein (in a free or bound form or carried by a vector) are considered elements of the same invention. Claims relating to other proteins of different biological activity (eg, an antibody to the original protein), or claims relating to DNA or nucleotide sequences incapable of encoding the original protein (eg, a hybridised nucleic acid such as cDNA or a sense sequence), are considered separate inventions.

Isolated stem cells for further differentiating or isolating into specialised cells or tissues can also be claimed.

3.4   Bioinformatics systems

The Israeli Patents Act does not specifically exclude methods for doing business, mental acts or computer programs from patentability. Yet, the definition of a 'patentable invention' requires the invention to be in 'any field of technology'. This served as the basis for excluding patent protection for software *per se*. However, computed systems making a 'technological use' of novel software to obtain an inventive 'technical result' will be patentable.[14] Despite the scarce case law, it therefore appears that computerised systems employed to discover the sequences of therapeutic genes and proteins can be protected by patents in Israel.

4.   Acts of patent infringement

The patentee is entitled to prevent any other person from exploiting the patented invention, either in the manner defined in the claims or in a similar manner which, in light of what is defined in the claims, constitutes the essence of the patented invention. 'Exploitation of an invention' is "any act that is one of the following – manufacture, use, offer for sale, sale, or importation for one of said acts". Export of patented goods from Israel also amounts to infringement, although it is not specifically listed as one of the acts exclusive to the patentee.[15]

If the invention is a process, the patent shall also apply to the 'direct product' of the process. When the defendant's product is identical to the 'direct product' of the process, the burden lies on the defendant to prove that it did not use the patented process. There will be a presumption of infringement of the patented process provided that:

- the patentee cannot determine, using reasonable efforts, which process was actually used for the manufacture of the defendant's product; and
- there is a 'high likelihood' that the defendant's product was manufactured using the patented process.[16]

In practice, it is difficult to prevail in an action for infringement of a process patent merely on the basis of the shifting of the burden of proof. Therefore, it is worthwhile to take experimental means in order to support infringement arguments. As an example of such approach, the analysis of the defendant's product to find chemical markers (such as impurities) attesting to the use of the patented process may be critical.

4.1   Contributory infringement

The Patents Act does not have any provisions relating to indirect infringement. The

---

14   MA (Jerusalem) 23/94 *United Technologies Corporation v PTO* Tak-District Court 1994(4), 1838.
15   MCM (Jerusalem) 814/05 *Orbotech Ltd v Camtech Ltd* Tak-Dist 2005(2) 2893.
16   CA 7614/96 *Zehori and Sons Industries Ltd v ' REGBA' Ltd* ILR 54(3) 721.

Supreme Court nevertheless incorporated this doctrine based on foreign case law, in particular US and UK law.[17]

According to the case law, a defendant will be found liable for contributory infringement if the following conditions are met:

- the defendant sold a component of a patented machine, combination or composition, or material or apparatus for use in practising a patented process, constituting a material part of the invention;
- the defendant knew, or should have known, that the article sold was especially made or especially adapted for use in infringement of the patent; and
- the article sold is not a staple article or commodity suitable for substantial non-infringing use.

In addition to contributory infringement, Israeli patent law recognises the doctrine of infringement by inducement.[18] Liability under the inducement-to-infringe cause of action is narrower in scope than liability under the doctrine of contributory infringement. Among other things, it must be shown that the defendants had a common design and there must be concerted action towards a common end.

4.2 National treatment of 'Bolar'-type provisions

Section 54A of the Patents Act provides a 'Bolar'-type defence. It applies to experimental acts for the purpose of obtaining a marketing approval prior to the expiration of the patent in Israel or in another country whose laws contain a Bolar-type defence. The provision reads as follows:

> 54A. An experimental act, which is part of an effort to obtain a license to market the product after the patent has expired, does not constitute 'exploitation of an invention', if the following two conditions are met:
>
> > (1) the effort to obtain a license is made in order to obtain a license in Israel, or in a country in which an experimental act on a patented invention for the purpose of obtaining a license is permitted before the patent expires;
> >
> > (2) any product produced under the terms of this section is not used – while the patent is in effect or thereafter – for any purpose other than obtaining a license as aforesaid;
>
> for the purposes of this section, 'license' means certification, permit or any other document required under law in order to market the product.

There has only been scarce litigation with regard to this defence and there still is some uncertainty with regard to its scope. Running bioequivalence studies for obtaining a marketing approval for a generic drug will clearly be exempt from infringement. On the other hand, stockpiling of products for marketing after the patent expiration or offers to supply the patented product will certainly not be

---

17 CA 1636/98 *Rav Bariach v Car Accessories Shop Havashush (1987) Ltd* ILR 55(5) 337; CA 7614/96 *Zehori and Sons Industries Ltd v 'REGBA' Ltd*, ILR 54(3) 721.

18 *Rav Bariach v Car Accessories Shop Havashush (Ibid)*; CA 817/77 *Beecham Group Limited v Bristol Myers Company* ILR 33(3) 757.

exempt. But there still remains a considerable grey area, such as the manufacture of pilot batches or the upscaling of manufacturing processes.

4.3 Experimental use exemptions

Any act that "is not on a commercial scale and is not commercial in character" is exempt. In order to qualify for the defence, both conditions must be satisfied. In addition, this defence was interpreted as a *de minimis* defence and is primarily relevant to provide protection for non-commercial academic research.[19]

The Patents Act provides that "an experimental act in connection with the invention, the objective of which is to improve the invention or to develop another invention" is not infringing. This experimental-use defence apparently applies to experimental acts with a commercial purpose. Comparative clinical trials between drugs assessing efficacy and safety profiles of the drugs are also covered.

In order to qualify for the experimental-use defence, the experiment's object must be the patented invention itself and it must be aimed at the testing of the essential features of the patented invention. Accordingly, the experimental-use defence is not intended to undermine the protection for diagnostic kits or research tools in biotechnology. In the same way, a person who wishes to test a new cure for cancer by applying it to a patented genetically modified mouse cannot rely on the defence.

5. Patent enforcement

Patent infringement proceedings can be brought by the patentee or its exclusive licensee and are initiated by submitting a statement of claim at the District Court. A defendant in patent infringement proceedings is entitled to challenge validity as part of its defence. The defendant may also file a petition for revocation of the patent at the Patent Office. In such case, the court will decide whether validity issues will be heard by the Patent Office or by the court. In most cases (unless the revocation proceedings at the Patent Office reached an advanced stage prior to the filing of the action for infringement) the court will hear both validity and infringement issues.

5.1 Obtaining information on the infringer and the infringement

Once the exchange of pleadings is completed, the parties may send each other demands for discovery of documents and interrogatories. Under the Civil Procedure Regulations, any documents containing information which may be pertinent to the subject matter of the proceedings must be disclosed. In practice, the approach adopted by the case law is restrictive and discovery under Israeli law is quite limited. It is by no means US-style discovery.

Documents containing trade secrets are in principle discoverable (possibly under appropriate protective orders), but it is often difficult to obtain court orders requiring such disclosure. There is no cross-examination or deposition of witnesses in discovery proceedings, which are essentially conducted by exchange of affidavits between the parties. However, if inappropriate discovery is made by one of the parties, it is possible to petition the court to order further discovery or additional response to interrogatories.

---

19   CF (Tel Aviv) 881/94 *Eli Lilly and Company v Teva Pharmaceutical Industries Ltd* Tak-Dist 98(3) 1586.

5.2 Interim relief

Claims for patent infringement are normally accompanied by a motion for a preliminary injunction. The motion for preliminary remedies must be supported by an affidavit and, where appropriate, expert opinions detailing all pertinent facts of the motion. The hearing of the petition for preliminary remedies will almost always involve live cross-examination of the affiants and experts. Depending on the complexity of the matter, a decision can be expected within a few weeks to a few months from the date of filing of the petition. In appropriate circumstances, it is also possible to file the petition *ex parte*, or even petition the court to grant *ex parte* search and seize (Anton Piller) orders. Attachment orders and Mareva injunctions are also available.

When deciding whether to grant preliminary relief, the court must consider whether the applicant demonstrated that there was a *prima facie* case and whether the balance of convenience justifies the granting of interlocutory remedies. Considerations such as laches, inequitable conduct and non-disclosure of pertinent facts are also relevant.

The defendant cannot challenge the validity of the patent in the preliminary injunction proceedings. The only exception is when the invalidity of the alleged patent 'screams to high heaven' and can be established without an in-depth review of the evidence.[20]

As a prerequisite for the granting of preliminary remedies, the applicant must submit a personal undertaking to compensate the defendant for any damages which it may sustain if and when the preliminary remedies are revoked by the court. In addition, if the preliminary injunction is granted, the applicant will be required to submit a bank guarantee as further indemnification for that purpose.

5.3 'Springboard'/post-patent expiry injunctions

When the patent is valid and infringed, the court will grant an injunction restraining further infringement of the patent. In appropriate circumstances, the court will also grant injunctions restraining the use of the 'poisoned fruit' of the infringement such as data accumulated in violating the patent. The use of the 'poisoned fruit' of the infringement can also be enjoined after the expiration of the patent.[21]

A patentee prevailing in the infringement proceedings is also entitled to monetary compensation. In practice, the patentee can claim either the actual damages that it sustained as a result of the infringement, or the profits that the defendant derived from the infringement. In order to determine the amounts due to the patentee, the court will order the infringer to submit detailed accounts of the extent of the infringement and of the profits that he derived therefrom. Most cases are divided into liability and damages phases, and the hearing of the damages phase commences only once liability has been established. Punitive damages (doubling the amount of damages ruled by the court) are available, but only rarely imposed.

---

20 MLA 920/05 *Hasin Esh Industries Ltd v Coniel Antonio (Israel) Ltd* Tak-Sup 2005(1) 4171; MLA 11964/04 *Zefi Profil Chen v Azulay* ILR 59(6) 350.

21 *Eli Lilly and Company v Teva Pharmaceutical Industries Ltd* (see note 19 above).

Ancillary remedies such as orders requiring the recall of patent-infringing goods, destruction of the goods etc are also available.

5.4 Unjustified threats

The making of groundless threats to sue for patent infringement is not an actionable wrong in itself (although it may possibly give rise to some other causes of action based on the law of torts). However, it is possible to seek a declaratory judgment that the product or process are not infringing. Such declaratory judgment will only be entered by the court if the applicant provided to the patentee the complete details of the product or process in question and the patentee declined to confirm that the product or process is not infringing within a reasonable timeframe. Moreover, in these proceedings, the applicant seeking a declaratory judgment of non-infringement may not challenge patent validity.

6. Compulsory licensing

The Patents Act originally contained rather draconic provisions with regard to compulsory licences. Among other things, as long as the patented product was not manufactured in Israel, compulsory licences could be easily obtained even when the patented product was imported to Israel, widely available and sold at reasonable prices. Moreover, compulsory licences were not limited to domestic sales and also covered manufacture in Israel for export markets. The royalty rates determined by the Patent Office were also extremely low.

In *Eli Lilly v Teva Pharmaceuticals*,[22] the Tel Aviv District Court brought an end to the profitable business of obtaining compulsory licences. The court held that compulsory licences are intended to ensure the supply of the product on the local market, rather than to grant competitive advantages to local generic manufacturers at the expense of the innovators of the drug. This decision in fact abolished the Patent Office practice of granting compulsory licences almost automatically when the drug was not manufactured in Israel. Thereafter, in 1999, the Patents Act was amended to bring the local legislation in conformity with TRIPS and compulsory licensing regained its original purpose – to provide adequate remedy in case of abuse of patent rights or shortage of the patented product on the domestic market.

Under the current legislation, an 'abuse of monopoly' giving rise to a compulsory licence will only occur if the entire demand for the product is not satisfied in Israel on reasonable terms or if the conditions imposed by the patentee with respect to the supply of the product or to the grant of the licence for its manufacture or use are unfair. It is specifically indicated that a compulsory licence must be granted primarily for the supply of the requirements of the local market. Overall, and contrary to the previous statutory mechanism, the current legislation reflects a balanced approach intended to safeguard legitimate public interests. In most cases, and in the absence of exceptional circumstances, importation of the patented product into Israel will be sufficient in order to defeat a petition for compulsory licence.

---

22   OS 6108/95 *Eli Lilly v Teva* (not published) August 31 1995.

7. Ownership, inventors and compensation

An employee must notify his employer in writing of any invention which he has developed as a result of the employment or during the period of employment, and as soon as possible after the date of invention. In the absence of an agreement to the contrary, any invention of an employee, developed as a result of the employment and during the period of employment, will be the employer's property. In determining whether an invention is an employee invention owned by the employer, courts will apply a flexible set of rules. Among other things, courts will analyse the relationship between the invention and the employee's duties, whether the invention was developed using the employer's resources and during working hours, etc. Most importantly, if an employee is of a senior position, he is deemed to be under special fiduciary duty to further the interests of the employer. For such employees, the scope of inventions belonging to the employer will be broader and it will be easier to establish the connection between the invention and the employment.

In the absence of an agreement between the employer and the employee on the employee's entitlement to compensation, a special Compensation Tribunal will determine whether the employee is entitled to compensation in consideration for the invention and the scope of such compensation. Among other things, the tribunal will consider the nature of the employer's position, the relationship between the invention and the employee's duties, the employer's initiative in developing the invention and the potential to exploit the invention commercially. In practice, proceedings before this tribunal have only rarely been invoked.

7.1 Application of the law of unjust enrichment

There has been an ongoing debate in Israeli jurisprudence as to whether imitation of technology can be enjoined on the basis of the laws of unjust enrichment in the absence of patent or design protection. The Supreme Court recognised this possibility, but formulated ambiguous guidelines which led to much confusion.[23] Thus, it was held that imitation *per se* is permitted but will be considered as 'unjust enrichment' if there exists an 'additional element' involving bad faith or unfair competition on the part of the defendant. Such 'additional element' may exist if the development of the imitated product required the investment of substantial resources by the plaintiff, if there was no functional justification for full-scale imitation, or when the defendant launched the imitation product before the original manufacturer had a fair opportunity to recoup the fruits of its investments etc.

In practice, low-end products whose development required minimal investments (such as toothbrush holders or towel hangers) were sometimes protected under the unjust-enrichment cause of action.[24] On the other hand, products whose development requires substantial resources, in particular pharmaceuticals, were not afforded protection under this cause of action. Thus, in *Merck v Teva*[25] it was held that the owner of an opposed patent application cannot invoke the laws of unjust

---

23   LTA 5768/94 *A.Sh.I.R v Forum Accessories* ILR 52(4) 289.
24   *A.Sh.I.R v Forum (ibid)*; CF (Tel Aviv) 1561/03 *Plasto-Wok (1990) Ltd v MAG for Plastic Ltd* Tak-Dist. 2008(4) 6565.
25   CF 2292/04 *Merck & Co Inc v Teva Pharmaceutical Industries Ltd ILR* (District Court) 2005(2) 4060.

enrichment to prevent the exploitation of the invention and it must await patent grant. Other than a certain degree of confusion, it does not seem that the unjust-enrichment cause of action contributed much substance to Israeli jurisprudence.

8. Branding and designs

Patents have been and remain the predominant means to protect pharmaceutical innovation. However, pharmaceutical branding has become more important in recent years. Branding is important both as a means to differentiate from competitors and also for the purpose of leveraging brand equity after patent expiry.

In addition to traditional word marks, branding can cover all aspects of the product – for instance, the trade dress of packaging and the shape, colour or colour combination of a pill. All these elements can be registered as trade marks in Israel. The qualification is that, for features such as the three-dimensional configuration of a pill, registration will be dependent on proof of acquired secondary meaning.[26] Even in the absence of registration, all the elements of the product that acquired secondary meaning can be protected on the basis of goodwill-related causes of action such as passing off. In order to prevail in a passing-off action, the plaintiff must prove secondary meaning and likelihood of confusion. In fact, the passing-off cause of action also encompasses instances of likelihood of association, attracting initial consumer interest and similar instances of taking a 'free ride' on the plaintiff's goodwill. The unjust-enrichment cause of action is also available in the absence of trade mark registration. As noted above, its boundaries are rather blurred and it serves de facto to enjoin imitation when the court finds that the defendant's conduct is inequitable.

The Israeli Trade-mark Office requires that the list of goods of pharmaceutical trade marks be limited to specific therapeutic indications. Broad definitions such as 'pharmaceutical preparations' without a specific therapeutic indication are not accepted. However, the Trade-mark Office may accept a trade mark for a broad 'pharmaceutical preparations' list of goods if the applicant undertakes to limit the list of goods to specified therapeutic indications within five years from the date of filing. A prerequisite for invoking this procedure is that the applicant proves that at the time of examination it has not yet determined for which drug the trade mark will be used. These restrictions do not apply to 'house marks' of pharmaceutical companies, which can be registered for a broad class of goods without limitation to specific therapeutic indications.

The Trade-mark Ordinance does not have any provisions relating to comparative advertisement. In Alonial v McDonald the Supreme Court, acting under the premise that as a general rule any unauthorised trade-mark use, even if not misleading, is an infringement, disqualified comparative 'puffing' or opinion statements, but nevertheless made positive comments about allowing fact-based comparative advertisements.[27]

Comparative advertisement of non-prescription pharmaceuticals is also governed at the regulatory level (any advertisement of prescription drugs is normally

---

26   CA 11487/03 August Storck KG v Alfa Intuit Foodstuffs Ltd (published in Nevo database).
27   CA 8483/02 Alonial v McDonald ILR 58(4) 314.

forbidden). The guidelines, issued by the Israeli Ministry of Health, allow comparative advertisements only between products with identical APIs. Any comparison must be based on works published in peer review scientific journals and the advertisement must not create an artificial advantage.

9.   Protecting valuable information

Israeli law with regard to trade secrets was codified in 1999 in the Commercial Wrongs Act. A 'trade secret' was defined as any business information which is not in the public domain and confers on its proprietor a competitive advantage. Unauthorised use of a trade secret is actionable. In addition to damages for breach of trade secret, the court may also order the infringer to pay the proprietor any profits it derived as a result of the use of the trade secret. Reverse engineering of a product for the purpose of 'extracting' trade secrets embodied in the product will not in itself be considered as a violation and is specifically permitted.

Restrictions on the freedom of occupation of an employee require an explicit contractual obligation. In the absence of such a covenant, non-compete obligations cannot be imposed. Even when there is an express non-compete obligation, the court has the discretion to intervene in the parties' agreement to ensure that the restriction will not be broader than reasonably necessary to protect the employer's legitimate interests in preventing the divulging of its trade secrets to competitors. In practice, courts often substantially shorten the non-compete period, or revoke altogether the non-compete covenant.[28]

10.   Counterfeiting

Israeli legislation was amended to incorporate the border measure provisions provided in TRIPS. The seizure procedure under Israeli law closely follows the path of Part III, Section 4 of TRIPS. In addition to the TRIPS border measure provisions, Israeli Customs has broad powers to seize and destroy shipments containing infringing goods. In accordance with these broad powers, Customs set up a simplified procedure whereby its staff are permitted to confiscate shipments of infringing goods without requiring the trade mark or copyright owner to take legal action or file a bank guarantee. Customs' border measures remain the most cost-effective means of combating piracy and creating a substantial barrier against the importation of counterfeit goods into Israel.

As the marketing of counterfeit goods amounts to a criminal offence, the Israeli police has set up an Intellectual Property Unit which monitors the markets, obtains intelligence and raids the premises of counterfeiters and businesses marketing counterfeit products. The unit is assisted by a second unit, set up by the Ministry of Health, which is dedicated to combating pharmaceutical piracy.

11.   Collaborative models – 'open innovation'

The idea of 'open-innovation' presents substantial advantages in the pooling and sharing of risks and resources and in fostering research. Despite the advantages, the

---

28   CA 6601/96 *AES Systems Inc v Saar* ILR 54(3) 850.

concept of collaborative innovation may not be 'natural' for an industry which relies so heavily on IP protection. On the other hand, the Human Genome Project is an example of highly successful life sciences innovation through an 'open' model. The success of the project demonstrates that there is room to consider and develop new 'open' innovation and licensing models also in the life sciences industry.

In addition to the cultural change that will be required, numerous legal issues will need to be regulated for a successful model to survive. These should include the allocation of proprietorship and access to the IP which will be developed under the open collaborative model, licence pools, and competition law issues. As such collaborative models are likely to be multi-jurisdictional, there will be legal challenges to make sure that the model will be valid and enforceable in all relevant jurisdictions. Yet, it seems that the main challenge would not be legal but cultural, as such open-innovation models defy more traditional life sciences innovation.

12. Hot topics

12.1 Patent term extension

As can be seen from the above, Israeli intellectual property laws closely follow the substantive and procedural framework in common-law countries and provide adequate protection to innovators. The exception is with respect to the system of 'patent term extension' (PTE). The PTE provisions were first adopted in 1998. They have since been subjected to extensive lobbying efforts by the local generic industry, which has resulted in subsequent statutory amendments curtailing PTE protection. In addition, attempts by the innovator companies to modernise certain obsolete aspects of the Israeli patent system – in particular the pre-grant opposition system – have so far been unsuccessful. These issues are likely to remain the focal point of the debate between different interest groups in the coming years.

PTE is available for a Basic Patent. The Basic Patent is a "patent that protects a pharmaceutical compound [API], a process for making the API, use of the API, or a pharmaceutical product containing the API, or a process for the making of the pharmaceutical compound containing the API, or a medical device that requires regulatory approval in Israel". Accordingly, numerous patents covering the approved product can qualify as a Basic Patent. The patentee can select which Basic Patent to extend, but only one Basic Patent can be extended with respect to each product.

The PTE petition must be filed within 90 days from the date of regulatory approval. A recent legislative change makes extensions more difficult to obtain. Yet, the Patent Office has so far liberally construed these new provisions.[29]

---

29 According to the revised Section 164A(1) of the Patents Act (Amendment No 7 of January 3 2006), the Registrar may extend the 90-day period for filing the petition for PTE only if the petition was not filed on time "as a result of circumstances over which the applicant and his counsel or agent had no control and which could not be prevented". In the Patents Registrar's Decision in connection with IL Patent No 112081, the Patents Registrar granted a two-week extension. See also unpublished decision in connection with IL Patent No 95572, where the regulatory approval was granted approximately eight years before the patent was issued. The patentee only filed the PTE petition when the patent was issued. The patentee's justification was that the scope of claims constantly changed during examination and the PTE petition was therefore filed after it transpired what the scope of monopoly would be. The PTO permitted the late filing on these grounds.

The Patent Office will extend the term of protection of the Basic Patent if the following conditions are met:

- The PTE petition was filed in good faith.
- The API, the process of its manufacture, its use, the pharmaceutical composition containing the API, or the medical device are claimed in the Basic Patent and the Basic Patent is in force.
- The pharmaceutical composition is registered in accordance with the Pharmacists' Ordinance.
- The registration is "the first regulatory approval permitting use of the compound in Israel for pharmaceutical purposes".
- If the pharmaceutical composition is registered in the United States, PTE in Israel will only be available if a Patent of Reference in the United States was also extended. The Patent of Reference is any patent that protects the pharmaceutical product or the medical device claimed in the Basic Israeli Patent. It is not required that the Patent of Reference be the corresponding patent of the Basic Israeli Patent.
- If the pharmaceutical composition is registered in any of the Countries of Reference, PTE in Israel will only be available if a Patent of Reference was extended in at least one of these countries. The Countries of Reference are Austria, Belgium, Denmark, Finland, France, Germany, Greece, Ireland, Italy, Luxembourg, the Netherlands, Portugal, Spain, Sweden and the United Kingdom.

The 'dependency' requirement (namely that PTE in Israel will only be available if PTE/SPC was granted in the United States and in at least one Country of Reference) is the distinguishing feature of the Israeli PTE system. The Patent Office interpretation of this requirement is that PTE in the foreign jurisdiction must be formally granted and issued before the expiry date of the Israeli patent.[30]

As indicated, PTE is only available with respect to the "first regulatory approval permitting use of the compound in Israel for pharmaceutical purposes". Therefore, PTE is not available for second or subsequent medical indications of a previously approved API. PTE is also not available for new formulations of previously approved APIs. In addition, salts, esters, hydrates or different crystalline forms of previously approved APIs have been specifically excluded in the legislation from PTE protection.

In the *Novartis* decision, the Patent Office held that PTE can be granted with respect to combination products only if at least one of the APIs was not previously registered.[31] In another restrictive decision it was held that the registration of a new enantiomerically pure product is not 'the first regulatory approval' permitting use of the compound. This decision attributed the earlier regulatory approval of the racemate to the single enantiomer product that was separated and developed much

---

30   Patents Registrar's Decision regarding petition for PTE of IL Patent No 88534.
31   Patents Registrar's Decision regarding petition for PTE of IL Patent No 97219. The decision was upheld on appeal.
32   Patents Registrar's Decision regarding opposition to PTE order for IL Patent No 90465. The decision was upheld on appeal.

later.[32] Nevertheless, it appears that this decision can be distinguished in the future, at the very least if the previously approved racemate and the single enantiomer have different pharmaceutical profiles.

The Israeli PTE will be granted for a period equal to the shortest duration of a PTE granted in any of the Countries of Reference. In addition, the extension period cannot exceed five years, and the overall patent and extension periods cannot exceed 14 years from the date of the earliest regulatory approval in any of the Countries of Reference. The calculation of the PTE period is another facet of the dependency principle of the Israeli PTE system. Its implications are a *de facto* PTE period of about two or three years.

In accordance with a recent agreement between Israel and the US government, the number of Countries of Reference that serve as a basis for calculating PTE period will be reduced to six (France, Germany, Italy, Spain, the United Kingdom and the United States). A legislative proposal in this regard is currently in the pipeline. However, this proposal also includes some potential pitfalls to patentees; one is that the current draft unjustifiably provides that an Israeli PTE order will expire if PTE in one of the Countries of Reference is revoked.

It is yet to be seen how the PTE legislation will develop in Israel in view of growing international pressure.

Italy

Gualtiero Dragotti
Roberto Valenti
DLA Piper

---

1.   Small molecules

Small-molecule compounds are litigated in Italy in traditional patent infringement actions. Italian law and case law regarding small molecules in both infringement and invalidity actions are discussed below.

1.1   Product and process claims

The Italian patent law[1] allows claims directed to products, methods[2] and uses.[3] A single patent may include multiple independent claims, including different types of claims, provided that they refer to the same inventive concept. The old approach, according to which product and process claims could not be combined in the same patent, has been superseded.

A claim to the use of a certain substance in the manufacture of a medicinal product for a certain therapeutic use, the so-called 'Swiss-type' claim, was common in the past in order to reflect the European Patent Office (EPO) case law in connection with first, second or subsequent medical use (or indication of efficacy) of a known substance or composition. Italian law did not mandate the use of Swiss-type claims in the past and, after European Patent Convention (EPC) 2000, they are no longer necessary before the EPO.

In an infringement action for a patented process, the patent owner must in principle provide evidence that the infringer's product has been manufactured according to the patented method. However, any product identical to that produced by the patented process shall be deemed, in the absence of proof to the contrary, to have been obtained by the patented process alternatively: (i) if the product obtained by the process is new; (ii) if there is a genuine probability that the identical product was produced by the process and the owner of the patent has been unable, in spite of reasonable effort, to determine what process was actually used. The provision under (ii) has been criticised and is not widely used by the Italian courts.

---

1   Section IV of Part II of the Italian IP Code (hereinafter IIPC), enacted by Legislative Decree No 30 of February 10 2005, and subsequent modifications.
2   Article 66 and 67 IIPC.
3   Article 46.4 IIPC. According to some commentators, claims directed to products must in any case be intended as limited to the uses disclosed by the inventor (Di Cataldo, "Sistema Brevettuale e settori della tecnica – Riflessioni sul brevetto chimico", in *Riv. Dir. Comm.* 1985, 277; Di Cataldo, "Fra tutela assoluta del prodotto brevettato e limitazione ai procedimenti descritti ed agli usi rivendicati.", in *Riv. Dir. Ind.* 2004, I, 111).

1.2 Scope of protection of claims and Markush formulae

Patent claims (generally, including claims for small molecules) are to be construed on the basis of the claims, to be interpreted according to the specification and drawings.[4] The older purposive construction, identifying essential and inessential elements on the basis of the 'core' of the invention and also taking into account the state of the art and the inventor's intent, is now deprecated.

Markush claims (which are not known by this name in the majority of Italian case law) are admissible, provided that the class shares common properties and effects; in contrast, the lack of unity of the invention can be invoked.[5]

In practice, the courts tend to consider each element of the group when evaluating the scope of a patent, and allow the patent owner to drop an element if necessary. Gaps between elements could be filled by invoking the equivalence doctrine, unless a certain element has been omitted due to an explicit choice made by the patent owner.

1.3 Metabolites

In principle, metabolites are patentable in Italy. However, there is no published case law on the subject matter.[6]

2. Second-generation inventions

In addition to protecting a pharmaceutical compound or device, there are other areas in Italian patent law where a secondary invention could provide additional patent protection. These include obtaining patents on combinations, an enantiomer (if one enantiomer is superior to the other), a selection from a group of compounds, a use of the invention, a method of treatment, claims to formulations and physical forms (such as crystals and salts) and, finally, using product-by-process claims to claim an unknown product made by a particular method.

2.1 Combinations

Italian case law differentiates between a new combination (which is patentable) and an aggregation of elements (which is not patentable). A new combination exists when the elements of the invention cooperate and interact together to create a new or unexpected result. In contrast, an aggregation is a sum of known parts without any synergy between the parts.

Once it is determined that the invention is a combination, and therefore patentable, the next step is to consider whether that combination is obvious and therefore not patentable on that basis.

2.2 Enantiomers

Enantiomer patents are valid as long as the technology would not have been obvious to a person skilled in the art. In at least one case, a patent related to a racemate was deemed infringed by an enantiomer, as the expert report confirmed that the

---

4   This principle is now provided by Article 52 IIPC.
5   A critical comment on the effects that Markush claims which are too wide may have on the scope of protection can be read in Di Cataldo, "Sistema brevettuale..." (see note 3) at page 297.
6   R Sgarbi, "La brevettazione del metabolite e delle pro-drugs", in *I Nuovi brevetti*, Milan 1995.

enantiomer was actually disclosed in the examples and at the time of the application it was not common to separate enantiomers.

2.3 Selection inventions

Selection patents are valid,[7] as long as the selected substance or group presents significant differences from the general class.[8] In principle, a patent on a selection invention is dependent on the patent on the general class, so that the use of the invention is subordinated to the consent of the owner of the principal patent,[9] according to the general discipline provided by Article 68.2 and 71 IIPC.

2.4 Methods of use and secondary indications

Claims directed to a new and unobvious use of a known product or compound are allowable[10] in all fields, including the life sciences. These uses cannot consist in methods for treatment of the human or animal body by surgery or therapy and diagnostic methods practised on the human or animal body, which per se are not patentable.[11] Claims to the discovery of a new property or advantage of a known substance are not patentable. Claims must cover a practical solution to a practical problem – something more than a mere discovery.[12] There may be some overlap between discovery and invention, particularly where biotechnological patents are concerned.[13]

2.5 Methods of treatment

As stated above, methods for the treatment of the human or animal body by surgery or therapy, and diagnostic methods practised on the human or animal body, per se are not patentable. This prohibition does not apply to microbiological methods and related products, or to products, substances or compositions used in these methods.[14]

2.6 Formulations and physical forms

In principle, formulation claims are patentable provided that the formulation is new and inventive. Dosage patents should be allowable as well, even if there is no established national case law. The scope of both formulation and dosage patents must be construed taking into account the prohibition on patent therapeutic methods.[15]

Claims to crystal forms or to a certain level of purity are patentable, as are claims to salt forms. However, the crystal form, the purity or the salt must meet the novelty and inventive-step requisites.

2.7 Reach-through claims

There is no case law in Italy on reach-through claims (ie, claims which define a

---

7   Luzzatto, "Brevetti chimici di base e di selezione", Riv. Dir. Ind. 1990, I, 299.
8   Milan Court of Appeal 13-7-1990, in GADI 1990, 640.
9   Sena G, I diritti sulle invenzioni e sui modelli di utilità, Milan, 2011, 145.
10 Article 46.4 IIPC.
11 Article 45.4 IIPC.
12 Article 45.2 IIPC.
13 Sena G, "Brevi note sulla brevettabilità delle scoperte e delle invenzioni biotecnologiche", in Riv. Dir. Ind, 2000, II, 364; Guglielmetti G, "La brevettazione delle scoperte-invenzioni", in Riv. Dir. Ind. 1999, I, 97.
14 Article 45.4 IIPC.
15 Dragotti, "I regimi di dosaggio alla prova del divieto di brevettazione", in Riv. Dir. Ind. 2009, I, 214.

method and then claim the products obtained using that method). However *product-by-process* claims are allowable, provided that the process is adequately disclosed[16] and that the resulting product has an industrial application.[17]

3.   DNA, biologicals and personalised medicine

3.1   Discoveries

Discoveries *per se* are not patentable. A patent must disclose (and may claim) a practical solution to a practical problem – something more than a mere discovery.[18] There may be some overlap between discovery and invention, in particular where biotechnological patents are concerned.[19] A mere scientific principle or theorem and laws of nature are not patentable.

The following discussion highlights some additional limitations on the scope of patentable subject matter in the area of biotechnology and medicine.

(a)   Gene patents

According to the new Section IV-*bis* of the Italian IP Code, enacted pursuant to EC Directive 98/44/CE on the legal protection of biotechnological inventions, gene sequences and partial gene sequences used to encode a protein or a partial protein are not patentable, unless the patent discloses and claims a useful function[20] so that the industrial application utility can be evaluated. Each sequence is deemed autonomous for patent purposes, and partial overlapping in non-active portions of the sequence does not affect the possibility of patenting a sequence. When evaluating the scope of protection of gene patents, omology could be taken into account.[21]

The patentability of genes does not extend to the entire organism, as this is not patentable, be it a plant or an animal.[22] It is, however, possible to patent an isolated element of the human body,[23] as well as an invention concerning plants or animals characterised by a specific gene or gene sequence, where the technical feasibility of the invention is not confined to a particular plant or animal variety.[24]

(b)   Stem cells

Human stem cells are not patentable. The prohibition is extended to any technical method which makes use of human stem cells.[25] As far as animal stem cells are concerned, Italian law, as well the EC Directive, does not allow the patenting of processes for modifying the genetic identity of animals which would be likely to

---

16   Article 51 IIPC.
17   Article 49 IIPC.
18   Article 45.2 IIPC.
19   See Sena and Guglielmetti at note 13
20   Article 81-*quinquies* IIPC.
21   According to Trib. Milano, November 11 1999, in *Riv. Dir. Ind.* 2000, II, 342, a claim to a sequence with a minimum omology compared to a disclosed sequence is valid.
22   Article 45.4 IIPC.
23   Article 81-*quater* IIPC.
24   Article 81-*quater* IIPC.
25   Article 81-*quinquies* IIPC.

cause them suffering without any substantial medical benefit to man or animal (and also animals resulting from such processes).

(c) Organic tissue

Human organic tissue could be considered a part of the human body, therefore being patentable.[26] It is, however, necessary that the tissue is produced by a technical method. Artificial organ- or tissue-like structures, created by combining cellular and/or inert components, should also be considered patentable.

(d) Bioinformatics systems

Despite the fact that information and mathematical formulae are not patentable *per se*, bioinformatic systems may, in principle, be patentable. However, they face the limits regarding biotechnological inventions,[27] combined with those regarding patenting information, algorithms and computer programs.[28]

(e) Copyright and sequence information

The Copyright Law[29] does not include sequence information among literary works. It is extremely dubious that a sequence may qualify for copyright protection. However, collections of sequences may enjoy the *sui generis* protection provided for databases,[30] and the presentation of sequence information may qualify for copyright protection. Copyright on computer programs may play a role, too.

3.2 Disclosure requirement

A patent specification must contain a disclosure of the invention, which must, on its own, be sufficiently detailed so as to enable a person skilled in the art to make the claimed invention.[31] The specification could include drawings or other illustrative materials useful to achieve a full understanding of the invention.

For some biotechnological inventions, there are specific disclosure requirements.

3.3 Sequence listing

Where a nucleotide or an amino acid sequence is disclosed, a 'sequence listing' in electronic format must be included.[32]

(a) Biological materials

Where it is necessary to comply with disclosure requirements, a deposit of biological material with an International Depository Authority will be required in accordance with the Budapest Treaty of April 28 1977.[33]

---

26   Article 81-*quinquies* IIPC.
27   See above.
28   Article 45 IIPC.
29   Law No 633 of April 22 1941 and subsequent amendments.
30   Database protection is disciplined by Section VII of Part IV of Law 633/1941.
31   Article 51 IIPC.
32   IIPC Regulations, Article 22.8. The electronic form requirements are set by guidelines of the Italian Patent and Trademark Office, but are not published on its website.
33   Article 162 IIPC.

(b)   Inclusion of examples

The specification must disclose at least a working example of the invention.[34] However, a violation of this provision will not lead to the invalidity of the patent unless the lack of examples affects the sufficiency of the disclosure. For biotechnological patents, where the industrial application must be explicitly disclosed and claimed, the need to include working examples could be more relevant.

4.   Acts of patent infringement

According to Article 66 IIPC, the patentee has the sole right to use and exploit the patented invention within Italian territory. In particular:

- It is an infringement of a product patent to make, use, exploit commercially, sell or import the patented product for this purpose.
- It is an infringement of a process patent to use the process or to exploit commercially, sell or import the product directly obtained by the patented process for this purpose. Where the product is new or a substantial likelihood exists that the product was made by the patented process and the patentee has made a reasonable effort to determine the process actually used by the prospective infringer, the burden of proof that the product is not made by the patented process is on the infringer.[35]

It is not an infringement of a patent to put the invention into practice for private and personal use for non-commercial purposes, or for research or experimental purposes.[36] Pharmaceutical patents do not prevent chemists from manually preparing the patented product at their premises, provided they do not use active principles industrially manufactured.[37]

Experimental use aimed at obtaining a marketing authorisation, in Italy or abroad, does not amount to an infringement.[38] However, the law as recently modified now provides that marketing authorisation applications for a medicinal equivalent to a patented product can commence no earlier than one year before the end of the patent life, including possible supplementary protection certificates.[39] According to some decisions, filing a marketing authorisation request before the one-year term may amount to patent infringement.[40]

Infringement is assessed by reference to the claims of the patent, interpreted according to the specification and drawings.[41] For the purpose of determining the extent of protection, due account must be taken of any element which is equivalent to an element specified in the claims (and this is known as 'infringement by equivalence').[42] According to the case law, an element should be deemed equivalent

---

34   Article 21.4 IIPC Regulations.
35   Article 67 IIPC.
36   Article 68.1(a) IIPC.
37   Article 68.1(c) IIPC. The latter part of the provision has been added recently, in order to better clarify (or restrict) the scope of the chemist's exception.
38   Article 68.1(b) IIPC.
39   Article 68.1-*bis* IIPC.
40   The majority of the decisions on these issues were granted before the introduction of the one-year term, so that there is no established case law on the correct construction of the current provisions.
41   Article 52.2 IIPC.

to a claimed one if the substitution was obvious to a person skilled in the art.[43]

It is also an infringement to supply essential elements of the invention (referred to as 'contributory infringement'), provided that the essential element is unequivocally aimed at infringing the patent. For medicinal products, statements made by the drug manufacturer in the product leaflet may play a role in assessing the infringement.[44]

5. Patent enforcement

Since 2003, IP cases have been decided exclusively by Specialised Sections on Industrial and Intellectual Property, instituted in 12 major *Tribunale* and Appeal Courts (namely: Bari, Bologna, Catania, Florence, Genoa, Milan, Naples, Palermo, Rome, Turin, Trieste and Venice).[45]

The Italian Patent and Trade Mark Office (PTO) has no judicial power; a refusal to grant a patent or a trade mark (although a refusal is not very common, as no substantial examination is carried out) can be appealed against to the PTO Board of Appeal, whose decision can be appealed against to the Supreme Court on points of law. The assessment of the validity of patents, as well as litigation on the infringement thereof, falls within the exclusive jurisdiction of the civil courts.

An action for infringement can be brought in either the IP specialised section of the territory where the defendant is located, or where the infringement is alleged to have occurred.[46] An action for nullity must be brought in the IP specialised section of the territory where the patentee elected domicile at the time of the application.[47]

In order to collect the evidence of the infringement, the patentee may seek a judicial description[48] order, (ie, the inspection of the machinery, process or goods suspected of infringing the patent). Judicial description is particularly useful for obtaining evidence of infringement of process patents and in all the cases where the purchase of goods is not sufficient to show evidence of the infringement. The description order may be granted without previous notice to the defendant (*inaudita altera parte*), when there is a chance that the defendant can prevent the successful execution of the order if informed in advance. The judicial description may also be extended to the means used to manufacture the infringing goods and to other evidence. The law requires that confidential information belonging to the alleged infringer must be kept confidential.

In terms of preliminary relief, a patentee may seek a seizure and a cease-and-desist injunction (*inibitoria*), as well as some auxiliary remedies.

The seizure[49] takes the infringing goods from the defendant's possession; generally, the seized goods are given into the custody of the defendant itself or a

---

42	Article 52.3-*bis* IIPC.
43	Supreme Court, January 13 2004, n 257, *Lisec v Soc Forel*, in *Giur. It.* 2004, 1680.
44	The issue was debated in a case before the Milan court, which was, however, settled before judgment.
45	The IP specialised sections were instituted by Law Decree No 168 of June 27 2003, in force since July 12 2003.
46	Article 120.6 IIPC.
47	Article 120.3 IIPC. According to some case law, it is still possible to bring the action in the IP specialised section of the territory where the defendant is located.
48	Article 129 IIPC.
49	Article 129 IIPC.

third party which is responsible for their custody until the end of the suit. The main result of the seizure is that the defendant cannot sell, exploit commercially or even use the seized goods until the seizure is lifted.

The *inibitoria*[50] injunction is a cease-and-desist order issued against the defendant until the end of the suit; it may include the order to withdraw the infringing goods from the market. An *inibitoria* injunction can be assisted by a penalty for any subsequent violation or for any delay in complying with the judge's order.

Both *inibitoria* and seizure injunctions can be assisted by an order to publish the decision in the press.[51]

Preliminary injunctions are granted after a summary and informal judgment, upon showing the following:[52]

- The injunction is necessary to avoid further damage to the petitioner, such damage being irreparable or not completely reparable by monetary compensation (*periculum in mora*). In IP cases, where market position and goodwill are at stake, the *periculum in mora* is generally assumed (even if extended tolerance may hinder the chances of obtaining the order).

- The case appears favourable to the petitioner (*fumus boni iuris*). The bare existence of an Italian patent may not be deemed sufficient to establish the *fumus*, as judges know that the Italian PTO does not carry out substantial examinations of patent applications. The petitioner should therefore reinforce his request with other indications as to the validity of the patent, such as affidavits by independent experts or by independent institutes, and/or copies of corresponding patents issued in countries where substantial examination is carried out. European patents that have already been granted, and which designate Italy, are generally deemed valid for injunction purposes. Nevertheless, most courts subordinate the grant of injunctions to a summary expert's report, and this practice significantly delays the procedure.

In the pharmaceutical drug context, there is often an overlap between proceedings before the regulatory authority (mainly AIFA, the Italian Pharmaceutical Agency) and before the courts. In particular, originators widely resort to access requests[53] in order to obtain, from AIFA, data and information on genericists' market authorisation requests, with the purpose of using them before the courts. Conversely, court proceedings are often used to legitimise access request information and proceeding before AIFA.

After the action on the merits, the following civil final remedies are available to the patentee:

- injunction;
- delivery-up or seizure of the infringing products;

---

50   Article 131 IIPC.
51   Article 126 IIPC.
52   For a more detailed explanation of procedural issues see M Scuffi, *Diritto processuale della proprietà industriale ed intellettuale – Ordinamento amministrativo e tutela giurisdizionale*, Milan 2009.
53   Administrative proceedings are regulated by Law No 241 of August 7 1990 and subsequent modifications.

- publication of the decision;
- damages;
- penalty for further infringements; and
- costs (generally limited to a portion of the costs actually paid).

As regards damages, in principle the patentee is entitled to compensation for the loss of profits directly and immediately attributable to the infringement.[54] This amount includes both the money the rights owner has lost due to the infringement (eg, a licence agreement was not reached as a result of the infringement) and the money the rights owner would have earned but for the infringement (eg, sales of patented goods made by the infringer instead of the patentee). Alternatively, the law allows the determination of damages on the basis of a fair royalty. There are no treble or punitive damages. The patentee is also entitled to the recovery of profits, either in the alternative to damages calculated on the basis of the negative economic consequences of the infringement, or as a supplementary request.[55]

Damages may be sought for infringing acts up to five years before the commencement of the litigation; interest is awarded until the payment takes place.

6. Compulsory licensing

In theory, compulsory licensing is an exception to the exclusive rights afforded by patents. The Italian IP Code provides two hypotheses of compulsory rights (dependent patents and insufficient exploitation of a patented invention) and some special provisions to stimulate licensing of patents for which a supplementary protection certificate (SPC) has been granted, limited to export purposes in countries where the patent protection has already expired. If patent rights are abused, antitrust laws may provide a further hypothesis for compulsory licences.

6.1 Dependent patents

When a patented invention amounts to an important technical advance with significant economic relevance compared with a previous patented invention, the owner of the second patent is entitled to request that the owner of the principal patent provides a licence, and on a compulsory basis.[56] The licensee is entitled to a cross licence.

The terms and conditions of the licence are set by the administrative authority and may be varied or revoked.[57]

6.2 Sufficient exploitation of the invention

If a patent is not sufficiently exploited in Italy within three years from the date of grant (or four years from the application date, whichever comes first), any

---

54   Article 125 IIPC.
55   The provision on recovery of profits was introduced in 2006 and it may be not fully consistent with Italian general principles on damages (G Dragotti, "The implementation of the Enforcement Directive in Italy", in *Italian Intellectual Property*, 2006, 105).
56   Article 71 IIPC.
57   Article 73 IIPC.

competitor may apply for a compulsory licence provided that he can show that the patent owner refused to license the patent under reasonable conditions. Exploitation in a country that is part of the World Trade Organization (WTO) can be considered exploitation in Italy under certain conditions.[58]

The terms and conditions of the licence will be set by the administrative authority and may be varied or revoked.[59]

6.3 Voluntary assisted licence

An entity willing to manufacture active principle ingredients protected by an 'old' Italian SPC[60] for exportation in a country where the patent rights have already expired may resort to a special procedure for 'voluntary' licences.[61] The procedure provides an assisted negotiation before a special commission appointed by the PTO. The commission may inform the antitrust authority in cases where the patentee refuses to grant the licence under fair-market conditions.[62]

6.4 Abuse of patent rights

The provisions regarding sufficient exploitation of patented inventions[63] contribute to reduce the need to resort to antitrust law to prevent the abuse of patent rights. In principle, the antitrust authority may order a compulsory licence as a remedy for patent abuses. It is, however, an established doctrine that the refusal to grant a licence does not amount to a patent (or other IP right) abuse.[64] The doctrine of essential facilities[65] and the need to keep open standards available to all the competitors may limit this principle.

6.5 Government use

A patent may be expropriated for public reasons, including military defence.[66] The expropriation may be limited to the use of the invention by the state,[67] thereby introducing a form of additional compulsory licence hypothesis. The patentee is entitled to an indemnification, to be established by the expropriation decree.

---

58   Article 70 IIPC.
59   Article 73 IIPC.
60   SPCs were introduced in Italy in 1991 by Law No 349 of October 19 1991. Their duration far exceeded the duration of SPCs later introduced by EC Regulation No 1768/1992. It was therefore necessary progressively to reduce the duration of the Italian SPCs (see Article 81.4 IIPC) and introduce some measures to limit the undesired effects of the extended duration of the national protection compared with the EC protection.
61   Article 81.5 IIPC.
62   At least in cases where the link between the voluntary assisted procedure and the antitrust authority led to the grant of a licence initially refused (AGCM February 8 2006 No 15175, *Glaxo v SFI*, in *Dir. Ind.* 2006, I, 1349).
63   See above, section 6.2.
64   ECJ October 5 1988 n. 238/1987, Volvo. For a more detailed explanation see Sarti, "Proprietà intelletuale, interessi protetti e diritto antitrust", in *Riv. Dir. Ind.* 2002, I, 550.
65   CE Mezzetti, "Diritti di proprietà intellettuale e abuso di posizione dominante: da 'Magill' a 'Microsoft' (Nota a Trib. I grado Comunità Europee, 17 Settembre 2007, n. 201/04, Microsoft Corp. c. Commiss. Ce)", in *Dir. Ind.*, 2008, 245.
66   Article 141.1 IIPC.
67   Article 141.2 IIPC.

7. Ownership, inventors and compensation

7.1 Inventorship

The author of the invention has the moral right to be acknowledged as such.[68] This right cannot be assigned, is not subordinated to the actual grant of a patent[69] and is transferred to the heirs of the inventor. Furthermore, the author of the invention has the economic right to file and obtain a patent; this right can be freely assigned.[70]

When the invention is the result of the combined efforts of more than one inventor (a common hypothesis when the research is carried out by teams of people), the rights are shared and managed according to the rules set out in the Civil Code for co-ownership.[71] Such a regimen, however, will be inadequate to deal with patent issues[72] (eg, it is a matter of debate as to whether each co-owner has the right independently to grant a licence on the patent), so that it always advisable to manage the patent though a juridical person, such as a company, where all the inventors own stakes corresponding to their contribution to the invention.

7.2 Ownership

The general rule is that the inventor is the first owner of an invention, even where it is made during the course of employment. In this latter case, however, the ownership of the patent rights is transferred *ex lege* to the employer, and the inventor or the inventors may qualify for compensation according to the following principles:

- Where the employee was hired for the purpose of inventing, the invention was made using employer's resources and the employees was paid also for the inventive activity, the employees keeps only the moral right.[73]
- Where the invention was made using the employer's resources but the employee was not paid for the inventive activity, the employee keeps the moral right and the right to receive compensation, calculated on the basis of the value of the invention. In the calculation of the compensation it is necessary to take into account the qualifications of the inventor, her or his salary, and the contribution of the employer's organisation.[74]
- Where the invention was made by an employee outside the above hypothesis, the employer has the right to acquire the rights to the invention if it pertains to the field of the employer's activity.[75]

Different rules apply where the employer is a university or a public research entity.[76]

---

68   Article 62 IIPC.
69   Vanzetti – Di Cataldo, *Manuale di diritto industriale,* Milan, 2011, 412.
70   Article 63 IIPC.
71   Article 6 IIPC; the Civil Code provisions on common goods are set forth in Articles 1100 to 1116.
72   Vanzetti – Di Cataldo, Manuale ... (see note 69) at page 414.
73   Article 64.1 IIPC.
74   Article 64.2 IIPC.
75   Article 64.3 IIPC.
76   Article 65 IIPC.

8.   Branding and designs

8.1   Trade marks in life sciences

Trade mark protection in the life sciences sector follows the general rules. Signs shall be new and have a distinctive character to be protected as a registered trade mark.[77] It should be taken into account that pharmaceutical trade marks often have a descriptive part, since they make reference to the active ingredients or the therapeutic effects, with some variations such as the use of a prefix or a suffix or the addition of some letters. Protection can also be extended to non-traditional trade marks such as colour, shape and size of pills or medical devices.

In both cases, the resulting marks will generally be considered weak. The general rule under Italian case law is that a weak mark may get protection only if the distinctive part of the mark is reproduced. Even so, the use of the descriptive part of the mark does not amount to trade mark infringement. Therefore, under Italian case law, minor modifications are often considered sufficient to exclude the likelihood of confusion.[78]

8.2   Extending protection using trade marks and designs

A trade mark cannot protect a functional structure or utilitarian feature of a product.[79] The shape, configuration, pattern or ornament of a pill or device may be protected through industrial design registration, provided that the design was not published more than one year prior to filing the application, is new (ie, not similar to any design already registered), has individual character and is not dictated solely by utilitarian function.[80] Industrial design registrations have a term of five years from the date of filing of the application, renewable up to a maximum of 25 years. Registration grants the owner the exclusive right to manufacture, offer to the market, import, export and use a product embodying the design, as well as designs not differing substantially therefrom for an informed user.

8.3   Advertising of pharmaceutical products (including comparative advertising)

The advertising of pharmaceutical products and curative treatments follows specific rules. Any marketing communication relating to medicinal products and curative treatments is forbidden in the absence of a marketing authorisation. Marketing communications relating to medicinal products and curative treatments are also forbidden in relation to medicinal products which can be distributed only under the prescription of a physician. In any event, it is forbidden to show a medicinal product or the name of a medicinal product in a contest (by way of comparative advertising) which may increase the use of the product.[81]

Moreover, marketing communications relating to medicinal products and curative treatments must take into account the sensitivity of the subject matter and

---

77   Article 12 and following, IIPC.
78   Court of Milan 5 February 2004 (*Allergan v Skelebert*).
79   Article 9 IIPC.
80   Article 31 and following, IIPC.
81   Article 113 and following, law no 206/2006.

display the utmost sense of responsibility. They must also accurately reflect the details contained in the fact sheet summarising the product specifications. Such marketing communication should draw the consumer's attention to the need for caution in using the product, explicitly and clearly encouraging consumers to read the package warnings and advising against the improper use of the product. In particular, marketing communications relating to over-the-counter products should include the name of the relevant medicinal product as well as the common name of the active ingredient. This latter information is not compulsory if the medicinal product contains more than one active ingredient or the communication is intended solely as a generic reminder of the product's name.

Moreover, marketing communications relating to over-the-counter medicinal products and curative treatments should not:

- suggest that the medicine is without side effects or that its safety or efficacy profiles are due to the fact that it is a natural substance;
- claim that the efficacy of the medicine or treatment is equal to or better than others;
- suggest that a medical consultation or surgical procedure is unnecessary or lead consumers to make an incorrect self-diagnosis;
- exclusively or principally address children or lead minors to use the product without appropriate adult supervision;
- make use of recommendations by scientists, health professionals or persons well known to the public, or refer to the fact that the medicinal product has been approved for sale, or improperly or misleadingly report certificates of recovery;
- compare the medicinal product with a foodstuff, cosmetic or other consumer product;
- suggest that the medicinal or the curative treatment can improve normal good health, or that avoiding a certain product or treatment can be harmful, unless the message refers to vaccination campaigns;
- use improper, misleading or frightening depictions of changes in the human body caused by disease or injury, or due to the effects of the medicinal product.[82]

Comparative advertising is permitted on an objective basis and must not contain false or misleading claims, statements, illustrations or representations. In addition to civil liability under the civil code, Italian law prohibits false or misleading statements tending to discredit the business or services of a competitor.

9.   Protecting valuable information

9.1   Preserving confidentiality
Business information and know-how, including commercial information, under the control of their legitimate owner, are protected as far as:

---

82   Article 25 of the Code of Marketing Communication Self-Regulation.

- they are secret in the sense that they are not wholly or partly generally known or easily accessible for persons operating in the sector;
- they are valuable as far as they are secret; and
- they are protected by measures which can be considered adequate to keep them secret.[83]

The acquisition, disclosure and abusive use of the confidential information is forbidden to everyone. The unfair competition rules may also apply.

Confidential information may also be protected through contractual means (ie, through non-disclosure agreements). To succeed in an action for breach of confidence, the plaintiff must establish that:

- the information used by the defendant was confidential;
- the plaintiff disclosed the information to the defendant in confidence; and
- the defendant misused the information.

(a) The information must be confidential

If the information is generally known to the trade or it concerns a type of product which is commonly known and widely available, it will not usually constitute confidential information. The view will be taken that diffusion to a limited number of persons may not destroy the confidentiality of the information.

(b) The information must be given in circumstances imparting a duty of confidence

The information is considered as imparting a duty of confidence when the document containing the relevant information is expressly stated to be 'Confidential' or 'Secret'. If this is not the case, whether the circumstances of disclosure imply an obligation of confidence is to be established on a case-by-case basis. Courts have placed significant weight on whether it would be industry practice to consider the information as being disclosed in confidence in the relevant circumstances.

(c) Unauthorised use

Before attracting liability, the confidential information must be misused such that the recipient obtains an unfair advantage.

9.2 Know-how and ex-employees: confidential information and trade secrets

Specific rules apply to employees. They are subject to confidentiality obligations under Article 2105 of the Italian Civil Code. Also, the employment contract will usually provide further obligations of confidentiality, based on which the employee must not reveal to competitors any confidential information such as customer names, addresses, times of delivery of goods and services, prices charged, etc.

---

83   Articles 98 and 99 and following, IIPC.

10.   Counterfeiting

10.1   Remedies against counterfeiters

Rights holders have a variety of remedies against counterfeiters. Civil remedies include interlocutory and permanent injunctions, damages, disgorgement of profits, and destruction of infringing products. The IIPC and the Criminal Code also provide criminal liability for trade-mark, copyright and patent infringement.

(a)   *Search orders*

Search orders are one of the most powerful remedies available against counterfeiters. These orders authorise the holder to attend a counterfeiter's premises, without notice, and request entry to search for and seize documents and objects relevant to the proceedings.[84] Under Italian case law, such orders will be granted only when there is a real possibility that relevant evidence will be destroyed or otherwise made to disappear; however, such orders are now fairly routinely issued in ordinary civil disputes and counteract the lack of disclosure mechanisms in Italy.

To obtain an order, the applicant must demonstrate:

- a *prima facie* case;
- that the alleged damage is serious;
- that the counterfeiter has evidence of the infringement in its possession; and
- that there is a real possibility that the counterfeiter may destroy the material.

The court has also to consider:

- whether the inspection would do harm to the alleged infringer or its case; and
- whether the interests of justice would be undermined by allowing the search order.

Before granting an order, the court may require the rights holder to post a bond to pay the defendant any damages it suffers if it is ultimately determined that there is no infringement.

(b)   *Border controls*

Italian law provides protection against counterfeit products entering Italy. The Italian Customs Agency, which oversees the importation of products, may seize counterfeit products destined for import. The enabling provisions require the rights holder to apply to the court for relief with appropriate evidence and potentially post security for damages, either before the counterfeit goods have been imported, or before they are released by Customs.

Customs officials, acting as public officers, may also seize goods based on intelligence and directives received from the police and in furtherance of police criminal counterfeit investigations, but also under indication of the rights holders.

---

84   IIPC Articles 128 and following.

11. Collaborative models – 'open innovation'

There are many types of collaborative innovation models, including participation in joint ventures with public institutions and strategic alliances between private entities. Patent owners are increasingly assigning all the patents owned to a central structure (so-called common licensing administrators) which will then implement a licensing strategy. While each model raises its own unique intellectual property and competition issues, several common issues can arise, including patent pooling and data privacy.

11.1 Patent pooling

Patent pooling agreements may raise issues from the viewpoint of anti-competitive behaviour, and may have an impact on patent enforcement, particularly where the patents enforced cover an industry standard. There is at least one precedent of the Court of Genoa in which the court adopted the so-called 'essential facility doctrine', stating that a preliminary injunction cannot be granted when:

- the patent owners have control over the essential facility;
- a competitor is prevented from obtaining a commercially viable alternative to such essential facility because of the patent;
- the patent owners have unreasonably denied the grant of a licence.

In that case, the subject matter was DVD technology and the court granted the requested injunction because there was no evidence that a licence had been unreasonably denied by the patent owner. Moreover, the patent owner had proposed the grant of a licence under the same conditions offered to other market players and the defendant had refused to enter into negotiations.

11.2 Data privacy

Many new cooperative models may also raise data privacy issues. The Italian Data Protection Code provides the rules on the processing of personal data. 'Processing' means any operation, or set of operations, carried out with or without the help of electronic or automated means, concerning the collection, recording, organisation, keeping, interrogation, elaboration, modification, selection, retrieval, comparison, utilisation, interconnection, blocking, communication, dissemination, erasure and destruction of data, whether or not the latter is contained in a data bank. Personal data means any information relating to natural or legal persons, bodies or associations that is or can be identified, even indirectly, by reference to any other information including a personal identification number.

As a general rule the data subject, as well as any entity from whom or from which personal data is collected, shall be preliminarily informed, either orally or in writing, as to, among other things, the purposes and practicalities of the processing, the entities or categories of entity to whom or which the data may be communicated, and the rights of data subjects to access information in the possession of the data controller. Whenever personal data is not collected from the data subject, the above information, also including the categories of processed data, must be provided to the data subject at the time of recording such data or, if communication of the data is envisaged, no later than when the data is first communicated.

As a further general rule, personal data may only be processed for the purpose for which it was specifically collected and with the consent of the data subject. Consent shall be given in writing if the processing relates to sensitive data which is – as is relevant in the context of the life sciences – personal data disclosing information relating to a party's health or sex life. Personal data that is no longer required for the purpose for which it was collected should be destroyed or otherwise made anonymous. Companies and research entities are required to formulate and implement policies and procedures relating to the processing and protection of personal data.

12.   Hot topics

12.1   The abuse of intellectual property rights in the pharmaceutical sector

On October 26 2010, the Italian Antitrust Authority commenced proceedings against Pfizer based on a generic drug company complaint, according to which Pfizer's conduct in the enforcement of its patent rights in relation to the active ingredient latanoprost (for the treatment of the eye glaucoma) was intended to prevent or delay the entry into the market of generic drugs and amounted to an abuse of a dominant position.

In June 2011, Pfizer offered a number of commitments to settle the case and avoid sanctions, including the grant of an irrevocable, royalty-free licence on the patent, a commitment not to request a pediatric extension, and the settlement of all pending litigation relating to this active ingredient.

12.2   Marketing authorisation under Article 68.1-*bis* IIPC

Another hot topic relates to the issue of the marketing authorisations requested by producers of generic drugs. According to Article 68-1-*bis* IIPC, companies intending to manufacture a pharmaceutical product may commence the registration procedure relating to the product containing the active ingredient one year before the expiry of the supplementary protection certificate or – in the absence of the latter – one year before the expiry of the patent on the active ingredient.

In a decision issued on February 14 2011, the Court of Turin indicated that Article 68.1-*bis* implements a general principle according to which preparatory activities for the marketing of an infringing product are *per se* infringements of patent rights. Accordingly, the filing of a marketing authorisation application – if carried out before the expiry of the patent – may be considered a preparatory act of marketing and result in an act of infringement of the patent rights.

The IP Court of Milan has, however, taken the opposite view on a number of occasions. For example, on October 21 2009 the Court of Milan concluded that patent rights do not extend to clinical trials conducted for a contested compound, even where the defendant in the case – an Italian biotechnology company – published on its website the news that phase III of the clinical trials had started and that researchers strongly hoped that the compound could become one of the first therapies to obtain marketing authorisation.

According to the Court, this was related "to studies and experimental situations, connected to the research of funding for that activity more than to commercial exploitation of the invention".

Similarly, the Court of Milan stated on July 11 2009 that neither the filing of an application for marketing authorisation nor the grant of a marketing authorisation can be considered *per se* acts of patent infringement. The exploitation of the patent requires some specific preparatory activities with reference to the production and commercialisation of the medicinal product such as "the purchase or production of active ingredients, the storage of the product, the organisation of the distribution activities in the territory or the launch of an advertising campaign".

Japan

Mami Hino
Abe, Ikubo & Katayama

1.   Small molecules

1.1   Product and process claims

Patent Law, Law No 121 of 1959 (Japan) (henceforth the Japan Patent Law) defines an invention as "the highly advanced creation of technical ideas utilising the laws of nature" (Article 2, paragraph 1). Compounds have been a patentable subject matter since the Japan Patent Law was amended in 1976.

Based on the above definition of an invention, a product of nature *per se* is not a patentable subject matter because it is not a 'creation'. A compound discovered in a soil can, however, be a patentable invention when it is isolated from the soil. Likewise, a gene can be patentable, even though it is a product of nature. A patent claim may distinguish such a patentable gene from a naturally existing gene simply by stating that it is 'isolated' or 'purified'.

Information *per se* is not patentable, because it is not a 'technical idea'. For example, a protein model defined by atomic coordinates, a pharmacophore and a DNA sequence are information *per se* and not patentable subject matters. However, it is mostly a matter of how to claim them. 'A protein defined by the following atomic coordinates...', 'a compound defined by a pharmacophore having the chemical formula below...,' and 'a DNA having the nucleic acid sequence described as Sequence No 1' are patentable subject matters.

A patent application claiming a compound defined by a pharmacophore will usually fail to satisfy the enablement requirement and/or the support requirement.

Japan Patent Law, Article 2, paragraph 3 provides three categories of invention:

- an invention of a product;
- an invention of a manufacturing product; and
- an invention of a process.

The provision defines 'practice of invention' differently depending on what category the invention belongs to as follows:

(i) in [the] case of an invention of product ..., producing, using, assigning, etc (assigning and leasing and, ... the same shall apply hereinafter), exporting or importing, or offering for assignment, etc (including displaying for the purpose of assignment, etc, the same shall apply hereinafter) thereof;

(ii) in [the] case of an invention of process, the use thereof; and

(iii) in [the] case of an invention of manufacturing a product, in addition to the action

as provided in the preceding item, acts of using, assigning, etc, exporting or importing, or offering for assignment, etc, the product produced by the process.

Thus, if a claim relates to an invention of a manufacturing method, the scope of protection of the claim reaches the product manufactured by the process. On the other hand, if it is a simple method other than a manufacturing method, the scope of protection of the claim is limited to the use of the method. In a case where a patentee contended that an invention of a method for checking quality during the manufacturing process was substantially an invention of a method for manufacture, the Supreme Court held that an invention of a simple process cannot be treated as an invention of a manufacturing method.[1]

Under Japanese practice, a 'system' is treated as a product, and a 'use' is treated as a method.

1.2 Scope of protection of claims

Each term in a patent claim is construed in view of the specification and drawings of the patent pursuant to Japan Patent Law, Article 70, paragraph 2. The prosecution history of the patent is also referred to. Since the claim term should be construed as to how a person of ordinary skill in the relevant art would understand it, descriptions of dictionaries, encyclopaedias and textbooks are also referred to.

When a claim recites 'a compound selected from a group consisting of a1, a2 and a3' (Markush claim), the scope of protection does not extend to A, a genus covering species a1, a2 and a3. Since it is not clear whether such claim covers a combination of a1 and a2, if necessary the claim should recite 'a compound selected from a group consisting of a1, a2 and a3 and combinations thereof'.

1.3 Product-by-process claims

When a claim defines a product by a process for making it, whether the scope of protection of the claim extends to the product made by another process is not always clear. The issue is being reviewed by the Grand Panel of the IP High Court. It is an infringement case[2] where a product-by-process claim of pravastatin sodium was at issue. The Tokyo District Court held as follows:

> *Since the scope of patent invention must be determined based on a claim (Patent Law, Article 1), when a product invention is purposely described by a method for making it despite the fact that it can be identified without using the method for making it, it is considered not appropriate to construe the scope of the patented invention excluding the description of method for making it. On the other hand, like the product at issue, sometimes it is difficult to identify and concretely describe a structure of product and cannot help but identify the product by the process for making it. In such case, it is considered not necessary to construe the scope of the patented invention limited to the product made by the method recited in the claim.*

In the case under review, the product-by-process claim in question was construed

---

1   Supreme Court 1998 (*o*) No 604, July 16 1999.

2   2010 (*ne*) No 10043, which is an appeal from a district court decision 2007 (*wa*) No 35324, March 31 2010.

to be limited to the product made by the process because there were claims reciting the product without the method, but cancelled because of the prior-art reference cited by the examiner. The IP High Court decision is expected in early 2012.

The author is not aware of any cases in Japan regarding metabolites. When a substance is publicly known and used as an active ingredient of a medicine, its metabolite is not necessarily publicly known and practised. A 'publicly practised invention' is one which is used in circumstances that can be known by a person without an obligation of confidentiality. Thus, a substance created in a human body would not be considered as being publicly practised, and a metabolite created in a human body cannot be prior art against a claim reciting the metabolite.

Also, when a claimed substance is created in a human body as a metabolite, such act is not considered as an act of infringement because creating it in a human body is not conducted as a business. A patent is infringed when an unauthorised person practises the claimed invention as a business.

2. Second-generation inventions

2.1 Combinations
To be patentable an invention must:
- be novel;
- have an inventive step; and
- be industrially applicable.

In order for a combination of two known components to have an inventive step, the combination must have effects which are more than would be expected from the prior art. In other words, effects which are superior to those of the prior art would not be sufficient for the combination to have an inventive step, if such improved effects are something that a person of ordinary skill would have expected. For example, when a combination of known active ingredients is claimed, the fact that the patients to whom the claimed medicine was prescribed continued taking the medicine longer than patients to whom two ingredients are prescribed separately (ie, improved compliance) may not establish the requirement for an inventive step even though the improved compliance is certainly a better effect than that of the prior art, when such effect would be expected by a person of ordinary skill in the art.

2.2 Enantiomers
When a compound is already known in its racemic form, it is not easy to show an inventive step of its optical isomer. This is because when a chemical structure of a compound is known, it is self-evident whether or not it has optical isomers. Also, because optical isomers generally differ from one another in their effects and side effects, a person of ordinary skill in the art would be expected to find which isomer is more effective than the other. Thus, even if a claimed enantiomer has greater effects than its racemic form, it would not be difficult to find such enantiomer.[3]

---

3   Tokyo High Court 2003 (*gyo-ke*) No 62 rendered on June 9 2004.

However, this decision stands on the premise that it was easy to resolve the compound in racemic form to optical isomers. If such optical resolution could not be achieved in the usual manner, an inventive step could be established.

When a compound is claimed reciting a chemical formula without specific disclosure of enantiomers, such enantiomers are encompassed by the claim. An enantiomer may be a selection invention as explained below. If a patent is granted to an enantiomer claim, the prior patent is a blocking patent for the enantiomer patent.

2.3 Selection inventions

When a prior-art reference describes genus or numerous options but does not specifically describe a species or an option within the genus or the numerous options, a claim of the species or option has novelty and can be patentable as a selection invention.

In the case of an invention whose effects are difficult to predict based on its structure, such as an invention of chemical compounds, the novel species/option has an inventive step over the prior-art reference describing genus or numerous options if it has an effect which is either not described in the prior-art reference or significantly superior to that which is described in the prior-art reference.

Even if a patent is granted to the species/option claim as a selection invention, the prior patent claiming genus/numerous options encompasses the selection invention as a blocking patent.

2.4 Methods of use and secondary indications

A known compound is patentable if it has a new and inventive effect. For example, if a new effect as an anticancer of Compound X which was known as an analgesic is discovered, a claim as follows would be patentable: 'An anticancer composition which comprises Compound X.'

Under Japanese practice, pharmacological data showing such effect or an equivalent in the original specification is mandatory in order to satisfy the enabling disclosure requirement. Generally, *in vitro* data is sufficient for this purpose, and clinical data is not necessary. If there is no such data in the original specification, such defect cannot be cured by submitting it later.

2.5 Methods of treatment

Methods of treatment are not patentable in Japan due to a lack of industrial applicability.

Thus, methods of treatment will usually be modified to secondary medical-indication claims. Such practice will now be easier, following revision of the Examination Guideline in 2010. Under the new guideline, a specific dose or dosage regimen is recognised as an element of a composition invention; therefore, a claimed composition for a medical indication with a specific dose and/or a dosage regimen will be found to be new, even if a prior-art reference describes the same compound and same indication without specifying the dose and dosage regimen.

2.6 Formulations and physical forms

A novel and inventive formulation or physical form of a drug is patentable. This usually occurs where a compound was already used as a drug in a certain physical form and is prepared in a new physical form. In order for the new physical form to be inventive, it must have effects which were not known in the prior form or which are significantly better than that of the prior form. When the effect is slightly better, the new form is considered mere optimisation and not sufficient for establishing an inventive step.

2.7 Reach-through claims

Reach-through claims seek to protect things which have not yet been discovered by an inventor, but which might be discovered in the future by making use of their invention, such as a compound defined by a method for screening. Reach-through claims are generally rejected on the ground of a lack of enabling disclosure.[4]

3. DNA, biologicals and personalised medicine

3.1 Discoveries

When a product of nature is discovered, such discovery *per se* is not a patentable subject matter. However, a newly discovered product, such as a biologically active substance, protein or gene, will be patentable (see section 1.1).

3.2 Gene patents and industrial application

Where a DNA sequence is newly determined, it may be difficult to demonstrate the requirement of industrial applicability if the utility of the sequence is not yet clear. In cases where a function of a protein encoded by DNA is deduced, or if the protein is expressed only in a specific cell of a patient with a specific disease, the DNA will generally be found useful and industrially applicable.

3.3 Stem cells and other organic material

Inventions of stem cells are industrially applicable. However, until recently it was not clear whether a method for handling stem cells would be industrially applicable since a method for handling biological materials collected from a person was not regarded as industrially applicable when such cells were intended to be brought back to the same person. The rationale behind this was that an invention which includes the human body as an element should for ethical reasons not be patented. However, the author is unaware of any restrictions placed on patenting biological material *per se* in Japan.

The Examination Guideline revised in 2009 explains that a method for inducing differentiation of human cells is industrially applicable, even where the cells are to be brought back to the original person. Therefore, it is now clear that the Japan Patent Office (JPO) will not reject claims of methods for handling stem cells as industrially non-applicable (although the courts have not yet decided this issue).

Likewise, the Examination Guideline explains that methods for making medical

---

4   IP High Court 2009 (*gyo-ke*) No 10170 rendered on May 10 2010.

materials from biological materials obtained from humans, such as artificial substitutes for human body parts (eg, artificial bone, cultured skin sheet), are industrially applicable.

3.4 Bioinformatics systems

Inventions which relate to bioinformatics are considered to be inventions relating to computers and software which deal with biological data such as gene sequences or amino acid sequences. Thus, the Examination Guideline for 'inventions related to computers and software' is usually applied to bioinformatics inventions.

The subject matter protected by Copyright Law, Law No 48 of 1970 (Japan) (hereafter the Japanese Copyright Law) is a "production in which the author's thoughts or sentiments are expressed in a creative way and which falls within the literary, scientific, artistic or musical domain" (Article 2, Paragraph 1, Item 1). Therefore, sequence information is not protected by copyright, since it is mere information rather than an expression of an author's thoughts or sentiments protected by copyright.

Data or software programs are not patentable subject matter. The Examination Guideline states that, in order to be patentable subject matter, data processing must be realised in a tangible form (ie, the data must be processed using a hardware resource).[5] The guideline provides two examples of patentable data processing inventions: (i) inventions to control equipment (eg, a rice cooker, a washing machine, an engine or hard disk), or to conduct processing in a tangible fashion that is associated with a level of control; or (ii) to process data based on physical properties or technical properties of a subject (eg, rotating speed of an engine, rolling temperature) in a tangible form.

A patent specification must contain sufficient disclosure for a person skilled in the art of the claimed invention (Article 36, paragraph 4). If the specification explains how the invention functions or is to proceed in the abstract, but does not explain how such function or process is realised by hardware or software, the patent application will be rejected for lack of enabling disclosure.

4. Acts of patent infringement

4.1 Infringement

When an unauthorised third party practises an invention claimed in a patent for commercial purposes, this constitutes direct infringement of the patent. (See section 1.1 above as to the implied meaning of 'to practise an invention'.)

In order to determine whether or not there is an infringement of a patent, the scope of the claimed invention (ie, 'technical scope of patented invention') must first be construed based on a description of the claims, the specification and drawings, the file history and the technical common knowledge of the area relevant to the invention. Next, the accused product or method is compared with the technical scope of the patented invention. If it falls within the scope of the technical scope of

---

5   See Tokyo High Court 1997 (*gyo-ke*) 206.

the patented invention, the accused product or method will constitute an infringement of the patent. Specifically, when the accused product or method includes all elements described in the claim, the claim is directly infringed; however, even if only one element is missing, there is no direct infringement.

If one element is replaced by an equivalent in the allegedly infringing product/method, there is an infringement under the doctrine of equivalents. The Supreme Court has held (in Supreme Court Case No 1994 (o) 1083, February 24 1998) that, for an infringement to be established under the doctrine of equivalents, the following five factors must be established:

- the replaced element is not an essential part of the claimed invention;
- the objective of the invention can be achieved with the replacement, and the same effects are obtained;
- it was easy to make such replacement at the time of infringement;
- the accused product/method was not easily conceived from the prior art; and
- the accused product/method was not intentionally removed by the applicant.

4.2   Indirect infringement

If one or some elements of a claim are missing in the accused product but the accused product still has no other use than to make the claimed product, then to make, transfer, import, or offer the accused product for sale, or to carry it for such purposes, will constitute an indirect infringement of the patent (Patent Law, Article 101, paragraph 1, items 1 and 3). When a component has other use(s) than the making of the claimed product (except a commodity that is widely distributed in Japan), but is essential for the claimed product to solve the problem, then to make, transfer, import or offer the component for sale will constitute an indirect infringement of the patent, knowing that it is used to make the claimed product (Patent Law, Article 101, paragraph 1, item 2).

In the case of method claims, the making, transfer, import or offer for sale of a product which has no other use than to be used for the method, or carrying it for those purposes, will constitute an indirect infringement of the patent (Patent Law, Article 101, paragraph 1, items 4 and 6). When the product has other uses than to be used for the claimed method (except a commodity that is widely distributed in Japan) but is necessary for the claimed method to solve a problem, the making, transfer, import or offer for sale of the product knowing that it is used for the claimed method constitutes an indirect infringement of the patent (Patent Law, Article 101, paragraph 1, item 5).

4.3   National treatment of 'Bolar'-type provisions

The courts used to be divided as to whether the making and use of a claimed compound for obtaining necessary data in preparation for an application for marketing approval constituted an infringement of the patent or an exemption provided by Patent Law Article 69 paragraph 1.

In April 1999, however, the Supreme Court held that such activities do not infringe the patent, because it is exempted as a practice of a claimed invention for an experiment or a study provided in Patent Law Article 69 paragraph 1.[6] In light of this decision, it is established that making and using a claimed compound to obtain data which is necessary for the preparation of an application for marketing approval will not infringe a patent (although there is no statutory provision to this effect).

4.4 Experimental-use exemptions

As noted above, Patent Law Article 69 paragraph 1 provides that the exclusivity of a patent right does not encompass a practice of the patented invention for an experiment or a study. All the cases so far decided by the courts on this issue relate to preparations for marketing approval, and therefore the meaning of 'experiment' or 'study' has not yet been clarified by the courts.

4.5 Submitting authorisations and offers to supply

As provided in Patent Law Article 2 paragraph 3, the offer for sale of a claimed product will be regarded as the practice of the relevant patent. Thus, the unauthorised offer for sale of a claimed product will infringe the relevant patent.

Filing an application for marketing approval or for pricing authorisation from the authority will not usually constitute an infringement.

It is possible to seek a preliminary injunction against an offer for sale provided that (i) such an offer for sale will constitute an infringement of the relevant patent, and (ii) the applicant demonstrates that there is a need for a preliminary injunction (see section 5.2 below). Given that the sale will not yet have occurred, it may be difficult to obtain information on the infringing product and to establish the necessity for a preliminary injunction unless the sale is imminent after the marketing approval.

4.6 Summaries of product characteristics (SmPCs)

Where a secondary indication of a known substance is claimed in a patent, the manufacture and sale by a pharmaceutical company of a medical product under the approval of the product containing the claimed effective substance for the indication will usually infringe the patent claiming the indication of the substance. When a patent claims a secondary indication of a known substance but does not claim the substance itself or the first indication of the substance, a court may grant an injunction to stop making a package and a package insert describing the claimed indication. However, the court cannot enjoin the defendant to make the substance itself, since the patent does not cover the substance or the first indication.

5. Patent enforcement

5.1 Obtaining information on the infringer and the infringement

There are no discovery proceedings in Japanese patent litigation. Although evidence preservation proceedings are available pursuant to Civil Procedure Law, Article 234, the requirements are extremely difficult to meet. A party who requests evidence preservation must specify the piece of evidence to be preserved, the fact(s) to be

---

6   Supreme Court Case No 1998 (o) 153, April 16 1999.

proved by the evidence, and the justification for the requested preservation of evidence (ie, the party must show that there is a *prima facie* case of infringement).

5.2 Interim relief

A patentee can seek an order for a preliminary injunction. In order to do so, the patentee must establish (i) an infringement of the patent and (ii) the necessity for a preliminary injunction (ie, by demonstrating that there is a possibility of irreparable harm if the alleged infringement continues). If neither the patentee nor any licensee practises the patented invention, it will be very difficult to demonstrate this necessity.

5.3 'Springboard'/post-patent expiry injunctions

Following the expiry of a patent, the former patentee will no longer be entitled to an injunction against an infringer (or one who was infringing the patent before its expiration). An injunction order will sometimes specify the expiration date of the order in accordance with the patent expiration date. In a preliminary injunction proceeding as to a patent at issue which is about to expire, the court may decide that there is no necessity for the preliminary injunction.

5.4 Unjustified threats

When an unjustified threat goes too far and reaches the level of criminal intimidation, it will be punishable as such. Criminal Law Article 222 paragraph 1 provides as follows:

> *A person who intimidates another through a threat to another's life, body, freedom, reputation or property shall be punished by imprisonment for not more than two years or a fine of not more than 300,000 yen.*

A warning letter concerning patent infringement will usually request a response within a certain time limit and state that if there is no response by the deadline, the patentee will take legal action. Unless such a letter has no legal basis and the letter's author is aware of this fact, the letter will not be treated as an unjustified threat.

5.5 Remedies

The main remedies for patent infringement are injunctions (Patent Law, Article 100), damages (Civil Code, Article 709) and attorneys' fees.

When seeking an injunction, the patentee must specify the nature of the activities of the infringer that the injunction is to restrain. As an additional measure for the prevention of infringements, patentees can also request the destruction of infringing products and half-finished products.

Japanese courts will automatically grant a permanent injunction where they find that a valid patent is being infringed. Even before a district court injunction decision becomes final and conclusive, it can be executed if it is accompanied by a preliminary execution declaration. Procedurally, however, the accused infringer may request a stay of injunction before a district court by depositing a sum of money as security and also by filing an appeal with the IP High Court. In such cases, the order will be stayed until the decision becomes final and conclusive.

Patent Law Article 102 paragraphs 1, 2 and 3 determine the calculation of damages, which can take three forms: (i) lost profits based on the infringer's sales and the patentee's profit rates; (ii) lost profits based on the infringer's profits; and (3) licence royalties. No punitive damages are available for wilful infringement. Reasonable legal fees may, however, be available.

6. Compulsory licensing

A person who needs to make use of a patent invention owned by a third party can seek a compulsory licence under the provisions of the Patent Law provided that (i) the third party's patent invention has not been practised for more than three continuous years (Article 83), (ii) the third party's patent invention is necessary to practice the person's own patent invention (Article 92), or (iii) the third party's patent invention is necessary in the public interest (Article 93).

The party requesting the compulsory licence must first ask the patentee or its exclusive licensee for a non-exclusive licence. When such negotiation fails, or the request for negotiation is turned down, the applicant can seek determination by the Minister of Economy, Trade and Industry.

In reality, however, no compulsory licence has ever been granted.

7. Ownership, inventors and compensation

When an invention is made, its inventor owns all rights to the invention including a right to file a patent application.

If the invention is within the scope of the inventor's job specification, the employer will have a royalty-free, non-exclusive licence to a patent granted on the employee's invention (Japanese Patent Law, Article 35, paragraph 1). In addition, the employee can assign rights to file a patent application as to future employee's inventions or a patent by employment regulation or an agreement, and in such case the employee is entitled to receive reasonable value under Japanese Patent Law Article 35 paragraph 3. The 'reasonable value' can be determined by a fair and reasonable process under Article 35, paragraph 4. Paragraph 3 was revised and paragraph 4 was added after the huge award (¥2 billion, though later settled for ¥3 million) to an inventor which was granted by the Tokyo District Court in 2004.[7]

8. Branding and designs

8.1 Trade marks in life sciences

There are no special trade mark rules or regulations that apply to trade marks on pharmaceuticals. However, an applicant must disclose the planned trade name of a medicine when filing an application for a marketing approval. When the planned trade name is similar to the name of another medicine so that it may cause confusion, the authority will instruct the applicant to modify the name.

8.2 Extending protection using trade marks and designs

Brand-name drug companies are making efforts to protect their trade marks and

---

7   Tokyo District Court, 2001 (*wa*) No 17772, January 30 2004.

designs so that their trade marks are not diluted by similar marks of generic companies.

Sankyo, the brand-name pharmaceutical company that developed and sells Mevalotin, commenced an invalidation proceeding against a trade mark registration for 'Mevalation' for medicine owned by a generic company Choseido. The JPO determined that the trade mark registration for 'Mevalation' for medicine was invalid because of the similarity with Mevalotin, and this would be likely to cause confusion among consumers. In its appeal case, the Tokyo High Court affirmed the JPO's decision.[8] Likewise, Merckhoei's trade mark registration for Mevastan for medicine was invalidated.[9]

Taisho Yakuhin's trade mark registration for 'Harnnat' for medicine was also invalidated, because it is confusingly similar to Astellas's trade mark 'Harnal'.[10]

However, the brand-name companies' efforts to protect their trade marks have not been particularly effective in terms of extending protection for their brand-name drugs. When patents and re-examination periods (data-exclusivity periods) expire, generic companies are able to obtain approval and to start sales of their drugs by choosing trade marks which are not too similar to brand-name companies' trade marks. For example, there are 32 approvals for pravastatin sodium (brand name Mevalotin).

8.3   Comparative advertising and advertising restrictions.
In Japan, prescription drug manufacturers have their own rule to comply with the Act against Unjustifiable Premiums and Misleading Representations. This rule provides that comparative advertising is admissible only when a company's own product is compared with another using the generic name and based on objective data. There is also a group of companies selling over-the-counter medication, and their rule completely prohibits comparative advertising.

9.   Protecting valuable information
Trade secrets and know-how are protected under the Unfair Competition Law, which prohibits the use and/or disclosure of another's trade secret. In particular, ex-employees are prohibited from using and/or disclosing the know-how of former employers under Article 2, paragraph 1, items 7 and 8 of the Unfair Competition Law. In order for information to qualify as a trade secret or know-how, it should be treated as secret. For this purpose, employers must ensure that employees' relevant terms of employment stipulate that trade secrets and know-how should be kept secret.

10.   Counterfeiting

10.1   Penalties
Counterfeiting a patented product or registered trade mark is a crime. Japan Patent Law Article 196 provides that the infringement of a patent right or exclusive licence

---

8   *Choseido v Sankyo*, Tokyo High Court 2004 (*gyo-ke*) 341.
9   *Merckhoei v Sankyo*, IP High Court 2005 (*gyo-ke*) 10418.
10   *Taisho Yakuhin v Astellas*, IP High Court 2007 (*gyo-ke*) 10427.

is punishable by up to 10 years' imprisonment or a fine not exceeding ¥10 million or a combination of the two. Trade Mark Law Article 78 stipulates the same penalties.

10.2 Customs

Pursuant to Customs law, an owner of a patent right or a trade mark registration can file with the Customs Office an application for a third party to cease and desist from the importation of infringing goods.

When an application is filed, the Customs Office will ask importer(s) to file an opinion within 10 days and will review the application as to whether it has merit. For patent infringement cases, the Customs Office summons three experts, and a hearing will take place. Based on the experts' opinions, the Customs Office will decide whether to accept, reject or suspend the application.

If the application is accepted, the Customs Office stops goods before importation and examines whether or not they infringe the patent or trade-mark registration at issue.

The application stays in effect for two years or up to the expiration date of the patent or trade mark at issue, and it should be renewed before expiration.

Customs proceedings are a strong measure in the fight against counterfeit drugs.

11. Collaborative models – 'open innovation'

'Open innovation' is "a paradigm that assumes that firms can and should use external ideas as well as internal ideas, and internal and external paths to market, as the firms look to advance their technology".[11]

The R&D costs of the pharmaceutical industry keep increasing, and there is always a risk that a drug candidate has to be dropped after clinical trials. In these circumstances, pharmaceutical companies, as in other industries, have been interested in, and now cannot avoid, open innovation.

Efforts have been made to modify the system to make it more useful for the open innovation paradigm:

- With effect from April 2009, an exclusive licence to a patent application can be registered with the JPO.
- A licence to a patent will be effective without licence registration with the JPO for a patentee who later acquires the patent without registration, following the Patent Law Amendment which will take effect in April 2012.
- A patent matured from an application filed for a misappropriated invention will be assigned to the rightful owner based on the final and conclusive court decision, under the amended Patent Law which will be effective in April 2012.

The current Japanese patent system still seems more compatible with the traditional paradigm that assumes that firms use internal ideas to advance their technology. For example, when a patent is owned in common, each co-owner can freely practise the patented invention without permission from the other owner(s),

---

11   *Open Innovation*, HW Chesbrough, Harvard Business School Press, 2003.

but each has to have the others' permission to license a patented invention to a third party. If one of the co-owners does not want to license the invention, the patented technology cannot be licensed.

12.   Hot topics

12.1   Patent term extension
A patent term extension of up to five years can be granted for the time taken to obtain a marketing approval, pursuant to Patent Law Article 67 paragraph 2:

> *67(2) Where there is a period during which the patented invention cannot be practised because an approval prescribed by relevant Act that is intended to ensure the safely, etc or any other disposition designated by Cabinet Order as requiring considerable time for the proper execution of the disposition in light of the purpose, procedures, etc, of such a disposition is necessary for practising the patented invention, the term of the patent right may be extended, upon filing a request for registration of extension of the term, by a period not exceeding five years.*

Under the current JPO practice, a patent term can be extended only once for a specific 'substance' for a specific 'use'. However, following the Supreme Court decision in *JPO v Takeda* of April 28 2011, the practice will soon change.

Patent term extension requests have been examined in terms of whether or not 'the patented invention cannot be practised because an approval is necessary' based on 'substance' and 'use' covered by the marketing approval and the patent at issue. If a combination of a substance and a use becomes the subject of a marketing approval for the first time, the substance and/or the use cannot be practised without marketing approval, and a patent term extension of a patent covering the substance and/or the use is allowable.

Later, when an additional marketing approval for another indication of the same ingredient was issued, a second patent term extension would be allowed for the same patent because the 'use' was different from the one for the prior patent term extension. On the other hand, if the second marketing approval related to a new formulation of the same ingredient and the same indication, the second patent term extension was not allowed.

In *JPO v Takeda*, Takeda requested an extension of the patent term, claiming a new sustained-release formula based on a marketing approval for the sustained-release formula of an active ingredient for an indication. However, since a patent term extension had previously been allowed for the same ingredient and the same indication (different formula), the JPO rejected the patent term extension request. The IP High Court overruled the JPO's decision, and the JPO appealed to the Supreme Court. The Supreme Court held that the patent term extension request should not be rejected based on the prior patent term extension, because the prior marketing approval did not fall within the scope of the patent claim at issue.

Netherlands

Bas Berghuis van Woortman
Mattie de Koning
András Kupecz
Simmons & Simmons LLP

Introduction

Being the domicile of many international headquarters, and with efficient and experienced courts, the Netherlands is regarded alongside Germany and the United Kingdom as a leading forum for intellectual property (IP) litigation in Europe. This chapter outlines the key features of intellectual property law and litigation in the Netherlands, and provides an overview of the most recent case law from the specialised patent court in The Hague. The case law in relation to community and Benelux trade marks will also be addressed, as well as other case law relevant to the field of life sciences.

1.   Small molecules

1.1   Product and process claims

Article 53(1) of the Dutch Patent Act (DPA) states:

> *Subject to the provisions of Articles 54 through 60, a patent shall confer on its proprietor the exclusive right:*
>
> *a. to make, use, put on the market, or resell, hire out or deliver the patented product or otherwise deal in it, in or for his business, or to offer, import or stock it for any of those purposes;*
>
> *b. to use the patented process in or for his business or to use, put on the market, or resell, hire out, deliver the product obtained directly as a result of the use of the patented process, or deal in any other way, in or for his business, or to offer, import or stock it for any of those purposes.*

The DPA distinguishes between patented products and processes. The protection conferred by Article 53(1) to products is 'absolute' (ie, a product claim gives the patentee exclusive rights for the product *per se*, no matter how the product is produced). A patented process gives the patentee an exclusive right to this process as well as any and all products which were manufactured by this process. If the product is not directly obtained through the patented process, for instance due to the fact that only one of the components in the chemical reaction which formed the product was made by the patented process, the product falls outside of the scope of a process claim.[1]

---

1   Dutch Supreme Court June 10 1983, NJ 1984, 32 (*Doxycycline*).

1.2 Scope of protection of claims

The scope of the patent and the exclusive right is determined by Article 69 of the European Patent Convention (EPC) and its national counterpart Article 53(2) of the DPA:

The exclusive right shall be determined by the content of the claims in the patent specification, the description and the drawings serving to interpret those claims.

Article 69 of the EPC states:

The extent of the protection conferred by a European patent or a European patent application shall be determined by the terms of the claims. Nevertheless, the description and drawings shall be used to interpret the claims.

The Protocol on the Interpretation of Article 69 EPC states:

Article 69 should not be interpreted as meaning that the extent of the protection conferred by a European patent is to be understood as that defined by the strict, literal meaning of the wording used in the claims, the description and drawings being employed only for the purpose of resolving an ambiguity found in the claims. Nor should it be taken to mean that the claims serve only as a guideline and that the actual protection conferred may extend to what, from a consideration of the description and drawings by a person skilled in the art, the patent proprietor has contemplated. On the contrary, it is to be interpreted as defining a position between these extremes which combines a fair protection for the patent proprietor with a reasonable degree of legal certainty for third parties.

Although the wording used by the District Court and the Court of Appeal is not always consistent, Dutch courts apply this European standard in determining the scope of protection of patent claims.[2] In order to prevent too strict an interpretation of the patent, the meaning behind the invention should be explored as a point of view. However, the court must also assert the patent by holding a reasonable degree of legal certainty for third parties, which in its turn could justify an interpretation which is stricter than the wording of the claim. According to the Supreme Court, therefore, a correct interpretation should hold the middle ground between a reasonable degree of protection and a reasonable degree of certainty for third parties.

1.3 Markush formulae

There is no recent Dutch case law particularly concerning Markush-type claims. Generally, the case law developed by the Boards of Appeal on novelty and inventive step in relation to selection inventions applies *mutatis mutandis*. As to novelty, the courts thus consistently apply the European Patent Office (EPO) 'photographic novelty test'[3] and where it concerns inventive step, the courts apply the problem–solution approach.[4]

One recent case which is noteworthy is *Ratiopharm v Eli Lilly*[5] (olanzapine).

---

2   Dutch Supreme Court September 7 2007, NJ 2007, 466 (*Lely/Delaval*) and Dutch Supreme Court January 13 1995, IEPT 19950113 (*Ciba Geigy/Oté Optics*).
3   Court of Appeal The Hague February 28 2008, IEPT 20080228 (*Ranbaxy/Warner-Lambert*).
4   See the *Glaxo/Sandoz* decision of the The Hague District Court, described further in section 2.1 of this chapter.
5   District Court of The Hague March 24 2010, H ZA 08-2126 (*Ratiopharm/Eli Lilly*).

Ratiopharm stated that the patent in suit lacked novelty, because the compound had already been disclosed in a scientific article. The article taught 4-peperazinyl-10H-thieno-benzodiapine analogues. The disclosed compounds had substituents R_1, R_2 and R_3, defining 12 different compounds. One of these compounds had R_1 = H, R_2 = CH$_3$ and R_3 = H, which are identical to the substituents of the patented compound olanzapine. However, the article disclosed a piperidine ring instead of the olanzapine piperazine ring (having a carbon atom instead of a nitrogen atom at the same position). The court, with Ratiopharm, held that the C atom instead of the N atom was a clear and unambiguous error in the publication which would strike the skilled person immediately. Furthermore, according to the court, the way in which the error should be corrected was also unambiguously clear. The court therefore invalidated the patent for lack of novelty. This decision deviates from decisions in many other countries, such as the United States, the United Kingdom and Germany, where the patent was considered valid, because Schauzu did not unambiguously disclose olanzapine.

2.   Second-generation inventions

2.1   Combinations

In the recent *Sandoz v Glaxo Group Lt*d decision, the court invalidated a patent relating to a combination of salmeterol and fluticasone propionate for lack of inventive step.[6] As mentioned, inventive step is usually assessed by the Dutch courts using the problem–solution approach. The District Court took a review article on asthma treatment as closest prior art (CPA). According to the court the document taught the skilled person the use of ß2 agonists, such as salmeterol, as a bronchodilator and the use of an inhaled corticosteroid as an anti-inflammatory component. This prior art also suggested combining the treatment components in a combined dose inhaler. After identifying the differences between the CPA and the patent, the court formulated the objective technical problem as finding a combination therapy as suggested by the CPA in the form of a combination inhaler with a longer-working ß2-agonist – for example salmeterol, as suggested by the CPA, in combination with a topical inhaled corticosteroid component that works better than the known agents. Because the CPA already suggests the use of a combination therapy with salmeterol and a better corticosteroid, the skilled person would, according to the court, search for an improved alternative for beclomethasone/ budesonide, and would select fluticasone propionate. Glaxo's arguments based on synergy failed because there was no such teaching in the patent specification. Based on the above, the court concluded that the combination patent lacked inventive step.

2.2   Enantiomers

The Dutch court, for example in the *Tiefenbacher v Lundbeck* case,[7] in assessing novelty, follows the consistent jurisprudence of the Boards of Appeal of the EPO, stating that disclosure of the racemate does not take away novelty of a specific

---

6   District Court of The Hague January 26 2011, HA ZA 09-2540.
7   District Court of The Hague April 8 2009, HA ZA 08-1827 (*Alfred E Tiefenbacher GmbH v H Lundbeck A/S*).

enantiomer; nor does the theory that it might be possible to separate the racemate into its enantiomers. Only the direct unambiguous disclosure of the specific compound in the form of an individualised technical teaching is novelty destroying (the concept of 'photographic novelty').

According to Dutch case law, establishing inventive step for enantiomer patents has proven difficult. The District Court of The Hague in the above-mentioned *Tiefenbacher v Lundbeck*, as well as the *Ratiopharm v Sepracor*[8] case, following T 296/87, judged that it must be assumed that testing enantiomers is the obvious first step in the search to improve a pharmaceutical containing one or more chiral centres. According to the District Court, a number of prior-art documents, for instance US FDA directives dating from 1987, gave strong pointers to research into the pharmacologic properties of the enantiomers of the racemate. Moreover, according to the court in the *Tiefenbacher v Lundbeck* case, the man skilled in the art could, with routine and systematic experiments and therefore without undue burden, solve the racemate. In the *Ratiopharm v Sepracor* case, the court judged that a man skilled in the art would know how to solve the racemate, in this case Cetirizine, into its enantiomers. Therefore in both cases the patents were invalidated for lack of inventive step.

2.3   Selection inventions

As mentioned above, the Dutch courts generally apply the EPO rules on selection inventions. As a general rule, a generic disclosure does not take away the novelty of a species. Inventive step will be assessed on a case-by-case basis using the problem–solution approach.

In the *Aventis v Apothecon* case,[9] the Aventis patent concerned "Use of terfenadine derivates as antihistaminics in a hepatically impaired patient". Aventis argued that the patent was a selection invention, based on the selection of a specific patient group, namely patients with liver diseases who could suffer from heart problems if allergy symptoms were treated with terfenadine. According to the Court of Appeal, fexofenadine (a terfenadine metabolite) was already known to the skilled person as an antihistaminic for the treatment of (amongst others) hay fever, from the compound patent which had expired in 2005. To successfully claim fexofenadine as a selection for the sub-group 'liver patients', there should be a "purposive selection/new technical teaching". According to the Court of Appeal, the selection in this case was arbitrary, because the patent did not describe a surprising therapeutic effect of fexofenadine in treating the sub-group 'liver patients'. The non-occurrence of heart problems was not deemed to be a surprising effect, because heart problems did also not occur in patients not belonging to the sub-group (ie, patients without liver disease). Therefore the patent lacked novelty.

2.4   Methods of use

In general, use claims are asserted by the Dutch Court in concurrence with the view

---

8   District Court of The Hague May 13 2009, IEPT 20090513 (*Ratiopharm/Sepracor*).
9   Court of Appeal The Hague March 16 2010, IEF 8683 (*Aventis/Apothecon*).
10 District Court of The Hague April 28 2010, IEPT 20100428 (*Ratiopharm/Sanofi-Aventis*).

of the EPO. In the *Ratiopharm v Sanofi-Aventis* case,[10] the District Court of The Hague for instance explicitly referred to G2/08, a case in which the Enlarged Board of Appeal of the EPO ruled that medical-use claims directed to new and inventive dosage regimes are allowable. In the *Ratiopharm v Sanofi-Aventis* case, however, the patent was revoked on lack of inventive step, because the patentee had not sufficiently argued why the claimed dosage regime of the combination of known products for use for the known indication was inventive in the specific case.

2.5 Formulations and physical forms

Dutch case law does not show a clear trend as to how the Dutch courts regard formulation patents. For instance in the two cases concerning a formulation patent in which judgment was rendered in 2010, in the case, *Mundipharma v Sandoz*[11] the formulation patent was upheld, while in the case *Teva v Aventis*,[12] the patent was invalidated. The latter case will be discussed in more detail next.

The patent comprised the use of compositions based on products of the taxane class of formula (I) in which R represents a hydrogen atom and R1 represents a tert-butoxycarbonylamino group, dissolved in a mixture of ethanol and polysorbate, for the production of perfusions for the treatment of tumours and leukemia.

The patent was revoked on lack of inventive step. Again, the District Court asserted inventive step by using the problem–solution approach of the EPO (see section 2.1 above). The District Court stated the objective technical problem as 'finding a composition containing docetaxel with an acceptable toxicity profile'. EP 738, which described docetaxel with Emulphor and ethanol, was seen as closest prior art. The question that needs to be addressed, therefore, was whether on the priority date the average skilled person, on the basis of EP 738 and seeking a solution for this problem, would have arrived at the docetaxel formulation (polysorbate as an alternative for Emulphor) according to claim 1 of the patent.

The most radical defence raised by Aventis was that the average skilled person would not have arrived at the solution described in claim 1 of EP 656, because on the priority date that person would not have been motivated to seek an alternative docetaxel formulation. This defence was dismissed by the court because in its opinion there will be a constant search for formulations having fewer and less serious side effects, even where those side effects are controllable. There is, and will be, an incessant ambition to achieve the best treatment results using the least possible medication, as any medication has inherent side effects.

Because the court furthermore found pointers in the prior art to the use of polysorbate as an alternative, the patent was invalidated for lack of inventive step.

2.6 Reach-through claims

In the *Fuel oil/EXXON*, T 409/91 decision of the Technical Board of Appeal (TBA) of the EPO, the Board had challenged unduly broad claims under Article 83 of the EPC which thus constituted a ground for opposition and for nullity. The Board stated:

---

11   District Court of The Hague April 7 2010, HA ZA 09/229 (*Mundipharma/Sandoz*).
12   District Court of The Hague September 1 2010, IEPT 20100901 (*Teva/Aventis*).

> *Thus, a claim may well be supported by the description in the sense that it corresponds to it, but still encompasses subject-matter which is not sufficiently disclosed within the meaning of Art 83 EPC, as it cannot be performed without undue burden or vice versa.*

and

> *In the Board's judgement, the disclosure of one way of performing the invention is only sufficient within the meaning of Art 83 EPC if it allows the person skilled in the art to perform the invention in the whole range that is claimed ...*

Recently, the District Court of The Hague invalidated a few patents for having too broad a claim and thus were not sufficiently disclosed within the meaning of Article 83 of the EPC. One of these cases is the *Novozymes v DSM*[13] case, which will be further discussed below in section 3.3, together with more general background on insufficiency. A second example is the *Bayer v Abbott*[14] case. In this case, the District Court of The Hague invalidated the Dutch part of Bayer's European patent covering anti-TNF-α human monoclonal antibodies. The Court considered that the patent covered high-affinity antibodies. According to the Court, however, such type of antibodies were not sufficiently disclosed in the patent description and therefore the patent was invalidated. This decision was different from the TBA decision. The TBA had considered the patent not to comprise high-affinity antibodies, and absent these high-affinity antibodies, the patent could be carried out by a skilled person without undue burden and therefore the patent was upheld by the TBA.

3.   DNA, biologicals and personalised medicine

3.1   Discoveries

Patentable inventions should be new, involve an inventive step and be useful. Things that already exist in nature, therefore, are mere discoveries and not inventions, since they are not novel. In the field of biotechnology, there is a thin line between discovery and invention.

In the last 30 years, increasing numbers of micro-organisms are being patented. There are both supporters and opponents of gene patenting. The supporters generally state that because they are able to explain the function and the industrial application of what was previously unknown, their discovery is an invention that should be patentable. Critics state that owing to new techniques it is becoming increasingly easy (and therefore routine) to isolate and purify DNA; identifying a DNA sequence should therefore not be patentable, because it does not involve an inventive step. Furthermore, allowing patents on DNA sequences based on a single function could even discourage further investigation of the often multiple functions of the patented strand of DNA.

The Biotech Directive (Directive 98/44/EC) gives some clarity on the patentability of biotech inventions. The Directive for instance gives guidance on the interpretation of *ordre public* in relation to Article 3(1) of the DPA. There still remain, however, many questions to be answered by the legislator or patent courts.

---

13   District Court of The Hague May 19 2010, HA ZA 09/3220 (*Novozymes/DSM*).
14   District Court of The Hague October 20 2010, IEF 9170 (*Bayer/Abbott*).

3.2   Gene patents and industrial application

In the recent ECJ *Monsanto* decision, a case which was referred to the ECJ by the District Court of The Hague, the ECJ has given its first clarification as to how the Biotech Directive should be interpreted. This case concerned Article 9 of the Biotech Directive, which states:

> *The protection conferred by a patent on a product containing or consisting of genetic information shall extend to all material, save as provided in Article 5(1), in which the product in incorporated and in which the genetic information is contained and performs its function.*

Monsanto holds a European patent to a genetic sequence which makes the soybean plant resistant to the herbicide glyphosate, used by farmers to kill off weeds without harming their soy crops. The technology, known as Roundup Ready, is not patented in Argentina. In 2005 and 2006 soymeal was imported from Argentina to the Netherlands which contained traces of Monsanto's claimed technology. This indicated that the soymeal originated from Roundup Ready soy beans. Monsanto claimed before the District Court of The Hague that the imported soymeal infringed the Dutch part of her European patent, because the patented gene sequence was present in the soy meal.

The ECJ ruled that Article 9 of Directive 98/44/EC is to be interpreted as not conferring patent right protection in circumstances in which the patented product does not perform the function for which it was patented. Furthermore, the ECJ ruled that the Biotech Directive precludes the national patent legislation from offering absolute protection to the patented product as such, regardless of whether it performs its function in the material containing it.

3.3   Sufficiency issues

Article 75 of the DPA states that the patent will be revoked by the court if it does not contain a description of the invention which is sufficiently clear and complete to allow a person skilled in the art to work the invention (ie, if the patent is sufficiently disclosed).

Until recently, the most interesting judgment concerning sufficiency issues was rendered in *Boehringer Mannheim v Kirin Amgen*.[15] The judgment addressed in detail the requirements of sufficiency of disclosure, also in relation to biotech patents.

The Court of Appeal held that patents relating to new substances often have a main claim that is directed at a class of chemical compounds meeting a certain structure formula. In such cases the inventor's merit is that he has ascertained that all compounds of that class have a certain therapeutic effect. A specific example may then sufficiently disclose an entire class of compounds as it will be clear to the skilled person that any compounds not included in the examples can be processed in the manner described and will have the intended therapeutic effect, which is directly related to the claimed structure.

This would likewise apply to all future embodiments that are part of that class but cannot be obtained via the method described in the patent. The Court of Appeal

---

15   Court of Appeal The Hague January 27 2000, IEPT 20000127 (*Boehringer/Kirin Amgen*).

acknowledged the question of whether a substance that could be obtained only with the use of an inventive method that was discovered later and showed the claimed therapeutic effect should no longer be covered by the scope of the patent. The court did not accept this point of view in instances where the compound in question proved to have the therapeutic effect that had already been made plausible on the submission date. After all, this would mean that the inventor had been right to claim this compound as the patent already described the inventive relation between the (structure of the) class of compounds and the therapeutic effect. Therefore, the Court of Appeal held that this relation, too, is to be considered part of the invention disclosed in the patent.

The Court of Appeal was of the opinion that if a substance is unambiguously characterised with the aid of a structure formula and the inventor discloses this knowledge, he is entitled to protection that extends to all compounds not specifically mentioned in the patent as long as the compounds satisfy the structure formula and evidence has been provided that they show the effect in question. The Court of Appeal's decision is supported by the *Genentech* decision (T 292/85).

Novozymes v DSM was the first case in which a patent was nullified by the Dutch patent court on account of insufficiency of disclosure. With a reference to the Guidelines for Examination in the European Patent Office, the District Court stated that lack of support/clarity is closely related to insufficient disclosure. In its assessment the District Court began by establishing that:

> *article 83 EPC requires that the patent application discloses the invention in a manner sufficiently clear and complete for it to be carried out without undue burden by a person skilled in the art within the whole area that is claimed (cf TBA 18 March 1993, T 409/91, Exxon). Thus, based on the patent description, the person with average skill in the pertinent art should be able at least to reproduce a substantial part of the embodiments of the invention.*

In short, the method claimed in the Novozymes patent related to adding to dough known lipolytic enzymes that have a known and specified (favourable) effect. The court held that the Novozymes patent covered innumerable variants and merely expressed the desire for these variants to have a certain functionality, without providing a sound, common, technical basis. In this respect the court held:

> *In order to be able to apply claim 1 within the whole area that is claimed, the Patent description has to teach the skilled person how to obtain the said enzymes. The Patent description describes those enzymes, however, not in terms of their chemical structure or composition. The skilled person is therefore unable to reproduce the chemical structure or composition. The patent description only defines the enzymes in terms of the desired activity to digalactosyldiglyceride, phospholipide and triglyceride, without disclosing a technical concept that teaches the skilled person how to obtain all enzymes that show the desired activity.*

As a skilled person would realise that not all alternatives would have the desired characteristics, he would have to resort to trial-and-error experimentation to identify the enzymes in question by testing every compound imaginable for the characteristics described in the patent. According to the court this meant that the patent description was in fact nothing more than an invitation to the skilled person to start a research programme, which imposes undue burden.

Based on the above, the court found that the mere circumstance that the description provided several (sufficiently disclosed) examples that resulted in enzymes covered by the claims was insufficient to assume that the patent was reproducible within the whole area that was claimed. In the absence of a technical concept suitable for general use, the skilled person will be unable to obtain on the basis of a few examples at least a substantial part of the other enzymes showing the desired activity. The court referred to the claim as a 'free beer' claim.

4. Acts of patent infringement

4.1 Infringement

Article 53 of the DPA sets out the exclusive rights of the patentee, as described above in sections 1.1 and 1.2.

Article 70 (1) of the DPA states the enforcement right of the patent holder in case of direct infringement:

> The proprietor of a patent may enforce his patent vis-à-vis any person who, without being entitled to do so, performs any of the acts referred to in Article 53, paragraph 1.

On the issue of direct infringement the *AGA v Occlutech* case is illustrative.[16] In this case, AGA had patented an intravascular occlusion device. One of the features of claim 1 was "clamps are adapted to clamp the strands at the opposed ends of the device". The alleged infringing device of Occlutech had only a clamp on one side of the device. Therefore, the court judged there was neither direct infringement nor infringement on the basis of equivalence – the latter due to the fact that the Occlutech device logically had a different closing mechanism on the side of the device that did not close with a clamp.

4.2 Contributory infringement

(a) Patents

Article 73(1) of the DPA states:

> The proprietor of the patent may institute the claims which are at his disposal in enforcing his patent against any person who, in the Netherlands or the Netherlands Antilles, or where a European patent is concerned, in the Netherlands, offers or delivers, in or for his business, the means for application of the patented invention, in respect of an essential part of the invention, in the Realm to persons other than those who by virtue of Articles 55 to 60 are empowered to work the patented invention, provided that that person knows or that it is evident considering the circumstances that those means are suitable and intended for that application.

So, for contributory infringement the two main conditions to be met are: (i) the infringer should offer or deliver the means for application of the patented invention, in respect of an essential part of the invention; and (ii) that person should know, or it should be evident that those means are suitable and intended for that application.

---

16   Court of Appeal The Hague October 19 2010, IPET 20101019 (*AGA/Occlutech*).

The Dutch Supreme Court[17] provided guidance on the interpretation of the condition 'an essential element of the invention', in a case related to Sara Lee's European patent for an "Assembly for use in a coffee machine, for preparing coffee". This assembly comprised a container and a 'pouch' (also: a 'pad'). The Court of Appeal decided that the sale of pads did not qualify as an act of contributory infringement. The Court of Appeal did not consider the pad to be an essential part of the invention. The Supreme Court upheld this decision. According to the Supreme Court, the fact that an element (the pad) is necessary to apply the invention does not necessarily make it an essential element. The Supreme Court further said that when deciding whether or not an element is an essential element, the judge may take into account whether or not the element concerned is an element which distinguishes the teaching of the patent over the prior art.

In *Safeway v Kedge*,[18] the supplier of a product prescribed specific installation instructions which would not infringe the patent. The president of the district court nonetheless judged that this was a case of indirect infringement due to the fact that the alleged infringer knew and that it was evident that customers would not follow the installation instructions but would make use of the patented process instead.

(b)   Other IP rights

Unlike Article 73 of the DPA, the other IP Acts in the Netherlands do not contain any provision with respect to contributory infringement. According to Dutch law, third parties who do not themselves infringe trademarks, copyright, design rights etc of the intellectual property right owner could commit a tort under the general tort clause of Article 6:162 of the Dutch Civil Code.

4.3   National treatment of 'Bolar'-type provisions

Article 10(6) of the Directive 2004/27/EC is implemented in Article 53(4) of the DPA, which states:

> *Carrying out necessary studies, tests and experiments bearing in mind application of Article 10, paragraphs 1, 2, 3 and 4, of Directive 2001/83/EC on the Community code relating to medicinal products for human use (OJEC L 311) or Article 13, paragraphs 1, 2, 3, 4 and 5 of Directive 2001/82/EC on the Community code relating to medicinal products for veterinary use (OJEC L 311) and the consequential practical requirements shall not be considered infringement on patents regarding medicines intended for human use, or medicines intended for veterinary use.*

An exact reference is made amongst others to Article 10(4) of Directive 2001/83/EC. Therefore biosimilar testing is explicitly excluded from the exclusive right of the patentee.

The explanatory memorandum with the amendment to the Dutch Patent Act to implement Directive 2004/27/EC and 2004/28/EC explicitly states:

> *All types of acts other than performing the necessary studies, tests and experiments for the purpose of applying Article 10, first to fourth paragraph, of Directive 2001/83/EC*

---

17   Dutch Supreme Court October 31 2003, IEPT 20031031.
18   President District Court of The Hague October 18 2010, IEPT 20101018.

and Article 13, first to fifth paragraph of Directive 2001/82/EC or acts covered by the scope of the research exception in Article 53, third paragraph, are not exempt from the exclusive rights of the patentee, under Article 53, first paragraph, of the Patent Act of 1995 and are therefore not allowed.

So, the exceptions to the exclusive rights of the patentee have to be strictly interpreted. The limitation of the exclusive rights of the patentee by the Bolar-type provision according to Dutch law is therefore (i) limited to the studies, tests and experiments and the consequential practical requirements necessary in seeking regulatory approval for a generic medicinal product; and (ii) necessary for the application of regulatory approval in the European Union and European Economic Area.

4.4 Experimental-use exemptions

Article 53(3) of the DPA states:

The exclusive right shall not extend to acts solely serving for research of the patented subject matter, including the product obtained directly as a result of using the patented process.

According to the explanatory memorandum to Article 53(3) of the DPA, the experimental-use exemption should be strictly interpreted. Only acts which are purely scientific in nature or aimed to further develop the patented subject matter are allowed.[19] An example is *Kirin Amgen v Boehringer*,[20] in which case clinical trials needed for finding a second medical use were judged to be allowed.

In *TeDe v Tetra Laval*,[21] TeDe had copied a machine which was patented by Tetra Laval in order to try to improve the patented product. The court judged that copying in order to search for an innovating improvement did not fall under the scope of the experimental-use exemption. There would only be a case of innovating research on an improvement if the machine was recreated with the improvement.

4.5 Submitting authorisations and offers to supply

According to standard Dutch case law, a mere application for a generic market authorisation or the granting of such market authorisation does not constitute an act of patent infringement. An 'offer for sale' does, however, constitute an infringing act under Article 53(1) of the DPA. The explanatory memorandum to Article 53(1) explicitly states that the term 'offer' in Article 53(1) of the DPA should be interpreted broadly.

The Hague Court of Appeal[22] has recently held that publication of a generic pharmaceutical in the Dutch 'G-standard' (a Dutch database for pharmaceuticals) prior to patent expiry amounted to an 'offer' under the Dutch Patent Act.

---

19   Dutch Supreme Court December 18 1992, NJ 1993, 735 (*ICI/Medicopha*rma).
20   Dutch Supreme Court April 21 1995, NJ 1996, 462 (*Kirin Amgen/Boehringer*).
21   District Court of The Hague November 17 2010, IEPT 20101117 (*TeDe/Tetra Laval*).
22   The Hague Court of Appeal November 2 2010 (*Glaxo Group/Pharmachemie*).

4.6 Summaries of product characteristics (SmPCs)

Because the mere application for a market authorisation does not constitute a patent infringement, the same goes for the SmPC, which must be completed and submitted to the Dutch medicines authority (CBG-MEG) before market authorisation is awarded. If a third party commits an infringing act, the infringement of the contents of the SmPC can, however, be used as a tool to prove patent infringement. This has especially been the case in matters of infringement of Swiss-type, second-medical-use claims. An example of this is the *Schering v Teva* case,[23] in which the court considered the Swiss-type claims not infringed due to the fact that the Swiss-type claims of the patent referred to a very specific group of patients, which patient group was especially carved out by Teva in the SmPC.

5. Patent enforcement

5.1 Obtaining information on the infringer and the infringement

Under Dutch civil procedural law, a plaintiff has the obligation to present its case in full in its introductory writ of summons, which means that all the facts including the evidence which substantiates those facts have to be presented at the start of the proceedings. Dutch law does not provide a discovery/fact-finding process, unlike the legal systems of the United States and the United Kingdom.

The Dutch Code for Civil Procedure (DCCP) nevertheless provides both the court and the litigants with instruments for obtaining information on infringement. Two of those, Articles 1019a and 1019b DCCP, are implementations of the Enforcement Directive (Directive 2004/48/EC).

Any party may, during the proceedings or outside proceedings, at its own cost, request the disclosure of certain specific documents. Three requirements have to be fulfilled. First, the claimant can only demand specific records, which are within the control of the opposing party. This excludes, or at least restricts, so-called 'fishing expeditions'. Secondly, there must be a 'legal relationship' between the claimant and the opposing party. Article 1019 a(1) of the DCCP explicitly states that infringement of an intellectual property right is such a 'legal relationship'. Thirdly, the claimant will have to show that a proper administration of justice requires that the identified documents must be disclosed. Article 1019a(2) however states that in the event of an infringement of intellectual property rights, disclosure of other documents that can be used as proof may also be requested.

When proceedings are pending between the patentee and the infringer, the court may at any time during the proceedings order any party to substantiate its arguments and to submit documents. If a party refuses to submit the documents, the judge may rule as he deems appropriate.

5.2 Interim relief

An interim injunction is a frequently used weapon for IP rights holders in the Netherlands. In general an interim injunction will be granted under Dutch law if:

---

23   District Court of The Hague November 10 2010, IEPT20101110 (*Schering/Teva Pharma*).

- the claimant can show that the urgency requirement is met. Generally, the courts hold that a continuing (or threatening) infringement of intellectual property rights creates an urgent interest;
- it is likely that the validity of the intellectual property right and the (threat of) infringement thereof will be established in proceedings on the merits. This is the substantive check; and
- the case is suitable for interim injunction proceedings.

Ex parte injunctions (ie, without the defendants being present) are also possible under Dutch law. For an *ex parte* injunction to be granted, the infringement must be made evident to the court. A hearing may be scheduled after the injunction has come into effect.

In practice, the courts are reluctant to deem a case unsuitable for summary proceedings. The same holds true for lack of urgency in the case of an ongoing infringement of IP rights.[24]

If the above-mentioned criteria are met, the court will give a decision weighing the interests of all parties and taking into account all circumstances of the case.

5.3 'Springboard'/post-patent expiry injunctions

Previously, post-patent expiry injunctions have not been awarded in Dutch case law. There are, however, examples in which parties have claimed post-patent expiry injunctions. In the *Angiotech v Sahajanand*[25] case, for instance, Angiotech had asked the court for a moratorium on the patented stent for a period of three years after patent expiry, due to the fact that Sahajanand had used infringing clinical trials data in the process for receiving a CE-permit. The court, which judged the patent to be infringed by Sahajanand, nevertheless did not award the three-year post-patent expiry moratorium, because Angiotech had not sufficiently proved causality between the infringing clinical trials data and the (early) receiving by Sahajanand of a CE-permit.

5.4 Unjustified threats

Where it is shown in court that threats are unjustified, these will constitute a tort under the general unlawful act of Article 6:162 of the Dutch Civil Code. The threatened party will have a right to claim damages suffered due to the unjustified threat. There is extensive Dutch case law on the unlawfulness of statements or announcements made to third parties regarding infringing actions, where such actions were judged not to infringe.[26] Provided the IP right is still valid, however, enforcement of the right is lawful, unless the IP right holder knows, or should have known, that its right was invalid.[27]

---

24   Dutch Supreme Court June 29 2001, NJ 2001, 602 (*Impag v Hasbro*).
25   District Court of The Hague May 3 2006, HA ZA 05/2016 (*Angiotech/Sahajanand*).
26   For example: President District Court of Arnheim September 30 2010, LJN: BN9748.
27   For example: President District Court of The Hague, September 8 2010, IPET 20100908 (*Wins/Hema*).

5.5   Remedies

If the Dutch court establishes that an intellectual property right is infringed, it will order an injunction. If claimed by the IP rights holder, the injunction is likely to be subject to penalty sums that will be due in the event of non-compliance with the injunction.

Damages, or surrender of profits, may be awarded by the court. Furthermore, the court may order the infringer to provide information about the amount of infringing products sold, and the identity of purchasers; and it can subsequently order a recall of the infringing products and order the infringer to destroy the infringing products recalled and still in stock. Lastly, the IP rights holder will be awarded full compensation for all judicial costs made.

6.   Compulsory licensing

6.1   Introduction

Articles 57 to 60 of the DPA provide a number of grounds for ordering a compulsory licence. Such a licence will be awarded:

- for reasons of general interest;
- due to non-use; and
- due to dependency.

It should be noted here that successful applications for compulsory licences are extremely rare in the Netherlands.

6.2   Ground A: In the general interest

Section 1 of Article 57 of the DPA relates to granting a compulsory licence in the general interest. It should be noted that this article has not yet been put into practice.

As prescribed by Article 31(b) of the TRIPS agreement, it should first be ascertained whether or not the holder of the patent is willing to grant a licence. The patentee will be provided with the opportunity to voice his opinion in writing and, if necessary, orally. If the patentee is not willing to grant a licence, the Dutch Minister of Economic Affairs can grant one if there is a valid claim on the basis of the general interest. Further, if the patentee and the future licensee appointed by the minister do not reach consensus on the amount of royalties, the judge will determine this amount to be paid to the patentee on the basis of Section 6 of Article 58 of the DPA.

According to the parliamentary history of the DPA, the term 'general interest' may be interpreted broadly. Moreover, what 'general interest' exactly stands for is not described by law; the general interest comprises the policy objectives the Dutch government pursues.

In the *Cook v Fujinon*[28] case, Fujinon argued that an injunction would be contrary to the interest of public health. However, the District Court of The Hague was of the

---

28   District Court of The Hague March 17 1995, BIE 1999 nr. 110, p 443 *et seq.*

opinion that the presence of acceptable alternative products justified an injunction despite the interest of a public-health argument.

6.3 Ground B: Due to non-use

Section 2 of Article 57 of the DPA relates to granting a compulsory licence due to non-use. The condition for the grant of such a compulsory licence is that the patentee has not implemented the patent for a period of three years starting at the date of the publication of the grant of the patent.

An invention comprised by a patent qualifies as 'implemented' if the invention is applied by an operating device, in which the concerned product, or the concerned method, is applied in good faith and manufactured to a sufficient extent. Furthermore, Section 1 of Article 58 of the DPA requires the infringer first to request a licence with the patentee, after which court proceedings may be initiated.

6.4 Ground C: Due to a dependent position

It may be possible to obtain a compulsory licence on the ground that a licence is required for the exploitation of a younger patent falling under the scope of an older patent, in so far as this patent embodies a key technical advance of economic significance in relation to the invention claimed in the older patent.

Similar to the non-use licence, this 'younger patentee' is required first to approach the 'older patentee' seeking a licence on commercial terms. If such licence is refused, the 'younger patentee' may make an application to the court for a compulsory licence on terms set by the court.

It should be noted here that pursuant to the relevant provisions, a cross-licence should be granted by the 'younger patentee' to the 'older patentee'.

7. Ownership, inventors and compensation

In principle the creator or inventor of a work or invention will be entitled to the copyright, design right, or patent right thereon. In an employment relationship, however, this may be different. Under Articles 12(1) of the DPA, 3(8) of the Benelux Treaty on Intellectual Property (BTIP) and 7 of the Dutch Copyright Act (DCA), the employer will in principle be entitled to the relevant IP rights when one of its employees:

- has filed a patent application, and the nature of his employment entails the use of specific knowledge;
- has made a design while carrying out his duties; and
- has made certain literary, scientific or artistic works within the scope of his duties.

However, pursuant to Article 12(6) of the DPA, the inventor of the claimed subject matter comprised by a patent may be rewarded appropriately for not being the holder of the patent. In such a case, the salary of the inventor must be such that the inventor can be considered as not being sufficiently rewarded for his invention. In determining the latter, the financial interest of the invention, and the circumstances under which such invention is devised, have to be taken into account.[29] However, in most cases the salary of the inventor is considered to be sufficient;[30] if the inventor is rewarded, it is likely to take the form of an award of a financial bonus. It should be noted here that a claim pursuant to Article 12(6) of the DPA expires three years after the grant of the patent.

There is no similar mechanism for the creator of a work protected by copyright. As the employer will be considered the creator, it is possible that the identity of the real creator will be unknown. This is particularly the case where works have also been made public by the employer.

Also under the BTIP, the employer will be considered the owner of a design which has been made in the course of the employee's normal employment. As such, the employee has no rights to any additional remuneration.

8.   Branding and designs

8.1   Trade marks in life sciences

When choosing a name for a new medicinal product, producers are required to take into account the provisions of the BTIP, as well as the provisions of European and national legislation on the authorisation of pharmaceuticals (such as EC-Regulation 726/2004 and the Dutch Medicines Act).

This has some important implications: even though applicants for a market authorisation are required to use one single name for their medicinal product in the entire European Union, this may not be possible because of (older) registered trade mark rights.[31] Article 6 of Regulation 726/2004 therefore provides that if a name is already registered as a trade mark, the producer is free to use a different name for the same product. Producers may also be faced with problems in relation to the risk of confusion, whereby different standards are applied under the laws on market authorisation and trade mark laws. An example of this different perspective is the case concerning the pharmaceuticals Fluralin and Fluvirin, where the Amsterdam Court of Appeal found no risk of confusion even though the names would be too similar under the rules of the Dutch Medicines Evaluation Board, which states that the names of pharmaceuticals having a different active pharmaceutical ingredient (API) should at least have three letters different.[32]

In addition to the rules described above, the general trade-mark law rules still apply. This means that a trade mark must fulfil the general criteria mentioned in Article 2.1 of BTIP, these being that the mark must be (i) a sign, (ii) capable of being graphically represented and (iii) capable of being able to distinguish the goods. Also, the trade mark must not be contrary to public order, be misleading, or be confusing as regards well-known trade marks.[33]

---

29   Dutch Supreme Court March 1 2002, NJ 2003, 210 (*TNO/Ter Meulen*).
30   Dutch Supreme Court May 27 1994, NJ 1995, 136 (*Van Ginneken/Hupkens*).
31   Regulation 726/2004, Article 6.
32   Court of Appeal Amsterdam, February 23 2010, *Solvay v Novartis*, JGR 2010/18.
33   Article 2.4 of BTIP.

8.2 Designs

Designs for the packaging of medicinal products, as well as the product leaflet, should also be included in an application for a market authorisation.[34] Consequently, a design may meet all requirements to obtain a market authorisation, but it may still be unable to be registered as a Benelux or Community Design since different standards are applied.

In order to obtain registration in the Benelux countries or the European Union as a whole, the design must be novel and have an individual character. When disclosing the design for the grant of a market authorisation, applicants should be careful not to lose the novelty of their design.

8.3 Extending protection using trade marks and designs

Producers of pharmaceutical products can extend the protection of their products by registering as much detail as possible in relation to the given product as a trade mark or a design. Protection is also extended considerably if producers are able to register different trade marks for their product in different countries, as this is likely to have an impact on parallel imports of their products.

8.4 Protecting against unfair competition

In the Netherlands, unfair competition is considered a species under the general doctrine of unlawful act provided for in Article 6:162 of the Dutch Civil Code.

The most important example of unfair competition is the so-called 'slavish copying' of a competitor's product. This entails copying of the elements which are not essential, causing unnecessary confusion amongst the public and damage to the competitor. This claim is not limited in time, and therefore provides an important tool when the producer no longer has any IP rights in the product.

8.5 Comparative advertising and advertising restrictions

The Dutch Medicines Act prohibits advertising aimed at children,[35] advertising aimed at the general public wherein a pharmaceutical is recommended by professionals,[36] advertising for pharmaceuticals which do not have a market authorisation, as well as advertising aimed at the general public for pharmaceuticals which are only available on prescription.[37] The rules applicable on advertising aimed at professionals in the industry is less strict, even though the advertising still needs to give information on the composition of the pharmaceutical, the therapeutic indications, the contra-indications etc.[38]

Comparative advertising is allowed, albeit within the applicable boundaries. In a case concerning a comparative advertisement used by Wyeth, the Advertising Code Commission (a non-governmental body set up by the advertising sector itself) referred to the prohibition contained in Article 88 of the Dutch Medicines Act,

---

34 Article 3.7(n) of the Dutch Medicines Act Regulations.
35 Article 89(a) of the Dutch Medicines Act.
36 Article 89(b) of the Dutch Medicines Act.
37 Article 84 and Article 85(a) of the Dutch Medicines Act.
38 Article 91 of the Dutch Medicines Act.

stating that it was not allowed to claim better or equal potency in relation to another registered pharmaceutical.[39]

8.6 Protection of online content

Online content is generally protected by trade marks and designs on the one hand (on the name of the product, the producer's logo, etc) and copyrights on the look of the website on the other.

9. Protecting valuable information

9.1 Introduction

In the Netherlands, the protection of valuable information is provided under civil law as well as criminal law. To date, no specific rules on unfair competition exist under civil law (apart from a specific labour law provision, the scope of which is limited).

9.2 Trade secrets

Unfair competition and the protection of trade secrets are not the subject of explicit provisions under the Dutch Civil Code. The Dutch Supreme Court's decision in *Lindenbaum v Cohen*[40] firmly established the concept of unfair competition including trade secrets and their protection under the general provisions in the Dutch Civil Code on unlawful acts. In subsequent case law, the protection of trade secrets was dealt with in more detail, and it has matured into a well-established right that can be invoked in court proceedings. Case law has defined a number of further circumstances that may constitute an act of unfair competition and an unlawful act, notably in situations whereby a third party wilfully takes advantage of such violations of trade secrets and misappropriation of confidential business information.

In *Philip Morris v State of the Netherlands*,[41] the court addressed the following cumulative criteria for accepting that information may constitute a trade secret:

- the extent to which the information is known outside the company;
- the extent to which the information is known among employees and others within the company;
- the measures taken by the company to maintain secrecy of the information;
- the value of the information for the company and its competitors;
- the endeavours of the company to obtain or develop the information; and
- the difficulty for third parties to develop or discover the information, without the help of the company and in a legal manner.

In this decision, the court broadly defined 'trade secrets' as all the information that a company keeps secret for considerations of competition.

---

39   Advertising Code Commission, July 28 2009, 'Wyeth-commercial', published in *JGR* 2010/29.
40   Dutch Supreme Court December 31 1919, NJ 1919, 161 (*Lindenbaum/Cohen*).
41   District Court of The Hague December 21 2005, LJN: AU 8410 (*Philip Morris/Staat der Nederlanden*).

9.3   Preserving confidentiality

Company information, comprising trade secrets as described above, can generally be protected by a non-disclosure agreement. Under the law of the Netherlands, there is no prescribed form for such an agreement. A non-disclosure agreement may *inter alia* contain several relevant terms on the (sharing of) confidential information and/or a penalty clause when the agreement is violated.

(a)   *Employment law*

Within the scope of employment law, a contract of employment may contain a confidentiality clause, setting out how the employee should handle confidential company information during employment, and after termination of the employment. Confidentiality clauses are regularly invoked to combat competition by ex-employees, as such competition is often deemed to be unfair competition.

On the basis of Article 7:678, Section 2(i) of the Dutch Civil Code, an employee who discloses trade secrets of the company where he is employed can be dismissed with immediate effect.

In addition to the above, the employer may – in specific circumstances – invoke specific provisions with regard to patents, designs and copyright.

9.4   Repackaging of medicines – trade mark issues

Case law in the Netherlands follows the case law of the Court of Justice of the European Union in which it is determined that the trade mark holder cannot contest the repackaging of medicines (which has to meet certain cumulative criteria) if the said repackaging is necessary to market the goods in the importing member state. The necessity of repackaging merely relates to the fact that the product is being repackaged, and not to the method or style of this repackaging. These criteria, which are further developed in case law of the Court of Justice, boil down to the following:

- the repackaging may not affect the product;
- the names of the producer and the 'repackager' must be specified;
- the presentation of the repackaged product may not harm the reputation of the trade mark; and
- the trade mark holder must be informed prior to the repackaging and, if it so requires, should receive a sample of the product to be marketed.

9.5   The use of copyright or database rights to protect genetic or protein sequence information

Under the Dutch Copyright Act (DCA), all works which have an original character are protected, except when the work is so trivial that no creativity whatsoever is required to create the work in question.[42] The scope of protection of the DCA is also limited by the fact that technical characteristics of a work are excluded from protection.

Even though there is no case law on the protection of genetic or protein sequences under the DCA, the writers feel that it is very unlikely that the sequence as such will be considered protectable in the Netherlands under the DCA. This is

---

42   Dutch Supreme Court, *Endstra v Uitgeverij Nieuw Amsterdam*, May 30 2008, NJ 2008/55.

primarily because of the fact that displaying a sequence requires a technical effort instead of a creative effort, meaning that the sequence does not qualify for copyright protection. This does not, however, exclude the protection of the database containing the sequence(s) under the DCA.

The chances of protection of the collection of sequences are greater under the Dutch Database Act (DDA). According to Article 1(a) of the DDA, every collection of works, data or other independent elements qualifies as a database if it has been organised systematically, if the elements are independently accessible, and if the collection, control or presentation of the elements has required a substantial investment. The producer of a database which meets these criteria owns the database rights to that database. As there are no requirements for the contents of any such database, the authors believe that it is possible to obtain protection for a database which contains genetic or protein sequences, provided that the database meets the above-mentioned conditions.

A database-right confers on the producer of the database protection against the following acts:

- downloading or reusing the whole, or a qualitatively or quantitatively substantial part of, the contents of the database; and
- repeatedly and systematically downloading or reusing a non-substantial part of the database, in so far as this is contrary to the normal exploitation thereof or this causes unjustifiable damage to the legitimate interests of the producer.

It should be noted that a database right only provides protection for the database as such, and not for the contents of the database. Strictly speaking, a database right cannot therefore protect a genetic or protein sequence.

10. Counterfeiting

10.1 Administrative and criminal penalties
Violation of the rules applicable to advertising as set out in the Dutch Medicines Act are punishable by a fine of up to €450,000.

The Dutch Criminal Code also stipulates penalties for persons who display, import, export, offer for sale, sell, deliver, distribute or store goods which bear false trade marks, false trade names, false indications of origin, or imitations of trade marks and designs. Persons who are found guilty of one of these actions can be sentenced to imprisonment for a maximum of one year, or a fine up to a maximum of €76,000.[43] Should the person act professionally, the maximum term for imprisonment is raised to four years.[44] These provisions are rarely applied, however, and normally only in the context of another criminal procedure.

10.2 Border detentions
In addition to the administrative and criminal penalties provided above, counterfeits

---

43   Article 337(1) of the Dutch Criminal Code.
44   Article 337(3) of the Dutch Criminal Code.

are also handled by a pan-European harmonised system for border detentions in Europe based on the so-called Anti-Piracy Regulation (Regulation 2003/1383/EC). Under this system, all EU customs authorities have the authority to detain goods which are suspected of infringing intellectual property rights *ex officio*, or following an application for action by the customs authorities which can be filed by the owner of the right concerned.

Dutch customs authorities are particularly active, and frequently act to detain shipments. As soon as Customs detains a shipment of goods, the addressee has the option of handing over the goods and offering them for destruction. If the addressee is not willing to do so, the owner of the right must start infringement proceedings in a court in the country where the goods have been destined for delivery.

11. Collaborative models – 'open innovation'

Increasingly in recent years, the Dutch government has encouraged 'open innovation' initiatives such as the Dutch 'innovatieplatform' (www.openinnovatie.nl) and Top Institute Pharma. Multinationals such as Philips are also increasingly trying to cover broader research fields through close cooperation with other private companies and universities.

When setting up open innovation initiatives, it is important to try to avoid possible problems, for example by agreeing to clear terms relating to IP licensing and sharing. The same goes for the choice of partners. Valuable technology does not always make a valuable partner.

Because open innovation is still in its infancy in the Netherlands, the writers refer the reader to the issues raised in other chapters of this publication relating to other jurisdictions.

12. Hot topics

12.1 *Ex parte* request under the Dutch Code of Civil Procedure – filing protective letters

Pursuant to Article 1019e of the Dutch Code of Civil Procedure, a party can ask for an *ex parte* interim injunction. This means that the interim-relief judge has the authority to issue an immediate injunction without the alleged infringer being heard. This far-reaching measure is in fact a violation of the system of 'checks and balances' and of the right to be heard in court. In order to address this problem, following the grant of an *ex parte* interim injunction the alleged infringer may file summary proceedings with the interim-relief judge to have the *ex parte* decision reviewed, and the alleged infringer will be heard at this stage.

Moreover, in order to mitigate the unfavourable position of the defendant, it is possible to file protective letters, based on the use of so-called *Schutzschriften*, as developed and generally accepted in Germany. Certain district courts have permitted a party to explain its objections against the allowance of an *ex parte* injunction in a protective letter. This option is not limited to patent law but covers all IP rights. Filing of a protective letter is confidential and will not be seen by the potential applicant.

Generally, a protective letter contains:
- a list of potential applicants and respondents;
- material grounds for non-infringement;
- material grounds for invalidity; and
- an overall consideration of all circumstances and the interests of the parties.

In this way, even though the court will not hear the defendant in *ex parte* proceedings, a defendant has the possibility of informing the court of his non-infringement and invalidity arguments. On the basis of this information, the court may then decide on the request for an *ex parte* injunction.

Russia

Alexander Christophoroff
Julianna Tabastajewa
Vladislav Ugryumov
Yulia Yarnykh
Gowlings International Inc

1. Small molecules

1.1 Product and process claims

Two types of patents coexist in Russia:

- Russian national patents that are granted under Part IV of the Civil Code (2008); and
- Eurasian regional patents which are effective in Russia and eight other contracting states that are granted under the Eurasian Patent Convention (1995).

Although in this chapter patentability issues are analysed in the context of the Russian Federation Civil Code, in general the provisions of the Eurasian law are very similar or even identical.

The legal rights for an invention in Russia are certified by a patent issued by the Russian Patent Office (hereinafter, the RUPTO), and are effective for 20 years from the application filing date. The term of a patent for a medicament, pesticide or agrochemical may be further extended for a period not exceeding five years. Under the Civil Code, an invention shall be granted legal protection if it is novel, involves an inventive step and is industrially applicable.

Current practice in the Russian Federation provides sufficiently broad opportunities for patent protection of inventions related to medical applications.

Claims relating to compounds with a defined structure should include the structural formula asserted by known techniques and physicochemical constants. The method by which the compound was obtained and its intended use should also be disclosed in the specification.

If the invention is directed to a composition, then the specification should disclose the ingredients included in the composition, their characteristics and quantitative content. The method of obtaining the composition should be disclosed. If a composition comprises a novel compound, then the process for obtaining such compound should also be disclosed. A claim drawn to a pharmaceutical composition should comprise an indication of an intended use.

Much attention in the RUPTO is paid to the conformity of pharmaceutical inventions to a requirement of industrial applicability. On the one hand, under the Civil Code the invention should be disclosed to an extent enabling it to be used for treating a human body. On the other hand, applications for such inventions are filed long before the clinical trials of the claimed products have been completed.

The general approach of the RUPTO is that if the invention is directed to known medicaments used for new indications (see below), then the specification should provide examples showing the usability of the product for the claimed indication and an explanation of the new effect of the product based on traditional medical knowledge and nosotropic or physiological processes in the body. If the invention is directed to new medicaments, then the specification should provide the information on the therapeutic activity of the product, its effects and selectivity using adequate models.

As regards the scope of a claim drawn to a manufacturing process, under item 2(2) of Article 1358 of the Civil Code, such claim would impart indirect protection to a product obtained directly by means of the patented process. Thus, if the product obtained by the patented process is new, an identical product shall *prima facie* be considered to be obtained by the patented process.

It is noteworthy that under Eurasian Patent Office (EAPO) legislation a claim drawn to a known product characterised through a method for manufacturing thereof is explicitly prohibited.

1.2   Scope of protection and Markush claims

Although the Civil Code, part IV, provides for admissibility of use of the doctrine of equivalents, in considering the case of an alleged infringement no clear-cut guidelines on how to apply the doctrine of equivalents exist. In the course of an infringement lawsuit, the question of whether a patent is used or not in an allegedly infringing product or process is to be answered by an expert appointed by the court. The expert is to address the issue of whether every feature of an independent claim or a feature equivalent thereto is used, or not. Generally speaking, while assessing equivalency of the features of a patent and those of an allegedly infringing product or process, the expert shall take into consideration such aspects as the function performed by an alleged equivalent, the technical effect attained, and the interchangeability of the allegedly equivalent features. Also, an equivalent feature shall not impart any new utility property or shall not offer any substantial advantage over the product or process being copied.

The RUPTO is familiar with applications that are directed to generic chemical compounds that have common essential structural elements, exhibit similar activity and are produced according to a common scheme. The first claim of the so-called Markush formula normally claims the whole group of chemical compounds covered by the general chemical structure, which may in principle encompass an unlimited number of individual compounds. One of the main requirements for Markush claims envisaged by the authors' practice is that all the values of the radicals should be defined and that all claimed substances should be novel and have the same biological activity. According to the practice, an application should contain examples of obtaining chemical substances encompassed by Markush claims as well as examples demonstrating the biological activity of the claimed compounds. The common approach to examining Markush formulae is that the values of at least the radicals having the same 'chemical nature' should be supported by at least one example in the specification. For example, to support C1-4 alkyl, it is sufficient to provide an example of the synthesis of the substance with any of methyl, ethyl, propyl or butyl.

1.3 Extension of pharmaceutical and agrochemical patents

Both Russian and Eurasian patents for pharmaceutical and agrochemical substances can be extended over their standard 20-year term for up to five years as a matter of compensation for time spent to get the first marketing authorisation.

It should be noted that Eurasian patents can be extended on a country-by-country basis to the extent that the national legislation of a given country permits this.

The conditions for the extension of a Eurasian patent differ slightly with regard to different Eurasian countries, but with regard to Russia they are the same as for Russian national patents, namely:

- more than five years must have elapsed between the filing date of an application for an invention relating to a medication, pesticide or agrochemical, the use of which requires authorisation, and the date of the first marketing authorisation; and
- the request for extension of the term must be submitted by the patent owner during the term of the patent's validity within six months after obtaining the first authorisation or of the patent grant, whichever comes last.

The term shall be extended for a period counted from the filing date of the patent to the date of the first authorisation minus five years, with the proviso that the term of the patent may not be extended for a period exceeding five years.

2. Second-generation inventions

Both the Russian and the Eurasian patent protection can be extended by patenting second-generation inventions such as combinations, selection inventions, methods of use and secondary indications, methods of treatment, formulations and physical forms.

All such second-generation inventions are treated as usual inventions and should be patentable (ie, they should in particular have an inventive step (and thus be non-obvious) over the principal invention in combination with other prior art. Therefore, enantiomers are normally not patentable over one another unless they demonstrate unexpected features.

If a second-generation invention is patented by a third party, it is still an open issue in Russia what rights such a patent will give *vis-à-vis* the respective principal patent.

2.1 Methods of use and secondary indications

Although lacking explicit support in the Civil Code, use claims (eg, German-type or Swiss-type claims) may also find protection in regard to pharmaceuticals.

Generally, first-medical-use claims are not allowable. That is, any claim to use a substance as a medicament should necessarily contain a medical indication, which may be expressed in various forms – most commonly in the form of a specific disease or condition. Otherwise, 'use of substance X as a medicament' would be considered as an undue generalisation lacking support in the specification, which would lead to rejection on the grounds of lack of industrial applicability.

Chemical compounds (including metabolites) and compositions for use in the treatment, diagnosis and prevention of a disease in humans and animals are considered as patentable. Where the compound is intended to be used for the treatment, diagnosis or prevention of a disease in humans and animals, reliable data should be provided corroborating its suitability for the claimed purpose and, in particular, should include information on the impact of the compound on certain steps of the physiological or pathological processes.

2.2   Methods of treatment

Methods of treatment or prevention of a disease in humans and animals and diagnostic methods, as well as products (substances or compositions) intended for this purpose, are all regarded as patentable inventions. In contrast to the EPO, the RUPTO recognises such methods as industrially applicable since they may be used in medical and veterinary applications, and thus are not directly excluded from protection under the Civil Code. The criteria for such 'industrial applicability' under Russian practice are wide-ranging and allow the possibility, in principle, for use in any field of activity and are not limited only to the industrial or commercial use of an invention.

An invention directed to a method of treatment, diagnosis or prevention of a disease in humans or animals should be supported by information on compounds used in the claimed method, factors exerting an impact on the etiopathogenesis of the disease or other authentic data corroborating the suitability of the claimed method for the treatment, diagnosis or prevention of the indicated condition or disease.

3.   DNA, biologicals and personalised medicine

3.1   Nucleic acids and proteins

The patentability of nucleic acids isolated from their natural environment or produced by means of a technical process (even where such products occur naturally) will generally be recognised in the Russian Federation.

The term 'natural environment' encompasses plants, animals and micro-organisms. Elements isolated from the human body or otherwise produced by means of a technical process are not regulated separately under the existing legislation, and are deemed to constitute a patentable invention.

In contrast to other chemical substances, for all nucleic acids and proteins the biological function defining the intended use should not only be fully described in the specification but should also be explicitly recited in the claims unless it follows in an obvious manner from the name of the claimed subject matter. The mere isolation of a nucleic acid or a protein and identification of its sequence without an indication of its function (which should be substantial, specific and credible) are not considered to constitute a patentable invention, due to failure to comply with the industrial-applicability requirement.

The Administrative Regulation currently in force provides for a list of the technical features that should be used to characterise isolated nucleic acids and proteins and recombinant nucleotide and amino acid sequences (including vectors and plasmids, fusions and chimeric proteins etc) in the most appropriate way. However, it should be noted that this list is neither obligatory nor exhaustive. The key point is to enable one to distinguish the claimed subject matter from prior art.

For nucleic acids and peptides and proteins with identified or partially identified structures, their nucleotide or amino acid sequences should be shown in the sequence listing. The sequence listing must meet the requirements of WIPO Standard ST.25 and must be presented in printed and computer-readable formats. Additionally, the applicant must present a declaration on the identity of the information in printed and computer-readable formats. Claims for such products should contain a reference to specific sequences from the sequence listing in the form of 'SEQ ID NO(s) ...'

In many cases, the RUPTO may recognise the validity of the claims not only for one specific nucleotide or amino acid sequence, but also for other sequences (eg, complementary or degenerate sequences or those possessing functional equivalence and structural homology). A group of such products may be protected under one independent claim if they share the same function and similar sequences. Such a group may be limited to several specific sequences, or the scope of protection may be broadened by including such features as 'variants', 'fragments', 'hybridising sequences', 'homologous and complementary sequences' etc, subject to the fact that examples supporting those features are disclosed in the specification.

As for genetic constructs, the most detailed technical characterisation thereof should include the description of all the nucleotide fragments (including nucleotide sequences), their interrelation, and the type of linkage. However, such precise characterisation is normally not required, taking into account that commercially available vector backbones are often used for creating genetic constructs. In such a case it is sufficient to indicate the name of the commercial vector and to define the inventive nucleotide fragment incorporated therein.

Claims related to antibodies should include reference to an antigen against which said antibody has been raised. In the case of monoclonal antibodies, a deposit of the hybridoma in accordance with the Budapest Treaty is generally required.

3.2 Stem cells and other organic material

While patentability of biotechnological inventions has generally been acknowledged, the Civil Code contains an exhaustive list of objects that cannot be subject to patent protection that includes: processes for cloning human beings; processes for modifying the germ line genetic identity of human beings; uses of human embryos for industrial or commercial purposes; and other solutions contrary to *ordre public* or morality.[1]

Additionally, the law contains objects that are not classified as inventions and these include: "discoveries, as well as scientific theories and mathematical methods,

---

1   Civil Code, Part 4, Art 1349 para 4.

proposals concerning solely the exterior appearance of manufactured articles and intended to satisfy aesthetic requirements, rules and methods of games, intellectual or business activities, computer software, ideas on presentation of information...".[2]

Furthermore, no legal protection as an invention shall be granted to "plant and animal varieties, and biological methods for producing therefrom except for microbiological methods and products produced by such methods".[3]

(a) Transformed cells, plants and animals

Stably transformed cells of micro-organisms, plants and animals that are classified as 'genetic constructs' have also been recognised as patentable inventions in the Russian Federation.

Since in most patent applications these subject matters are characterised by the presence of the transforming genetic element, a description of such genetic element is required. Typically, the description includes characteristics of the nucleotide sequence and its encoded products. Other features that may be used for a transformant include: features or properties acquired thereby; its origin; taxonomic properties; mutation of the natural genome; and cell culturing conditions.

As for a cell, the protection may be provided to a cell as such, eukaryotic or prokaryotic cell, micro-organism (eg, bacteria, yeast), family, genus or species of the micro-organism or specific cell strain (line).

3.3 Copyright and sequence information

Sequence information is not protectable by copyright in Russia.

3.4 Sufficiency issues

The breadth of the scope depends on the intended use of the invention and the support in the specification.

Depositing micro-organism strains can often be considered to be necessary. Under the Civil Code, the possibility of carrying out the invention for a cell strain, or a method in which it is used, shall be confirmed by means of a description of the method of producing the strain. If such description is not sufficient for producing the strain, a deposit of the strain in accordance with the Budapest Treaty is required, the date of which should precede the priority date of the application.

Patentability of transgenic plants and animals is acknowledged under the Russian Civil Code if the technical feasibility of the invention is not confined to a particular plant or animal variety. Varieties are protected under a separate section of the Civil Code (Chapter 73), under the title 'Right for a selection achievement'.

4. Acts of patent infringement

4.1 Infringement

Patent infringement for both Russian and Eurasian patents is determined on the

---

2 Civil Code, Part 4, Art 1350 para 5.
3 Civil Code, Part 4, Art 1350 para 6.

basis of either literal use of every feature of a particular claim, or the application of the theory of equivalents. Only independent claims can be used for determining the infringement. The specification can be referred to for interpretation of the independent claim.

According to the Civil Code, acts of infringement are the production, use, sale, offer for sale, importation, other introduction into circulation, or storage with one of those purposes of a product that falls under a product patent claim, or automatically performs the process that falls under a process patent claim, or performing the process that falls under a process patent claim.

Product-by-process protection is provided. If the product of the patented process is new, there is also a presumption that the defendant's product is obtained by means of the patented process.

4.2 Contributory infringement

There are no provisions on contributory infringement and, therefore, only those who add the last feature to a product or carry out the last operation of the process are normally considered infringers. Others might hypothetically be considered as culpable participators under the criminal law if applicable; however, no such cases are known.

4.3 National treatment of 'Bolar'-type provisions

For both Russian and Eurasian patents, limited exceptions from patent rights are provided. Principal exceptions relate to:

- transportation vehicles temporarily or accidentally entering the country (applicable on a reciprocity basis);
- scientific research or experimental use ('Bolar'-type protection);
- emergency situations (natural calamities, catastrophes, accidents), with payment of reasonable compensation to the patent owner;
- non-commercial use for private purposes;
- occasional preparation in pharmacies based on physicians' prescriptions; and
- exhaustion of rights (on a national basis).

4.4 Experimental-use exemptions

Submitting for authorisations and offers to supply are deemed to be infringing acts, but preparation for submission for authorisation is deemed to be scientific research or experimental use according to a recent case and can be conducted before the expiration of a patent, while actual submission is only allowed after expiration of a patent. There are no known cases on the other possible transitional situations (eg, an offer to supply after patent expiry made before patent expiry).

4.5 Summaries of product characteristics (SmPCs)

A statement of intended use in SmPCs will normally be considered as a feature of a claimed purpose of the patented product or process, but not as a claimed operation in the patented process.

5. Patent enforcement

Patent enforcement litigation in Russia (for both Russian and Eurasian patents) is generally quick, fair and equal for both Russian and foreign parties. However, the burden of proof for the plaintiff in litigation is high, as the plaintiff must prove that the infringement took place. The defendant, on the other hand, does not need to prove the opposite and, in cases of reasonable doubt, can simply deny the plaintiff's allegations.

In addition, or as an alternative to civil litigation, there are also other enforcement possibilities through administrative agencies (eg, Customs or antitrust authorities) and/or the criminal law. Administrative agencies concentrate on injunctions and do not award damages, and their decisions can be challenged in the court.

5.1 Obtaining information on the infringer and the infringement

Some limited assistance in obtaining information on the infringement is available from the court. This includes court orders to produce information or documents, which are usually addressed to third parties but sometimes to the defendant. In order to obtain such orders, the plaintiff must demonstrate that the information or documents are necessary to resolve the current case and impossible to get without the court's involvement. Usually a previously unsuccessful attempt to obtain the requested information or documents will be required.

5.2 Interim relief

Interim relief is available in the form of temporary injunctions; however, the burden of proof is of necessity quite high, so the courts have often been reluctant to go down this route.

In order to obtain an interim injunction, the plaintiff must demonstrate that failure to order such an injunction may lead either to unenforceability of the prospective decision by the court, or to gross damage to the plaintiff's business. In practice, judges tend to rule that an interim injunction should not replace the outcome of the case. Petitions for interim injunctions are heard *ex parte* within one day of being filed.

Plaintiffs seeking an interim injunction may be required to post a bond.

'Springboard'/post-patent-expiry injunctions are not dealt with in the legislation, and there are no known cases on this issue.

5.3 Unjustified threats

Russia has not adopted any legislation in relation to unjustified or unfounded threats by a patent owner, although such threats may hypothetically be considered as unfair competition if made between actual competitors in the market. There are no known cases on this issue. However, if unjustified threats are sent to business partners (eg, distributors), rather than to the source of the product, this will be considered unfair competition in relation to the source of the product.

5.4   Remedies

The principal remedies available in Russia are injunctions and damages. On receiving a decision of the court confirming patent infringement, it is also possible to cancel the medical registration of the infringing pharmaceutical product.

6.   Compulsory licensing

The Russian Civil Code provides that compulsory licensing will be granted:

- by the court, at the request of a person willing and ready to use an invention when it is not used, or insufficiently used, by the patent holder for a period of four years after the date of the patent issue and if such non-use or insufficient use results in a lack of goods, works or services in the market and if the patent holder has refused to grant a licence on terms commonly used in practice;
- by the court, at the request of an owner of a second patent which cannot be used without simultaneous use of the first patent;
- by the Government of the Russian Federation, in the interests of national security.

All compulsory licensing should be accompanied by fair remuneration.

No such compulsory licences have been issued to date. However, courts tend to give some protection to the owners of second patents where they are sued by the owners of first patents, and this is currently an open issue and a hot topic.

7.   Ownership, inventors and compensation

Unless otherwise provided for by a contract between the employee and the employer, the arrangements set out next apply.

Primary ownership of a prospective patent right for an invention created by an employee in the course of his employment (based on general responsibilities and/or specific orders) and thus considered an in-service invention will belong to the employer. The employee must notify the employer in writing of the creation of such invention.

Where the employer, within four months from the date of notification by employee, fails to file a patent application, or to transfer the right to another person, or to inform the employee on keeping the relevant subject matter as a trade secret, the right to file a patent application and to obtain a patent moves to the employee. In such cases, the employer retains a limited right to use the invention.

Where the employer obtains a patent, or transfers the right to another person, or decides to keep the relevant subject matter as a trade secret, the employee inventor is entitled to payment.

An invention created by an employee with the use of financial, technical or other material assets of the employer, but not in the line of his usual responsibilities or specific orders given by the employer, is not considered an in-service invention and belongs to the employee. However, the employer will have limited rights to use such an invention, or to obtain compensation for the resources used.

When an invention is made in the course of a contract other than an employment contract, the distribution of rights is generally defined by the contract, with some rules being provided by law in cases where the parties have failed to agree on this.

8. Branding and designs

8.1 Trade marks in life sciences

Similar to other jurisdictions, trade marks are very often registered with the RUPTO to extend protection of patents for inventions and designs. Since, on expiration of the patent protection, a trade mark remains the only IP right of the owner, civil disputes involving trade marks are quite common in Russian court practice. There have been several court cases in recent years where the producers of medicines have tried to obtain cancellation of a competitor's trade mark on a loss-of-distinctiveness basis in order to start volume production of the medicine under a famous trade mark without paying any royalties to the trade mark owner.

In *Bayer AG v The Russian Patent Office*[4] the Russian Supreme Arbitration Court overturned the decisions of the lower courts which had upheld the decision of the RUPTO by which the legal protection of the trade mark, Aspirin, had been cancelled. The Supreme Arbitration Court found that the lower courts had erred in finding that the trade mark had lost its distinctive character and become genericised.

8.2 Protection using unfair competition

Use of third-party IP rights is regulated by the law 'On Advertising' and is monitored and supervised by the Federal Antimonopoly Service of the Russian Federation. In general, the law prohibits unfair and inaccurate advertising which may mislead consumers, damage a competitor's reputation or otherwise constitute an act of unfair competition. Comparative advertising is not permissible, as it is regarded as being unethical, containing as it might incorrect comparisons of the advertised products with the products in circulation manufactured or sold by other parties. Another type of unfair advertising not admissible by the law is when the product, the advertising of which is prohibited by a certain means, in a certain place or at a certain time, is advertised under the image of another product whose trade mark is identical or confusingly similar to the trade mark which is subject to the advertising restrictions.

The law 'On Competition Protection' prohibits unfair competition, including:
- disseminating false or inaccurate information which may impair a legal entity's business reputation or inflict damage thereon;
- misinformation with regard to the manufacturer, the nature, method or place of production or the product's usability, quality or quantity;
- incorrect comparison between the legal entity's products with the products manufactured by other legal entities;
- sale, exchange or other placing in the market of the products associated with the use of another legal entity's intellectual property or means of individualisation of the products, works or services; and

---

4   Ruling of the Supreme Arbitration Court of the Russian Federation No 6567/97 dated August 11 1998.

- illegal acquisition, use or disclosure of information comprising commercial, trade or other proprietary secrets.

Unfair competition related to the acquisition and use of the exclusive right to the legal entity's means of individualisation is not admissible. Legal protection of an IP right can be challenged and further annulled if it was obtained in violation of the requirements of the law 'On Competition Protection'.

8.3 Protection of online content

Under the Russian Civil Code, online content constitutes the object of copyright, which subsists in any qualifying works upon its creation without a requirement to obtain a registration. The basic term of protection is 70 years and the start of the term depends upon the particular circumstances.

9. Protecting valuable information

9.1 Preserving confidentiality

The law 'On Commercial Secrets' provides that the proprietor of the information is entitled to consider such information as information comprising a commercial secret. The owner at his own discretion can define the list and composition of such information. The law further defines 'know-how' as information of any type (industrial, technical, economic, organisational, etc) without restriction as to subject matter.

For information to be protected:
- it must have actual or potential value due to the fact that it is unknown to third parties;
- it must not be freely accessible;
- the holder must have taken all required measures to preserve its confidentiality; and
- its confidentiality would enable its holder to increase revenue and avoid unnecessary expenses.

There is no obligation or possibility to register know-how, unless the owner wishes to obtain alternative protection in the form of a patent for an invention, a utility model or such like.

This right survives for as long as confidentiality is maintained.

The holder of information comprising a commercial secret must take the following measures to preserve its confidentiality:
- draft a list of commercial secrets;
- restrict access to the commercial secrets;
- keep a record of persons having access to such information;
- include provisions regulating commercial secrets in employment contracts and civil agreements; and
- mark hard copies (documents) containing information which comprises a commercial secret with the words 'Commercial Secret'.

9.2   Know-how and ex-employees

As part of the employment relationship, an employer must ensure that its employees are:

- familiar with the list of commercial secrets;
- aware of the commercial-secret regime and any liability for violation thereof; and
- provided with the conditions necessary for observing the commercial-secret regime.

The Civil Code stipulates that an employee who has obtained access to the employer's commercial secret must keep such information confidential until the exclusive right thereto terminates. Therefore, labour contracts will usually provide for the employee's obligation to preserve the confidentiality of any information which becomes known to him in the course of discharging his work responsibilities. Such provision will remain enforceable during the term of the labour contract. In accordance with the Labour Code in Russia, an employee can be held liable for disclosing information which comprises a commercial secret (full financial liability and dismissal). In certain cases, disclosure of confidential information results in criminal liability.

Labour contracts can also provide for an obligation of confidentiality for a period of time following termination of the employment contract. However, in practice such provisions of the contract are rarely enforceable and, in fact, employers are not adequately protected by applicable law against the disclosure of confidential information by former employees. The only exception to this practice is where an employee is held liable under the Criminal Code, but this is relevant only where the employer is able to demonstrate large-scale damage arising from the disclosure of confidential information.

10.   Counterfeiting

Counterfeiting of medicines is one of the most serious and dangerous problems in Russia's pharmaceutical market. The Government of Russia provides weak enforcement against producers of counterfeit medicines. Counterfeit products currently represent 12% to 15% of the market, the vast majority of which is produced by local manufacturers. Russian law does not specifically criminalise pharmaceutical counterfeiting and injunction measures are not applied. A definition of a 'pharmaceutical counterfeit' was introduced in the Law on Medicines in August 2004 and was later kept in the Law on Medicine Circulation enacted in September 2010; however, no related prosecution articles are available in the criminal and civil legislation. In the same way, there is no procedure for evidence gathering or acceptance by the courts to facilitate court proceedings in counterfeiting cases.

The main article of Russian legislation currently applicable in cases of pharmaceutical counterfeits is that which addresses trade mark infringement. However, the Criminal Code applies only in cases of numerous violations, or those involving significant damages – and even in those cases where the Criminal Code applies, the penalties are inadequate ($4,000 to $33,000 maximum). The penalty set in the Administrative Violations Code is even lower ($1,300 maximum). The Russian parliament has been debating a potential increase in criminal and administrative liabilities for several years, but nothing has been done so far.

11. Collaborative models – 'open innovation'

There are no specific legal instruments for open innovation in Russia.

12. Hot topics

12.1 The 2010 law on the circulation of medicines

Law No 61-FZ 'On Circulation of Medicines' (the Law) was adopted on April 12 2010 and came into effect on September 1 2010, replacing Federal Law No 86-FZ 'On Medicines' of June 22 1998 (the 'Law on Medicine'). Since its adoption a little more than a year ago at the time of writing, the Law has already undergone three amendments.

According to the Minister of Healthcare and Social Development (the MoH), the adoption of the new Law was prompted by the prevailing economic situation, which differed significantly from that which existed 11 years previously when the Law on Medicines was enacted.

In addition to the detailed regulation of drug approval, the Law establishes the procedure for clinical trials and expert examination of the medicine to be conducted for the approval of drugs. The Law introduces a uniform state duty payable to the federal budget, and state control over prices for essential medicines.

The process of marketing approval is transparent, with the relevant information published on the website of the MoH which is now responsible for medicine approval. The Federal Service for the Supervision of Public Health and Social Development (*Rosdravnadzor*), which was in charge of the approval of medicines under the Law on Medicine, is now responsible, *inter alia*, for the supervision of conducting medicine clinical trials and organisation of medicine safety monitoring.

The Law imposes a duty on the MoH to keep state registries of approved medicines, medical organisations accredited to conduct clinical trials, issued authorisations for clinical trials, investigators qualified for conducting clinical trials, and maximum prices of medicines. As at the time of writing (mid-2011), no such registers have been placed on the MoH's website. According to the MoH representative, the registers will be made available to the public early in 2012.

The Law establishes the general framework of legal requirements applicable to the circulation of medicines including development, pre-clinical and clinical trials, examination, marketing approval, compliance, quality control, advertising, production, preparation, export/import, distribution, application and destruction of medicines. The Law also regulates the circulation of narcotic medicines, psychotropic substances and radioactive medicines in accordance with specific regulations.

The Law sets out the most fundamental principles in the pharmaceutical area, while it is envisaged that certain regulations (such as approval procedures, conduct of clinical trials, etc) will be developed in greater detail by way of adoption of corresponding resolutions by the Russian Government or MOH's orders, administrative rules and similar legal acts. Although certain subordinate legal acts have already been adopted, other regulations (eg, the Rules of Good Clinical Practice) are still to be drafted and/or enacted.

Although the new Law was intended to create an adequate solution to the issues of the safety of medicines, price regulation and control, and procedures for the introduction of medicines to the pharmaceutical market, as well as to close the legal loopholes with regard to the approval of medicines, it lacks clarity in many respects. Unfortunately, there remains much ambiguity which, together with the lack of detailed regulation of certain issues, the absence of court practice and the inability of the government authorities to provide clear answers to questions arising in the course of application of the Law, makes it difficult for the pharmaceutical companies to operate smoothly.

Spain

Miquel Montañá
Clifford Chance

1. Small molecules

1.1 Product and process claims

The First Transitory Provision of Law 11/1986, of March 20 1986, on Patents (the Patent Act) established that inventions of chemical and pharmaceutical products would not be patentable before October 7 1992. When Spain deposited the Instrument of Adhesion to the European Patent Convention (EPC), it formulated a reservation stating that European patents, in so far as they confer protection on chemical or pharmaceutical products as such, will not have any effect in Spain. As a result, for years the level of protection of chemical and pharmaceutical products in Spain lagged behind neighbouring countries such as France, Germany or Italy, which had accepted the patentability of pharmaceutical products in 1968 and 1978.

The case of Italy is particularly interesting, since in 1978 the Constitutional Court annulled the provisions of the law that prohibited the patentability of pharmaceutical products on the grounds that those provisions were discriminating against research and development in the pharmaceutical sector. In Spain, this historical grievance was corrected by the Supreme Court (Administrative Chamber) in four judgments handed down on November 4 2010, in which the Supreme Court admitted the addition of product claims to patents filed before October 7 1992.

In addition to product claims, process claims and 'Swiss'-type claims are also patentable, although the latter are not explicitly mentioned in the Patent Act. Historically, this created some confusion, for the Spanish Patents and Trade Marks Office (SPTO) took the view that Swiss-type claims were also prohibited by the First Transitory Provision of the Patent Act. This interpretation was corrected by the High Court of Justice of Catalonia (Administrative Chamber) in a judgment dated March 4 1998, where it concluded that only product claims fell within the scope of that provision. This was further confirmed by the Court of Appeal of Madrid (Commercial Chamber) in a judgment dated October 26 2006.

1.2 Scope of protection of claims and Markush formulae

The scope of protection of patent claims is governed by Article 60.1 of the Patent Act, which establishes that "the scope of protection conferred by the patent or by the patent application is determined by the contents of the claims. The description and the drawings serve, however, to interpret the claims." Prior to the entry into force of the Revision Act 2000 of the EPC, it was argued that Article 60.1 of the Patent Act was further away from literal interpretation than Article 69 of the EPC. The reason was that whereas the former uses the words "contents of the claims", the latter stated "the terms of the claims". However, this subtle difference has blurred since December 13 2007, when the aforementioned Revision Act came into force, as it has deleted the words "the terms of" from the text of Article 69 of the EPC. This, coupled with the introduction of Article 2 in the Protocol of Interpretation of Article 69 of the EPC, which now requires the interpreter to take equivalent elements into account, has prompted the balance to turn a little further away from literal interpretation.

In some cases, there has been debate as to the scope of protection of Markush formulae. For example, in the ciprofloxacin saga, the Court of Appeal of Barcelona took the view that in the event of contradiction between Markush formulae and a compound named in the patent, the Markush formulae must prevail (eg, judgment dated January 2 2000 in *Bayer AG v Chemo Ibérica et altri*). Another question that has arisen is what is the scope of protection of Markush formulae that may cover chiral compounds when flat lines are used (ie, when the stereochemistry is not depicted). In its decision of November 30 2010 handed down in the rivastigmine case (*Proterra AG et altri v Cinfa et altri*), Commercial Court Number 1 of Barcelona took the view that unless the claim disregards a specific form, Markush formulae using flat lines would cover all possible stereochemistry forms.

The array of rights conferred by the patent is governed by Article 50 of the Patent Act:

> *The patent confers on its proprietor the right to prevent any third party who does not have the proprietor's consent from:*
>
> *(a) Manufacturing, offering, introducing on the market, and using a product which is the subject matter of the patent or importing or possessing such product for any of the purposes mentioned.*
>
> *(b) Using a process which is the subject matter of the patent or offering such use when the third party knows or the circumstances make it obvious that the use of the process is prohibited without the consent of the proprietor of the patent.*
>
> *(c) Offering, introducing on the market, and using the product directly obtained by the process which is the subject matter of the patent or importing or possessing such product for any of the purposes mentioned.*

Article 61.1 of the Patent Act is also relevant in the case of process claims:

> *When a product for which there is already a corresponding patent for its manufacturing process is introduced into Spain, the owner of the patent shall, in respect of the product introduced, have the same rights as those granted under the present Law for products manufactured in Spain.*

Although the Patent Act does not include provisions similar to paragraph (c) of Article 50 and Article 61.1 for Swiss-type claims, it is expected that Spanish courts would also apply these provisions to Swiss-type claims since, like process claims, they form part of the so-called 'activity' claims.

1.3   Metabolites

For the time being, the echoes of the landmark US case *Schering v Geneva*,[1] which applied a broad inherent anticipation rule that would make it difficult to patent metabolites, have not reached Spanish courts. Taking into account the narrower 'implicit' disclosure concept applied by the Spanish Supreme Court in its judgment of April 27 2011 (*Lek Pharmaceuticals v Warner-Lambert Company LLC*), when it confirmed the novelty of calcium atorvastatin (EP 409281), we would expect Spanish courts to find that a metabolite is new provided it had not been identified in the prior art.

Another point discussed in academic circles, not in court for the time being, is what amount of a patented metabolite would be required to be found to conclude infringement. In principle, Spanish courts would probably neglect *de minimis* quantities which do not have any technical effect.

2.   Second-generation inventions

Since, as mentioned, Spain has been a contracting party to the EPC since 1986, the criteria for admitting the patentability of combinations, enantiomers, selection inventions, methods of use and secondary indications, formulations and methods of treatment are very much driven by case law from the European Patents Office (EPO) Boards of Appeal.

The dearth of domestic case law on these issues has prompted Spanish courts of appeal to look for guidance in the case law from the EPO's Boards of Appeal. Although such so-called 'case law' is neither case law in the proper sense nor binding on national courts, the Spanish courts will treat it as authoritative doctrine.

2.1   Combinations

As regards combinations, the authors would expect Spanish courts to hold combination patents valid as long as the combination presents synergies and those synergies would not have been obvious to person skilled in the art.

2.2   Enantiomers

The leading case dealing with the patentability of enantiomers was *Lek v Warner-Lambert*, where, in its judgment of October 18 2007, the Court of Appeal of Barcelona (Commercial Chamber) dismissed the revocation action that Lek had filed against patent EP 409281, which protected the calcium salt of atorvastatin. The judgment was based on the 'two-list principle' enshrined by the EPO's Board of Appeal in its Decision 12/81 – which, ironically, was the legal basis that Lek had used to question novelty.

In its judgment of April 27 2011 the Supreme Court upheld the judgment of October 18 2007 issued by the Court of Appeal of Barcelona (Section 15). For the time being, this judgment is the main reference for selection inventions and, in particular, for enantiomers.

2.3   Selection inventions

Other relevant judgments confirming the validity of claims protecting selection inventions include the judgment of June 12 2001 from the Court of Appeal of Barcelona (Commercial Chamber), which confirmed the novelty and inventive

---

1   US Court of Appeals for the Federal Circuit, October 28 2003.

activity of the besylate salt of amlodipine (EP 244944) over a former patent that described amlodipine and other salts (ES 520389).

2.4 Methods of treatment and Swiss-type claims

Moving on to other types of claims, such as 'Swiss'-type claims, accepted by the EPO's Boards of Appeal to address the exclusion from patentability of methods of treatment of humans or animals, there are several cases where the scope of such claims has been discussed from different perspectives. For example, during the last few years several manufacturers of generic medicines filed revocation actions against Swiss-type claims alleging that they *de facto* protect a pharmaceutical product and that, therefore, they were contrary to Spain's Reservation to the EPC. According to this reservation, European patents would not have effects in Spain in so far as they protect chemical or pharmaceutical products as such. In its judgment of October 26 2006 (*Ratiopharm v Warner-Lambert Company LLC*), the Madrid Court of Appeal (Section 28) rejected this view after noting that Swiss-type claims, like 'process' claims, are so-called 'activity' claims and therefore they do not protect a product as such. The Barcelona Court of Appeal (Section 15) reached the same conclusion in its judgment of October 18 2007 (*Lek Pharmaceuticals v Warner-Lambert Company LLC*).

The admissibility of these types of claim has, of course, been further clarified after EPC 2000 came into force on December 13 2007. The exclusion formerly contained in Article 52.4 has been moved to Article 53(c) to reflect that methods-of-treatment claims are excluded from patentability due to ethical and public-health reasons, rather than lack of industrial applicability. For the avoidance of doubt, new Article 53(c) makes it clear that such prohibition will not apply to products, in particular substances or compositions, for use in any of these methods.

2.5 Formulations

As regards formulation patents, as in other countries, they are normally subject to obviousness attacks based on the argument that trying this or that formulation would only require routine work. In a judgment of December 30 2010 handed down by Commercial Court number 1 of Madrid in *Actavis v Novartis*, the court dismissed a revocation action filed by Actavis against patent EP 948320, which protects a delayed-release formulation of fluvastatin. Among other reasons, the court took into account that the expert used by Actavis to question the inventive activity of the patent was not a formulation expert.

2.6 Reach-through claims

The debate sparked by the US case *University of Rochester v GD Searle*[2] regarding the patentability of so-called reach-through claims has not yet reached Spanish courts. The interests of research institutions in trying to patent research tools through these types of claim to ensure future profits on downstream products identified and/or developed using such tools sometimes encounter problems of an insurmountable

---

2   US Court of Appeals for the Federal Circuit, February 13 2004.

lack of enablement (ie, insufficient description).

Although these types of invention are normally the result of valuable research efforts that should be rewarded, the format of the claims filed in relation to these patents is often too speculative. It is likely that Spanish courts will also follow the EPO Boards of Appeal in this area, rejecting, for example, patent applications to compounds defined solely by a function that the inventors have discovered. According to the EPO Boards of Appeal, these types of claim would not meet the sufficiency requirement, as a person skilled in the art would have to embark in a trial-and-error crusade to identify which compounds have the alleged new effect.

3. DNA, biologicals and personalised medicine

The legal framework of inventions in the field of biotechnology derives from Directive 98/44/EC of the European Parliament and of the Council of July 6 1998 on the legal protection of biotechnological inventions (the Biotech Directive), which was implemented through Law 10/2002, dated April 29 2002, and in force since May 1 2002. Therefore, broadly speaking, the legal regime of inventions in the field of biotechnology mirrors the principles followed throughout the European Union.

Taking into account the relatively new nature of these inventions, at present there is a dearth of case law in this field. One of the few relevant cases was decided by the Court of Appeal of Barcelona (Commercial Chamber) in a judgment dated April 10 2000 (*Kirin Amgen Inc v Boehringer Manheim*). In a way, this decision was a forerunner *avant la lettre*, since it applied the principles of the Biotech Directive before the law that implemented it had come into force.

All in all, the stance that Spanish courts will take with regard to some of the most disputed topics surrounding patents in the field of biotechnology, such as the sufficiency requirement, is still unclear. This is also the case for patents which constrict the trend towards personalised medicine, such as patents which protect diagnostic methods. According to the legal regime currently in force, it is unclear whether the use of patented diagnosis methods might benefit from the experimental use or from 'Bolar'-type exemptions, whether third parties accused of infringing these types of patents would have an *ordre public* defence, or whether they would be able to obtain a compulsory licence from the patent owner. For the benefit of the future development of personalised medicine, it would be desirable for the Spanish parliament to enact rules clarifying the legal regime. It would also be desirable to modify the rules applicable to compulsory licences, as the rules currently in force are so complex that they have hardly ever been used.

4. Acts of patent infringement

4.1 Infringement

As mentioned in section 1 above, according to Article 50 of the Patent Act the so-called *ius prohibendi* (ie, right to prevent others) conferred by the patent entitles its owner to prohibit:

- the manufacturing, offering, introducing on the market, and using a product which is the subject matter of the patent or importing or possessing such product for any of the purposes mentioned;

- the use of the patented process or the offer of such use when the third party knows or the circumstances make it obvious that the use of the process is prohibited without the consent of the proprietor of the patent; and
- the offering, introduction on the market, and use of the product directly obtained through the process that is the subject matter of the patent, or the import or possession of such product for any of the purposes mentioned.

Pfizer v Airfarm et altri[3] raised the interesting question as to whether or not the patentee is entitled to prevent others from placing products 'in transit' in a Spanish bonded warehouse. As the Patent Act did not contain an explicit answer to this question, in its judgment of June 12 2001 the Court of Appeal of Barcelona (Commercial Chamber) reached an affirmative conclusion relying on Article 31.2(c) of Law 32/1988, of November 10 1988, of Trade Marks, which entitled trade mark owners to prevent others from placing products bearing the trade mark in transit in the Spanish territory.

4.2   Contributory infringement

'Contributory' infringement is governed in Article 51 of the Patent Act:

1. The patent also confers on its holder the right to prevent any third party from supplying or offering to supply means for the putting into practice of the patented invention related to an essential element of it to persons not entitled to exploit it, without its consent, when said third party knows, or it is obvious in the circumstances, that such means are suitable and intended for putting the invention into effect.

2. The provisions of the foregoing paragraph are not applicable when the means to which it refers are products which can normally be found on the market, unless the third party induces the person to which it makes the supply to commit any act forbidden under the foregoing Article.[4]

On June 12 2001,[5] the Court of Appeal of Barcelona (Commercial Chamber), handed down the first judgment in which a Spanish Court applied Article 51 and concluded that, in order to find contributory infringement, Article 51 of the Patent Act requires:

(1) a delivery or offer of products, plans, information, or other elements which do not directly constitute an offence against the patent; (2) that the products, plans, etc serve to put into practice an essential element of the patent; (3) that the offer is directed to persons who are not authorised to exploit the patent; (4) that the offeror or transferor knows or the circumstances make it evident that the means are suitable for putting the invention into practice; (5) that it knows that they are destined for this purpose; and (6) that the cause for exception envisaged under number (2) of the same provision is not met.

Neither Spanish courts nor academic authorities have dealt extensively with what should be considered as "means to put the invention into effect"; nonetheless,

---

3   Court of Appeal of Barcelona, (Commercial Chamber), June 12 2001.
4   Article 50 of the Spanish Patent Act refers to direct infringement.
5   As note 3 above.

some indications exist. In the handful of cases relating to indirect infringement dealt with by the Spanish courts, the subject matter of the dispute was whether a particular physical product could be considered a "means to put the invention into effect". In the judgment dated June 12 2001, when assessing the requirements for contributory infringement the Court of Appeal of Barcelona mentioned that the supply or offer of "plans, information and other elements which do not directly constitute infringement" may be considered as means to put the invention into effect.

Spanish courts will consider, on a case-by-case basis, whether or not the means supplied or offered to be supplied is an 'essential element' of the patented invention. The Spanish Patent Act provides no guidance on this. For this purpose, Spanish courts will interpret the claims of the alleged infringed patent in accordance with Article 69 of the EPC and the Protocol for its interpretation in order to determine which elements are necessary to prepare, create and/or implement the patented product, or are necessary for its normal functioning.

In accordance with Article 51 of the Patent Act, the 'knowledge' required for contributory infringement is either actual knowledge (the person supplying or making the offer to supply 'knows') or constructive knowledge ('it is obvious in the circumstances' that the means are suitable for putting that invention into effect), taking into account the means supplied or offered and the person to whom it is delivered or offered.

In its judgment dated June 12 2001, the Court of Appeal of Barcelona stated that there are two required items of knowledge:

- that the means supplied or offered are suitable for putting the invention into practice; and
- that the means supplied or offered are intended to put the invention into practice.

The patent owner has the burden of proving that the defendant had actual or constructive knowledge that the supplied means are suitable and intended for putting the invention into practice.

In the cases in which Article 51 of the Patent Act has been applied, the Spanish courts decided on the basis of constructive knowledge. The Court of Appeal of Barcelona concluded that the fact that the pharmaceutical market is highly specialised and those that operate in it are very qualified was sufficient to deduce that the defendant company's managers knew the product offered and supplied was suitable to manufacture the product directly obtained through the patented process.

Contributory infringement does not exist in cases where the means supplied or offered are staple commercial products. The exact wording used by the Spanish Patent Act is "products which are generally available on the market", so a Spanish court will construe Article 51.2 of the Patent Act in the sense that the offer and supply of products, which are generally available on the market, irrespective of whether they relate to an essential element to put a patented invention into practice, is exempted from contributory infringement, unless the offeror induces the person to whom the offer is addressed to commit acts of infringement.

As far as the authors are aware, Article 51.2 of the Spanish Patent Act has never been applied in Spain, so it is uncertain how the Spanish courts will interpret it. We would apply by analogy the legal doctrine that has interpreted the inducement to contractual breach prohibited by the Spanish Unfair Competition Act, which establishes that, when considering whether there has been an inducement to infringe, the relevant issue is the "objective ability that the behaviour may achieve this end, independently of whether it is effectively achieved". In these circumstances, when the means supplied are staple commercial products, Spanish courts might consider that there is an indirect infringement, provided that the patent holder can prove the existence of behaviour objectively addressed to make the supplied person infringe the patent, even if no direct infringement is committed.

An interesting question, which has not received a definitive answer for the time being, is whether assigning a marketing authorisation of a drug to a third party may be considered an act of 'contributory' infringement, the rationale being that a marketing authorisation forms part of the 'means' necessary for working the invention.

4.3   'Bolar'-type provisions and experimental-use exemptions

Moving on to the exemptions most relevant to life sciences, the 'experimental use' exemption stands out. The original text of Article 52 paragraph (b) of the Patent Act stated that the rights conferred by the patent do not extend to: "acts carried out for experimental purposes relating to the subject matter of the patented invention". Before Directive 2004/27/EC of the European Parliament and of the Council of March 31 2004, amending Directive 2001/83/EC on the Community Code relating to medicinal products for human use, introduced the 'Bolar' exemption into Community law, several cases discussed whether use of a patented product or process for the purpose of obtaining a marketing authorisation of a generic drug fell within the scope of application of the 'experimental use' exemption. In its decisions of September 26 2002, December 13 2004 and July 25 2006, the Court of Appeal of Barcelona reached a negative conclusion. The court found that filing an application for a marketing authorisation, as such, was not an act of infringement. However, the court also highlighted that using a patented product or process for the purpose of obtaining a marketing authorisation (eg, submitting samples to the health authorities) would be an act of infringement.

Only three days after the last decision was published, Law 29/2006, dated July 26 2006, on guarantees and the rational use of drugs and health products (known as the 'new Medicament Law') came into force, which implemented the 'Bolar' exemption into Spanish law. The SPTO had initially proposed to add a new paragraph (g) including this new exemption in Article 52 of the Patent Act, which includes the catalogue of exemptions to patent rights. However, as a result of the lobbying efforts deployed by the manufacturers of generics, which were interested in trying to obtain the retrospective application of the new exemption, the Ministry of Health finally introduced the exemption in paragraph (b) of Article 52 – that is, mixed with the 'experimental use' exemption. As a consequence, since July 28 2006 paragraph (b) of Article 52 reads as follows:

The rights contained in the patent do not cover:

(...)

   (b) Acts carried out for experimental purposes relating to the subject matter of the patented invention, in particular studies and trials carried out for the authorisation of generic drugs, in Spain or abroad, and the corresponding practical requirements, including the preparation, obtaining and use of the active principle for such purposes.

In addition, in the Recitals of Law 29/2006 the Ministry of Health inserted a confusing sentence suggesting that the 'Bolar' exemption was introduced for 'clarification' purposes.

The odd way of implementing the 'Bolar' exemption created a great deal of confusion among Courts of Appeal. For example, the Court of Appeal of Barcelona, which, as mentioned, had traditionally maintained that these types of preparatory acts were not covered by the 'experimental use' exemption, in its judgment dated March 17 2008 changed its position after noting that the legislature had 'clarified' that these acts were indeed already excepted by the 'experimental use' exemption. The Court of Appeal of Madrid had reached the same conclusion in its judgment of 16 September 2008.

However, the Supreme Court reversed this line of interpretation in its judgment of June 30 2010, highlighting that the 'experimental use' exemption on the one hand, and the 'Bolar' exemption on the other, had different origins, rationales and scope of application. According to the Supreme Court this was very clear before Law 29/2006, which notwithstanding its alleged 'clarification' goal had in fact obscured the matter. The Supreme Court added that it would be absurd to argue that the 'Bolar' exemption was already comprised within the scope of application of the 'experimental use' exemption of the Patent Act, taking into account that in 1997 Spain, together with the other EU member states, had filed a complaint against Canada before the World Trade Organization (WTO) for introducing a 'Bolar' exemption, arguing that it was contrary to the TRIPS Agreement.

In conclusion, this important judgment made it clear that the 'experimental use' exemption and the 'Bolar' exemption are two different exemptions, and that the latter may not be applied retrospectively. Nonetheless, many grey areas will need further clarification in future judgments, such as whether the 'Bolar' exemption applies to acts carried out for obtaining marketing authorisation of generics only, or whether it may also be applied to acts conducted in relation to marketing authorisations for innovative drugs.

4.4   Offers to supply

A related question is whether storing product in preparation for launch (ie, stockpiling) or making offers to supply prior to the expiry of the patent would be an act of infringement. The first part of the question was answered by the WTO Dispute Settlement Body in its report of March 17 2000, where it clarified that stockpiling would be an act of infringement. In relation to offers, in its judgment of May 9 2008 (*Pfizer Limited et altri v Stada*) the Court of Appeal of Barcelona concluded that the publication of an advertisement of amlodipine, as such, was an act of infringement.

5.   Patent enforcement

5.1   Obtaining information on the infringer and the infringement

Articles 129 *et seq* of the Patent Act entitle patent owners to file an application for an *ex parte* inspection (called *Diligencias para la comprobación de hechos* or *Diligencias*) of the premises of the potential defendant. The process is roughly equivalent to the French *saisie-contrefaçon* type of inspections. However, in Spain the applicant must file some evidence which may allow the judge to suspect that the company to be inspected may be infringing the asserted patent, and shall justify that no means other than the *Diligencias* are available to verify the presumed infringement.

If the *Diligencias* order is given, the judge will set a bond which the applicant must deposit with the court to cover any damage which the inspection may eventually cause to any entity inspected, such as damages for disruption of production.

The inspection is conducted by a court committee comprising the judge, the court secretary, who will be present to record minutes of the inspection, and one or more independent technical experts appointed by the judge *ex officio*. Although in principle the powers of the court committee are limited to inspecting the plant and describing the machines, equipment and tools found at the premises, the Court of Appeal of Barcelona has construed the law as establishing that the court committee has the power to make photocopies of any documents, take samples and/or pictures of any products or substances and conduct any other type of activities which may be necessary in order to determine whether or not the applicant's patent has been infringed, with the limitation that the inspection not be used as a means to violate any trade or industrial secrets or carry out acts which represent unfair competition.

If the judge concludes from the results of the inspection that infringement can be presumed, the judge will provide the claimant patentee with a copy of the minutes of the inspection and any document, sample or picture taken during the inspection, so that the latter can then use them to prepare its claim in the main action and as evidence of infringement in such action. No other collateral use may be made of the information in the certificate by the claimant, which may not even disclose it to any third party.

5.2   Interim relief

According to Spanish law, the patent holder whose rights are being infringed, or might be infringed imminently, can apply for the interim injunctions necessary to secure the future enforcement of any favourable judgment obtained in the main infringement proceedings. Interlocutory injunctions must be applied for, in normal circumstances, when the main claim is filed, and applications filed before the main action will only be accepted if the applicant proves reasons of urgency or necessity which justify the request at such time.

The interim proceedings progress independently of the main proceedings, with their own self-contained procedure. The application for an interlocutory injunction is normally made *inter partes* (ie, with notice to the other party). However, when the applicant proves reasons of urgency or that hearing the other party might frustrate the purpose of the preliminary injunction, the court can order such interim injunctions within the following five days, without hearing the defendant, which is then entitled to file an opposition against the interim injunctions ordered.

A court order containing the interlocutory injunction, if granted, may order from the defendant, *inter alia*: the cessation of any act which infringes the applicant's patent rights, or its prohibition where there are reasonable indications that an act is about to begin immediately; the impounding of infringing items; and the depositing by the defendant of a bond guaranteeing payment of the damage already caused to the patent holder.

In order to obtain an interim injunction, the applicant must justify:

- industrial exploitation of the patented invention in either Spain or in the territory of a member state of the WTO in such a way that the exploitation is sufficient to satisfy the demand of the Spanish market;
- *fumus boni iuris* (good evidence of infringement): the applicant must have strong *prima facie* evidence supporting its application. Normally, this is proved by obtaining expert reports arguing the reasons that lead an expert in the field to conclude that the defendant's product falls within the scope of the allegedly infringed patent;
- *periculum in mora* (danger from a delay): the applicant must show that circumstances could occur during the main action which, unless an interim injunction is ordered, could prevent or make it more difficult to enforce a favourable judgment; and
- the offer of a deposit of a bond to compensate the defendant for any damages which the interlocutory injunction might cause and specify the amount of such bond.

5.3 'Springboard'/post-patent-expiry injunctions

There has been only one case where a 'springboard' preliminary injunction was ordered. On September 1 2004 Pfizer filed an application for a 'springboard' preliminary injunction based on patent ES 520389 (which was due to expire on December 19 2004) against all companies which had obtained authorisation to market generics of amlodipine. Since for the purpose of obtaining marketing authorisation the 'notice to applicants' published by the EU Commission required the submission of samples to the Spanish authorities, Pfizer assumed that the companies that had obtained marketing authorisation would have submitted samples. So, Pfizer requested the court to order a preliminary injunction preventing the defendants from launching, after patent expiry, during a period of time identical to the time elapsed between the filing of their applications and the date on which the patent expired. The case was assigned to Commercial Court number 3 of Barcelona, which denied the application for the preliminary injunction requested. Pfizer filed an appeal before the Court of Appeal of Barcelona which, on December 13 2004, handed down a landmark decision ordering the 'springboard' preliminary injunction requested. Although the injunction was lifted some months later, as it turned out that the defendants had not submitted samples, this case shows that, as in England and Germany, the concept of 'springboard' injunctions is not alien to Spanish courts.

5.4   Unjustified threats

According to Article 64 of the Patent Act:

1. Any person who, without the consent of the patent holder, manufactures or imports from abroad articles protected by a patent, or uses a patented process, will in every such case be liable for any damage so caused.

2. Persons performing any other act of exploitation of an article protected by a patent will only be required to compensate for any damages so caused where they have been warned by the patent holder as to the infringement thereof, accompanied by a request that they cease such infringement, or where they have acted deliberately or negligently.

Therefore, under Spanish law, sending so-called 'warning' letters is not only a legal right but, in some circumstances, a legal obligation. Like all legal rights, it may not be exercised in an abusive way.

In July 2003, Ratiopharm filed an unfair-competition action against Pfizer on the grounds that the latter had sent hundreds of 'warning' letters to the wholesalers that were planning to distribute generics of amlodipine in Spain. According to Ratiopharm, these letters were an unjustified threat contrary to the Unfair Competition Act. In its judgment of July 12 2010, the Supreme Court rejected the complaint, noting that the facts mentioned in the letter sent by Pfizer were accurate and that, therefore, the unfair competition claim could not be upheld.

5.5   Remedies

Article 63 of the Patent Law includes an illustrative list of the types of relief that may be sought in the main proceedings by a patent holder whose rights have been infringed. These include, amongst others:

- an order for cessation of the infringing acts (ie, an injunction);
- an award of damages;
- an order for seizure of infringing items and the means exclusively used for their production or for carrying out the patented process;
- an order for assignment of the ownership of the seized infringing items to the patent holder whenever possible, in which case the value thereof will be deducted from the amount of damages awarded and, if the value exceeds the amount of damages awarded, the patent holder will pay the excess amount to the infringer;
- adoption of any other measures necessary to prevent the patent infringement continuing and, in particular, alteration or destruction of infringing items when this is essential to prevent infringement of the patent; and
- publication of the judgment against the patent infringer, at the infringer's expense, by means of advertisements and notification to other interested persons.

6.   Compulsory licensing

According to the (1986) Spanish Patent Act, when it is not possible to exploit a patent without infringing a previous patent, the owner of the later patent is entitled to obtain a compulsory licence from the owner of the previous patent through the payment of 'adequate' compensation, as long as the owner of the later patent proved that his invention represents "significant technical progress of considerable economic importance". The application for this type of compulsory licence must be filed before the Spanish Patent and Trade Marks Office (ie, it is an administrative procedure).

In practice, these types of application are very rare.

7.   Ownership, inventors and compensation

According to Article 10 of the Patent Act, the right to a patent shall belong to the inventor or to his successors in title and it shall be transferable by any of the means recognised in the law. Where an invention has been made jointly by a number of persons, the right to obtain a patent shall belong to them jointly. Where the same invention has been made independently by various persons, the right to a patent shall belong to the person whose application bears the first date of registration in Spain. As far as the owner of the patent application or of the patent is concerned, the inventor shall have the right to be mentioned in the patent as being the inventor.

Under its Title IV (Articles 15 to 20) and Title XIII (Articles 140 to 142) the Patent Act regulates so-called 'employees' inventions' (inventions arising from a labour relationship), distinguishing three types of inventions. The most significant aspects of the regulations can be summarised as follows:

- Title IV applies to inventions created by those individuals (employees) who are linked to the company concerned through an employment agreement.
- In order to classify any invention as of a labour nature, it is necessary for the employment agreement to be in force at the time the inventions are produced, even though the Patent Act provides for certain exceptions to this criterion.
- The Patent Act establishes the remuneration or compensation to be paid, if any, to the employee inventor and conciliation proceedings to solve any dispute arising in relation to an employee's invention.

As a general comment, it is important to note that all provisions established under the Patent Act for employees' inventions are of a mandatory nature and hence it is not possible for the parties to an employment agreement to agree on any aspect arising in relation to employees' inventions except where such agreement aims to ameliorate and/or increase employees' rights arising therefrom.

The provisions of the Patent Act, when dealing with employees' inventions, refer to three different categories of inventions, which are relevant in order to ascertain the compensation to be paid to the employee inventor. These are discussed further in the remainder of this section.

7.1   So-called 'service inventions'

'Service inventions' are those which result from research activity, which explicitly or implicitly constitutes the object of the employment agreement.

In these cases, the invention belongs to the employer, since the objective of the employment agreement is precisely the development of a research activity for which the employee has been hired and is remunerated. Hence, the employee shall have no right to any additional remuneration, unless his personal contribution to the invention and the significance of such invention clearly exceeds the purpose of the employment agreement ('additional remuneration').

Article 15.2 of the Patent Act establishes that the employee could be entitled to so-called additional remuneration if (i) the personal contribution of the employee and (ii) the significance of the invention for the company clearly exceed the purpose of the employment agreement. Again, it should be noted that on the few occasions where existing case law has analysed potential requests of additional remuneration, the courts have come to the conclusion that in each particular case the referred invention was within the scope of the employment agreement and did not clearly exceed the purpose of the employment agreement.

One of the most important elements to consider in any given employment relationship is that those employment agreements of individuals who render services who may entail the creation of an invention should clearly determine this purpose within the scope of the employment agreement, or even in the definition of the individual's job, in order to prevent disputes on whether or not potential inventions exceed the scope of the services rendered under the employment agreement.

The Patent Act establishes that in cases where there is no agreement on the amount to be paid as additional remuneration, the parties must ask for special conciliation proceedings before starting any legal action before the courts.

Thus, only in the event that these parties do not reach an agreement in the conciliation proceedings would the employee be allowed to take legal action against the employer. The courts would have to consider the circumstances in which the employment agreement was entered into, particularly whether or not the invention clearly exceeds the main purpose of the employment agreement.

7.2   So-called 'mixed inventions'

Article 17 of the Patent Act refers to 'mixed inventions' as those inventions that, without fulfilling the requirements to be classified as service inventions (ie, the employee has not been specifically hired to develop research activities), arise as a result of the employee's professional activity in the company and have been obtained under the predominant influence of know-how acquired in the company or using means provided thereby.

Hence, given that the main purpose of the employment agreement does not constitute research activity, the result obtained by the employee is of an extraordinary nature. This circumstance is fundamental in order to distinguish this type of invention from service inventions.

In the case of mixed inventions, the employer has the right to choose either to assume the ownership of the invention or to retain the right to use it. In such event, the employee has a right to 'fair economic compensation' ('economic compensation'). Otherwise, should the employer not exercise its right, the employee will be entitled to have the ownership of the invention, even though he will lose his right to be compensated.

The economic compensation in cases of mixed inventions applies to cases where the employee's invention arises from the employee's professional activity in the company and has been obtained under the predominant influence of know-how acquired in the company or using means provided thereby, even though in most cases the employee has not been specifically hired to develop research activities (Article 17.2 of the Patent Act).

Article 17 of the Patent Act establishes that economic compensation should be "a fair economic compensation, fixed according to (i) the industrial and commercial relevance of the invention and (ii) the value of the means or knowledge provided by the company and the employee's own contribution".

The Patent Act does not describe the process for calculating compensation, so, again, the agreement between the employer and employee should be considered. Besides, legal scholars have interpreted that the amount to be paid could take into account the criteria established under the Patent Act to determine the damages caused to the owner of a patent in a case of a breach of patent rights, which normally takes into consideration (i) the gains the owner of the patent would have obtained and/or (ii) the unfair benefits obtained by the company for the use of the invention.

7.3 So-called 'free inventions'

These are deemed to be any invention different from service inventions or mixed inventions. Article 16 of the Patent Act provides that these inventions belong to the employee. In this case, given that the invention belongs to the employee, the matter of compensation is not relevant.

However, it must be noted that the right of the employee could be limited by his labour obligation of non-competition with the business activity of the company, according to Article 5(d) of the Spanish Workers' Statute.

Finally, ownership of inventions made by an academic as a result of his research responsibilities in a university and which come within the scope of his teaching and research functions shall belong to the university. The academic shall nevertheless have the right to participate in the benefits gained by the university from working or assigning its rights in an invention. The university may assign ownership of an invention to the academic who is its author, in which case it may keep a non-exclusive, untransferable and gratuitous licence. Where an academic derives benefits from working an invention, the university shall have the right to participate therein. Where an academic makes an invention as a result of a contract with a state or private body, the contract shall specify which of the contracting parties owns the invention.

8.   Branding and designs

8.1   Trade marks in life sciences

With the promotion and establishment of generic drugs in Spain, the protection of the branding and design of the reference products or innovator-branded drugs has become increasingly important in the last few years.

Article 85 of Act 29/2006, of July 26 2006, on guarantees and the rational use of drugs and healthcare products (the Drug Act) states:

The health administrations will promote the prescription of drugs identified by the active principle on the medical prescription. In those cases where the prescribing specialist indicates simply the active principle, the pharmacist will dispense the drug with the lowest price and, where the price is the same, the generic drug, if there is one.

Therefore, we should take into account that at present in Spain innovator-branded drugs are facing (i) the incentive for the doctor to prescribe by active principle (without identifying the brand) and (ii) the pharmacist's obligation, upon receiving a prescription stating the active principle, to dispense the cheapest drug (which is not usually the innovator-branded one) and, where the price of the innovator and generic drug are the same, the generic drug if there is one. That is, if the innovator drug and the generic drug cost the same, then provided the prescription is for the active principle, the pharmacist will dispense the generic drug.

8.2 Protection using trade marks and designs

Relevant in this regard is the judgment handed down by the Supreme Court (Civil Division, Section 1, number 204/2010 dated April 7 2010, by Judge Rapporteur Mr José Ramón Ferrándiz Gabriel), which upheld a ruling delivered against a company producing generic drugs for unfair competition and trade mark infringement, given that the company: (i) "copied the design and colours of the tablets" of the innovator-branded drug; (ii) reproduced the trade mark of the innovator drug in the brochures edited by the company to promote its generic drug; (iii) reproduced identical drawings to those used for the innovator-branded drug in the brochures edited to promote the generic drug, as well as certain scientific data obtained, not from experience gained from the use of the generic drug but from the use of the innovator-branded drug; and (iv) copied, committing an infringement of the trade mark rights of the owner of the innovator drug, the registered graphic trade mark of such drug, which consisted of the design of the tablet and the colours that identified such brand, leading to confusion.

The facts of the judgment are based on (i) the use in the generic drug of capsules of the same colour and size as the innovator-branded drug (ie, green-and-cream-coloured capsules); (ii) the placing of advertisements in brochures and medical publications by means of a drawing that implied an almost exact copy of the drawing employed by the owner of the innovator-branded drug; (iii) also using a graph relating to the bioequivalence of both products not accredited directly but, rather, through a third drug belonging to the owner of the innovator-branded drug; (iv) providing, for advertising purposes, the misleading reference to more than 50 million treatments that were not carried out with the generic drug but with the innovator-branded drug; and (v) presenting, in short, the generic drug as a product identical to the innovator-branded drug. The court of first instance already concluded that the conduct used in the advertising and design of the generic drug represented an infringement of the trade mark rights, not only because the use of the same graph and a design and colour identical to the compared products entailed a risk of association between both products for consumers, but also because this confusion was generated deliberately, taking advantage of the reputation of one of the drugs (the innovator-branded drug) to launch the other (the generic drug) onto the market.

8.3   Comparative advertising in pharmaceuticals

In Spain, a distinction is made between advertising aimed at healthcare professionals and advertising targeting the general public. Advertising for healthcare professionals is subject merely to simple communication requirements to the autonomous community where the company is based (Spain has 17 autonomous communities, which are territorial entities that, under the Spanish constitution, have legislative autonomy and executive powers, as well as the authority to govern themselves through their own representatives). Meanwhile, advertising for the general public may only deal with those medicines that are not financed and do not require a prescription (ie, are non-prescription or over-the-counter medicines) and such advertising must be authorised in advance by the Ministry of Health in Madrid, which has nationwide authority, in the event that the advertising is for the entire country, or for an autonomous community if the advertising is to be limited to a particular region. Advertising of prescription-only medicines to the general public is prohibited in Spain.

Comparative advertising in Spain is allowed provided that certain requirements are met. According to Spanish legislation (transposing Directive 97/55/EC of the European Parliament and of the Council of October 6 1997 amending Directive 84/450/EEC concerning misleading advertising so as to include comparative advertising, performed by Spain in 2002), public comparison, including comparative advertising, by means of an explicit or implicit reference to a competitor shall be permitted, in the case of those requirements relevant to pharmaceutical products, when the following conditions are met:

- it compares goods or services meeting the same needs or intended for the same purpose;
- it objectively compares one or more material, relevant, verifiable and representative features of those goods and services, which may include price;
- it does not present goods or services as imitations or replicas of goods or services bearing a protected trade mark or trade name; and
- the comparison will not contravene the rules on unfair competition applicable to acts of deception, denigration or exploitation of another party's reputation.

Furthermore, the Spanish Code of Good Practices for the Promotion of Medicines and Interaction with Healthcare Professionals of Farmaindustria (the Spanish Pharmaceutical Industry Association) establishes that comparative advertising must always comply with the regulations on fair competition. It cannot be disparaging, and comparisons must be based on relevant and comparable aspects. At any rate, and particularly as regards comparative advertising, care must be taken to ensure that the sources on which statements are based are valid and immediately accessible to the competitor. Any information, statement or comparison included in promotional materials should be well founded. Such foundation (or rationale) should be provided to physicians and all other healthcare professionals on request. In particular, any comparison made between different medicines must be scientifically verified. Such rationale need not be provided, however, for statements related to the indications approved in the summary of product characteristics.

In addition, the ANEFP, the Spanish industry association of over-the-counter medicines, regulates comparative advertising in their "Code of deontological rules for the promotion and advertising of non-prescription medicines and those not financed by the National Health System and other self-care healthcare products" as follows:

(a) *Public announcements of non-prescription medicines cannot suggest that the effects of said medicine are greater than or the same as those of another treatment or medicine.*

(b) *Announcements that contain comparisons must be designed in such a way that the comparison is not misleading and must be performed objectively in relation to one or more essential, pertinent, verifiable and representative characteristics, in accordance with the principles of fair competition.*

(c) *Announcements should not implicitly or explicitly denigrate or slight other companies, activities, products or services.*

(d) *Announcements will not be based on information stating that a certain product does not contain a component used in competing products, giving the impression that said component is unsafe or noxious.*

In addition, the guidelines issued by some autonomous regions in Spain clearly acknowledge the acceptance of comparative advertising in pharmaceutical products if certain conditions are met. For example, the "Guide for the advertising of medicines for human use" of the Department of Health of the Generalitat de Catalunya (third edition, October 2009) states:

- *When comparing the effectiveness, safety or other properties of different medicines or active principles as an advertising instrument, the source must be cited and a reference copy must be made available to the health authorities so that they can confirm that it is an accurate reproduction of the original. Moreover, no information should be omitted such as the statistical significance of the results, for example, and the results of different studies or clinical trials will not be compared in the same table or diagram, unless the source is a meta-analysis.*

- *Comparative advertising between medicines that have comparable safety and effectiveness and an equivalent therapeutic effect is not prohibited. According to the provisions of the General Advertising Law, this comparison must be confirmed scientifically.*

- *Unfair advertising is that which causes direct or indirect discredit, denigration or contempt for a person or company or its products or activities, or leads to confusion regarding the same. Comparative advertising is also unfair when it is not based on the essential, similar and objectively demonstrable characteristics of the products.*

Therefore, comparative advertising in pharmaceuticals products has been permitted and widely used by the industry in Spain, and its implementation has been duly controlled by the competent bodies of the Spanish Pharmaceutical Industry self-regulation systems (applying the Codes of Practice), by the competent health authorities and by the Spanish courts.

However, it is worth mentioning the "National Health System guidelines on the advertising of medicines for human use aimed at the general public" (the

Guidelines), dated April 2011 (published July 7 2011), which were prepared by the Ministry of Health, Social Policy and Equality through the Directorate General for Pharmacy and Healthcare Products, including a "Code of good practice on the advertising of medicines aimed at the general public". These Guidelines clearly establish the following with respect to the principle of comparative advertising:

- Comparative advertising is not admissible when it suggests that the effect of the medicine is the same or greater than that of another treatment or medicine. Article 8.3 of the General Advertising Law states that the administrative authorisation for advertising messages will have to respect the principles of fair competition, meaning that it must not be damaging to competitors.
- Medicines may only be compared if they belong to the same laboratory. In the event that one of the medicines possesses a different active principle, this active principle may be highlighted provided all the compositions of the other medicines involved in the comparison are listed.
- The inclusion of market share, sales or other data is unacceptable as it is always comparative.

The Guidelines state that:

The Ministry of Health, Social Policy and Equality, as the authority responsible for 'the advertising of medicines aimed at the general public' through the Directorate General for Pharmacy and Healthcare Products intends, in the very near future, to update said rules, adapting them to the new European norms and the new technologies in advertising that have emerged in recent years, as well as developing the tools necessary to facilitate the work of all the agents involved in this function (public, industry and administration). One of the first steps we wanted to take in this direction is the presentation of these 'Guidelines' which introduce a 'Code of good practice on the advertising of medicines aimed at the general public'.

As we have seen, the declaration of the "Principle of non-comparative advertising" in the Guidelines is based on the fact that "Article 8.3 of the General Advertising Law states that the administrative authorisation for advertising messages will have to respect the principles of fair competition, meaning that it cannot be damaging to competitors." The text quoted by the Guidelines was amended by Law 29/2009, of December 30 2009, which amends the legal regime governing unfair competition and advertising to improve consumer protection (and which entered into force on January 1 2010); the previous rules no longer exist, having being replaced by Article 5.3, which states equivalently that "the granting of authorisations will respect the principles of free competition, meaning that it must not cause damage to competitors". In fact, that section is included in the rules governing advertising of certain goods and services, establishing that:

1. *The advertising of healthcare materials or products and those subject to technical-healthcare regulations, and including products, goods, activities and services liable to generate risks for the health or safety of persons or their possessions, or the advertising of gambling, betting or lottery activities, may be regulated by special rules or subject to a prior administrative authorisation. Said system may also be*

> *established when the protection of the constitutionally recognised values and rights so requires.*
>
> 2. *The regulations implementing the provisions of the foregoing paragraph and those that, in regulating a product or services, contain rules on advertising of the same, will specify:*
>   - *The nature and characteristics of the products, goods, activities and services whose advertising is being regulated. These regulations will establish the requirement that the advertising of these products contain the risks associated with the normal use of the same, as the case may be.*
>   - *The form and conditions of the dissemination of the advertising messages.*
>   - *The requirements of authorisation and, if applicable, registration of the advertising, when subject to the prior administrative authorisation system.*
>
>   *When drafting these regulations, the business organisations representing the sector, the associations of agencies and advertisers and the consumers' associations must be consulted, where applicable, through their institutional representation bodies.*
>
> 3. *The granting of authorisations will respect the principles of free competition, meaning that no harm will be caused to competitors.*
>
> *Any refusal of applications for authorisation must be reasoned.*

In the author's opinion, the introduction of the prohibition by the recently published Guidelines (which seems to restrict comparative advertising to those cases in which the medicines compared belong to the same laboratory) in the current legislative context (which is permissive, as we have seen, provided certain requirements are met) does not quite fit with the legal framework currently in force.

9. Protecting valuable information

A company wishing to prevent a third party from using valuable information not protected in the form of an intellectual property right would typically rely on the Spanish Unfair Competition Act 3/1991 of January 10 1991, as amended by Law 29/2009 of December 30 2009 modifying the Legal System Governing Unfair Competition and Advertising to Enhance Consumer and User Protection (the Unfair Competition Law). In addition, Spanish Organic Law 10/1995, of November 23 1995, which approves the Spanish Criminal Code, protects trade secrets under Articles 278 and 279, since disclosing trade secrets is a criminal offence.

Article 278 of the Spanish Criminal Code, referring to obtaining data to discover a trade secret, establishes a prison sentence of two to four years and a fine payable over 12 to 24 months (ranging from €2 to €400 a day) for anyone who obtains through any means data, written and electronic documents, computer programs or other objects for the purpose of discovering a trade secret. Conviction will be higher – in particular, a prison sentence of three to five years and a daily fine payable over 12 to 24 months – when the person found guilty spreads, discloses or passes on the discovered secrets to third parties.

Article 279 of the Spanish Criminal Code, on the spreading, disclosure or passing-on of trade secrets, establishes that anyone legally or contractually bound to maintain the confidentiality of a trade secret who is found guilty of spreading, disclosing or passing on the trade secret will be sentenced to two to four years in prison and a daily fine payable over 12 to 24 months. In that respect, it should be noted that the sanctions would be applied at a rate of 50% when the guilty person uses the secret for personal profit.

According to Article 13 of the Unfair Competition Act:

1. *The disclosure or exploitation, without the authorisation of the owner, of industrial secrets or any other kind of business secrets to which access has been gained legitimately, but subject to reservations, or illegitimately, as a consequence of any conduct foreseen in the following section or in Article 14, shall be considered unfair.*

2. *The acquisition of secrets by means of espionage or similar procedures will also be considered unfair.*

3. *The prosecution of the violation of secrets established in the above sections does not require the existence of the requirements established in Article 2. However, the violation must have been carried out with a view to the author or a third party profiting therefrom, or in order to negatively affect the owner of the secret.*

Interestingly, the Spanish regulations governing non-disclosed valuable information or know-how are so scarce that Spanish courts have looked for additional guidance to Article 39 of TRIPS, which has been the main legal basis applied in the cases where Spanish courts have been called to draw the contour of what types of information may be covered. This was the case, for example, in judgments of June 25 2009 [WL AC 2010\1299] and January 13 2009 [WL JUR 2009\142860] from the Court of Appeal of Barcelona (Section 15) and judgments of October 15 2010 [WL JUR 2011\37159] and January 22 2010 [WL AC 2010\1383] from the Court of Appeal of Madrid (Section 28).

10. Counterfeiting

The increasing awareness regarding the need to protect intellectual property to foster creativity and technological progress has allowed counterfeiting to find its way into the Spanish Criminal Code. Wilful infringement of a trade mark, a patent, a copyright or a trade secret, for example, is categorised as a criminal offence. However, due to the lack of specialisation of criminal courts and their other priorities, successfully enforcing an intellectual property right before criminal courts continues to be an uphill battle – although in recent years a lot of progress has been made, particularly in clear-cut trade mark cases.

The so-called Customs Regulation (Council Regulation (EC) No 1383/2003 of July 22 2003 concerning customs action against goods suspected of infringing certain intellectual property rights and the measures to be taken against goods found to have infringed such rights) has been very useful, as it provides a one-stop shop for preventing the importation of goods that may infringe intellectual property rights registered in Spain. A simple application filed before the central customs authorities will be sufficient to prompt the seizure of suspicious goods held in Spanish Customs.

11. Collaborative models – 'open innovation'

As a result of the difficulty of inventing new molecules and the increasing safety requirements set by health authorities, R&D costs in the life sciences sector have skyrocketed over the last few years. This has caused the life sciences industry to join forces through mergers and acquisitions, but also through less formal collaborative models.

As in all Western countries, one of the hot topics has been whether traditional models should be replaced or complemented by 'open innovation', and how 'open' the new forms of collaboration should be. From a legal perspective, the new model has called for a detailed crafting of licence agreements built on the legal regime currently in force (particularly, the provisions on licence agreements of the 1986 Patent Act), and for a careful review of this type of collaborative model from the competition law angle. Although this type of model and, in particular, the 'open innovation' model, cuts across all industry sectors, the recent interest expressed by the EU Commission in the pharmaceuticals sector has prompted companies to be particularly careful.

12. Hot topics

The main hot topics encountered by the life sciences industry over the last few years have been the bounds of the 'Bolar' provision, and the effect of Articles 27.1 and 70 of TRIPS on patents filed before October 7 1992, the date when the old transitional period preventing the enforceability of 'product claims' came to an end.

With regard to the first question, the Supreme Court (Civil Chamber), in a landmark judgment of June 30 2010 reversed the view of the Courts of Appeal, which had interpreted that the 'Bolar' provision introduced by Law 29/2006 was a mere 'clarification' of the experimental-use exemption present in the Patent Act since 1986. According to the Supreme Court, the two exemptions have different origins, rationales and scopes of application, which caused the Court to conclude that the new 'Bolar' exemption may not be applied to acts carried out before it came into force (ie, July 28 2006).

As regards the second topic, on November 4 2010 the Supreme Court (Administrative Chamber) handed down four important judgments, concluding that Articles 27.1 (non-discrimination principle) and 70 of TRIPS allow owners of chemical or pharmaceutical process patents filed before October 7 1992 to add 'product claims'. These judgments have reversed several decisions handed down by the Spanish Patents and Trade Mark Office in 2006, which had rejected the publication of revised translations of European patents where 'product claims' had been added relying on Articles 27.1 and 70 of TRIPS.

United Kingdom

Catherine Drew
Sarah Innes
Gareth E Morgan
Winston & Strawn

Introduction to UK patents legislation

In the United Kingdom, patent law is based on the European Patent Convention (EPC) which, in turn, borrows many concepts and principles from the Community Patent Convention (CPC) that never entered into force. The main source of statute is the Patents Act 1977 (the Act). This has been amended on numerous occasions since its enactment.

As a basic requirement for patentability, the Act requires that an invention be novel, be inventive, be capable of industrial application and also be disclosed sufficiently such that a skilled person can put the invention into effect from the information contained in the patent specification.[1] The lack of any of the above properties forms the basis for grounds either for refusal of grant of a patent application or for the revocation of a granted patent.[2] Any person can bring actions for the revocation of a granted patent in the United Kingdom; there is no need to have been threatened with a claim of infringement prior to bringing such an action. In addition, a patent must be framed with sufficient clarity that it can be understood by a skilled person. Lack of clarity is a ground for the refusal of a patent application, but not a ground of revocation of a granted patent.[3]

Patents directed at new compounds provide *per se* (or product-based) protection for the patentee. A patentee of such a patent can, if the invention is the first time that the patented product has been available to those in the art, prevent any use of the patented product, not just any use that is disclosed in the patent.[4] There is one important exception to this rule and this is discussed in section 3.2 below.

Processes can also be protected by claims in a patent and the protection conferred by such claims extends to uses of the process and also to direct products of the patented process.[5]

1.   Small molecules

Metabolites can be protected by patent claims in much the same way as any other chemical entity. However, litigation has demonstrated that such inventions are

---

1   Sections 1 to 4 the Act.
2   Sections 18 and 72 of the Act.
3   Section 14 the Act.
4   *Generics [UK] and Others v H Lundbeck A/S* [2009] UKHL 12.
5   Section 60(1)(c) of the Act; also *Pioneer Electronics Capital v Warner Music Manufacturing and Another* [1997] RPC 757; and *Monsanto Technology LLC v Cargill International SA and Another* [2008] FSR 8.

vulnerable to being attacked on the basis of previous disclosures concerning the 'parent' molecule. The leading case in which a metabolite claim was analysed is the *Merrell Dow* terfenadine case.[6] This judgment provides guidance as to the circumstances under which metabolites would be patentable. The *Merrell Dow* case was significant as the arguments raised the possibility that although an anti-histamine called terfenadine had been known in the art and its use well understood, the later discovery (and patenting) of a metabolite that was formed in the liver of patients taking terfenadine could have prevented companies marketing terfenadine itself for fear of infringing the later metabolite patent. Lord Hoffmann gave the lead judgment in the House of Lords (as it then was) and held that the previous disclosure that one could take terfenadine in order to exert an anti-histamine effect on the human body sufficiently disclosed the metabolite of terfenadine made in the liver following administration of terfenadine. It did not matter that the metabolite was not given a name, nor its chemical formula deduced.

Remarkably, later filed 'use' patents relating to the use of the same terfenadine metabolite were also the subject of litigation.[7] In this case, the use of the metabolite was said to reduce side effects in certain patients over the use of terfenadine itself. Here, given knowledge of the side effects of terfenadine was not common general knowledge at the priority dates of the 'use' patents, these were anticipated and/or obvious over the disclosures of the ways to produce and use the terfenadine metabolite as an anti-histamine.

2. Second-generation inventions

Article 27 of the TRIPS agreement requires that patents are available for products or processes in all fields of technology, provided they are new, inventive and capable of industrial application. All inventions must comply with the same levels of novelty, inventiveness etc in order to qualify for patent protection, and so whilst 'second-generation inventions' is a useful way to refer to inventions made during the development of the delivery and use of an new active substance after launch, it should be remembered that patents that relate to them are patents in their own right granted according to the requirements of the relevant legislation.

What is not in doubt is that patents for second-generation inventions are valuable assets for innovator pharmaceutical companies extending the monopoly on some of their most valuable products.

However, in the United Kingdom the manner in which the English courts scrutinise patents in the course of litigation tends to mean that patents that are perhaps close to the boundary to what is and is not obvious will be more difficult to defend than in other jurisdictions in Europe where the courts are more deferential to the decisions of the EPO (both the Opposition Division and the Technical Boards of Appeal (TBAs)) than are the English courts. Put simply, a patent challenger has a far greater toolkit available in revocation actions before the English courts than it would

---

6   *Merrell Dow Pharmaceuticals v H N Norton & Co Ltd and Another* [1996] RPC 76.
7   *Teva Pharmaceutical Industries and Another v Merrell Pharmaceuticals Inc and Others* [2007] EWHC 2276 (Ch).

in other European jurisdictions. By way of example, the reformed Patents County Court which is designed to be a streamlined and low-cost recovery jurisdiction set up to hear simple, lower-value patent cases still contemplates trials of one to two days in length, which is longer than any other first-instance court in the European Union.

In reviewing recent cases in the English Patents Court, it is noticeable that because inventions often consist of the selection of known formulations, constituents to purify or actives to combine, there is usually a significant discussion to determine whether there was a motive for the skilled person to seek to improve the original product in the manner that the patentee has achieved in the patent. The reason for this is that, if a motive to improve can be made out, the challenger has an easier job to convince the court that a selection of one option from a list of known and obvious alternatives did not amount to an invention, rather to simply routine product development.

2.1 Combinations

The mere placing side by side of two known agents in their known formulations, strengths and presentations would not fulfil the requirements for patentability. This is simply demonstrated in the 'sausage machine' case.[8] In this case, the patentee sought protection for a machine to mince meat and fill the minced meat into skins to make sausages. Whilst the resulting machine was no doubt useful, it comprised two already existing machines and thus there was nothing novel or inventive about the machine – the placing together of the two machines had not resulted in a new or improved result over what would have been produced had the two original machines been used in sequence.

The same must of course be applied to pharmaceutical preparations. A patent for a combination would not be valid if the patented invention comprised the presentation of two known pharmaceutical agents, presented in their known formulations, strengths and doses and each acting in the same manner and as efficaciously as was already known (ie, no surprising interrelation between the agents is demonstrated). The novelty of a combination resides in the new or improved result brought about by the agents acting together.[9] However, a surprising result – a so-called 'golden bonus' – will not render a combination inventive if it was nonetheless obvious to combine the elements,[10] and this holds true also for unexpected combinatorial effects in combination pharmaceutical products.[11] The former point was highlighted in *Teva v Merck*,[12] concerning the product dorzolamide in combination with timolol, for the treatment of glaucoma. At first instance Floyd J found the patent to be invalid, lacking novelty and inventive step over a prior disclosure of the concomitant administration of both agents. Starting from this reference, the skilled team would have embarked on a development programme which would have taken some time and effort, but they would have been highly

---

8   *Williams v Nye* [1890] 7 RPC 62.
9   *British Celanese Ltd v Courtaulds Ltd* [1935] 52 RPC 171.
10   *Hallen Co v Brabantia (UK) Ltd* [1991] RPC 195.
11   *Glaxo Group Ltd's Patent* [2004] RPC 43.
12   *Teva UK Ltd v Merck & Co Inc* [2009] EWHC 2592 (Pat).

motivated and would have entertained a fair expectation of success and thus would have reached their goal without invention. An appeal against this decision was rejected[13] and leave to appeal to the UK Supreme Court was refused.

Often, patents claiming a combination product are drafted in the form of method-of-use patents, as the novelty of these combinations resides in the dosage regimen utilised.

Infringement of combination patents is committed by a third party, without consent, carrying out any of the infringing acts[14] in relation to the combination product – that is, all agents together or in relation to any individual agent of the combination (provided of course that all other integers of the claim, relating to for example the dosage and pharmaceutical form, are made out).

2.2 Enantiomers

The view of the English courts is that the disclosure of the racemic mixture in the prior art is no bar to the novelty and inventiveness of a claim to the single active enantiomer.

In the escitalopram[15] case, the High Court found the patent claiming the (+) enantiomer of citalopram to be novel and inventive, but insufficient. In relation to novelty the court held that whilst the (+) enantiomer is comprised in the racemic mixture, the existence of pure (+) enantiomer was not enabled by the basic citalopram patent and isolating the (+) enantiomer was inventive. At first instance Kitchin J concluded that the product *per se* claim was too broad as Lundbeck had a monopoly over all methods of isolation of the enantiomer, which was not commensurate with the technical contribution to the art (ie, one method of isolation of the enantiomer).

The Court of Appeal and House of Lords subsequently disagreed with this analysis, leading to clarification of the *Biogen* insufficiency principle (see note 25) and its limitation to claims to a product partially defined by its process of manufacture, rather than product claims *per se*. However, the judge's finding on novelty and obviousness was upheld and was subsequently followed in the *Daiichi* case.[16]

Of course the novelty and inventiveness of each claim to an enantiomer is not guaranteed. In the atorvastatin case,[17] for example, the Court of Appeal confirmed the finding of anticipation of Pumfrey J, where an earlier patent application had provided instructions to formulate three preferred enantiomers, of which one was claimed in the patent in suit. Interestingly, a trial is due in early 2012 where a challenge of insufficiency (based on the findings in escitalopram and levofloxacin) is to be brought against the earlier atorvastatin patent, claiming that it did not teach the skilled person how to obtain each of the different enantiomers.

---

13   *Merck Sharpe and Dohme v Teva UK Ltd* [2011] EWCA Civ 382.
14   Section 60 of the Patents Act 1977.
15   *Generics (UK) Ltd and Others v H Lundbeck A/S* [2009] UKHL 12.
16   *Generics (UK) Ltd v Daiichi Pharmaceutical Co Ltd* [2009] EWCA Civ 646.
17   *Ranbaxy Ltd & Arrow Generics Ltd v Warner-Lambert Company* [2006] EWCA Civ 876.

2.3   Selection inventions

The English law on selection patents was clarified in the olanzapine case.[18] In this case the Court of Appeal rejected the view that the disclosure of a class of compounds necessarily disclosed each member of that class, regardless of whether the class was designated by a general formula or a list of each member. To anticipate a selection invention, what is instead required is 'individualised description' of the relevant selected members of the class – be it an individual compound or a smaller class of the earlier disclosed compounds. In relation to inventive step, the mere arbitrary selection of a member of a class does not provide a technical contribution; there must be some particular advantage or technical effect which attaches to the selected compound. If the patent claims a class of selected compounds, then the particular technical effect must be produced by substantially all of the selected compounds.

2.4   Methods of use, method of treatment and secondary indications

Prior to implementation in 2007 of amendments to the EPC 2000, second-medical-use claims could only be worded in the 'Swiss-type'[19] form 'use of substance X in the manufacture of a medicament for the treatment of disease Y'. The Patents Act 2004 amended the Patents Act 1977, introducing a new Section 4(A), of which subsection (4) enables protection of second medical uses by a direct claim (ie, 'the use of X for the treatment of disease Y'). Furthermore, the European Patent Office (EPO) Enlarged Board of Appeal (EBA) in case G02/08 indicated that Swiss-type claims are no longer allowable. (This decision is not retrospective.)

Regardless of the form of the claim, it is established law that second-medical-use claims relating to the use of a substance for the same therapeutic application as the prior art, but claiming a different technical effect or new information about how the therapy works, will be found to lack novelty.[20] The Court of Appeal[21] in the *Bristol-Myers Squibb* case went further and stated that a claim which defined the new use of the known therapeutic agent in terms of a dosage regimen were disguised method-of-treatment claims and lacked novelty over the use of the agent to treat the same condition at a different dose.

In the finasteride[22] case, the Court of Appeal again considered claims to a new dosage regime for using the same therapeutic agent to treat the same condition (androgenic alopecia). Although this case came before the court before the EBA had given its ruling in G02/08, the TBA had referred questions to the EBA and thus the court was aware that this was a live issue at the EPO. The Court of Appeal allowed the appeal, ruling that dosage regimen claims were novel and inventive, bringing English practice in line with the practice of the EPO. In doing so, Jacob LJ indicated that the court was able to depart from its own precedent on the grounds that in the earlier *Bristol-Myers Squibb* case there was no clear rationale in relation to novelty and

18   *Dr Reddy's Laboratories v Eli Lilly* [2010] RPC 9.
19   Originally allowed by the EPO Enlarged Board of Appeal in G 05/83 EISAI/*Second Medical Indication* [1985] 3 *OJEPO* 64 and followed by the Patents Court in *John Wyeth's and Schering's Applications* [1985] RPC 545.
20   *Bristol-Myers Squibb v Baker Norton Pharmaceuticals* [1999] RPC 253.
21   *Bristol-Myers Squibb v Baker Norton Pharmaceuticals* [2001] 1.
22   *Actavis v Merck* [2008] RPC 26.

thus the court could instead follow the 'settled view' of the EPO. In conclusion, both the EPO and the English courts now accept that where it is already known to use a medicament to treat an illness, Article 45(5) of the EPC does not exclude that this medicament be patented for use in a different treatment by therapy of the same illness. Furthermore, patenting is not excluded where a dosage regime is the only feature claimed which is not comprised in the state of the art; claims to dosage regimes, where the dosage regime is the only 'new' feature, can be novel.

2.5   Formulations and physical forms

Patents may be granted for different polymorphs, metabolites, compound analogues or formulations. The validity of each patent will depend upon each new compound or form of compound being new and inventive (together with a sufficient description of such compound). There are no generally broad statements that can be made about such patents, save that given that the compound itself will already have been discovered, it may be that patents relating to crystal forms, for example, are more liable to be revoked by the courts than the basic compound patents.[23]

2.6   Reach-through claims

The view of the UK Intellectual Property Office (IPO) on reach-through claims is that such claims are insufficient and unsupported by the patent description.

If the inventor has identified a new biological receptor or enzyme which is implicated in a disease or collective group of diseases, the patentee would wish to have a claim to compounds which activate or interact with this receptor and thus are useful in the treatment of the disease conditions. The problem with such a claim is that to identify the downstream compounds being claimed, the skilled person would need to undertake extensive experimentation, and thus the claimed invention cannot be performed without undue trial and error (ie, the patent is insufficient). Further, some of the compounds which are claimed may inevitably be compounds that have already been identified (after all, the skilled person would normally start testing compounds that were already known) and the claim would lack novelty.

Thus whilst there is no absolute bar to claims such as these, they will be liable to a myriad of attacks as to their validity, as demonstrated in the *Lilly v HGS* case.[24] HGS had a patent on a protein, neutrokine-α, which claimed numerous different applications for the protein. The Court of Appeal upheld the judgment of Kitchin J that claims to the uses of the protein were purely speculative and thus the invention was not susceptible of industrial application. Following the overturning of the industrial-applicability ruling in the UK Supreme Court in *Human Genome Sciences Inc v Eli Lilly and Co* [2011] UKSC 51, the case will now return to the Court of Appeal in order to hear the HGS appeals against Kitchin J's (as he then was) findings on inventive step and insufficiency of the reach-through claims in the patent.

---

23   See *Les Laboratoires Servier v Apotex Inc* [2008] EWCA Civ 445 – a patent to a crystalline form of perindopril was held invalid. The exception to this suggestion is *Leo Pharma A/S v Sandoz Ltd* [2009] EWHC 996 where Floyd J upheld a patent claiming a crystalline hydrate of a known vitamin D analogue, calcipotriol, and its use in a topical preparation for the treatment of psoriasis.

24   *Eli Lilly & Co v Human Genome Sciences Inc* [2010] EWCA Civ 33.

3. DNA, biologicals and personalised medicine

The key intellectual property rights protecting biotechnology-derived inventions (and so the key incentive for innovation in this sector) are patents. Perversely, the development of patent law in Europe relating to biotechnology-derived inventions has not been satisfactory. One finds that both statute and case law have been subject to compromise positions being taken to fudge the legal position, often to accommodate the diverse range of moral and ethical positions taken across the EU member states. This has resulted in a number of biotechnology-driven areas suffering from legal uncertainty as to what is patentable subject matter and/or what granted European patent claims can be enforced (as a matter of principle) in the national courts around Europe.

In the United Kingdom, a number of biotechnology patents have been litigated through the court system.[25] Issues surrounding claim scope, sufficiency, obviousness and industrial applicability have come to the fore when biotechnology patents have sought to be enforced. When reviewed, at the heart of most of the cases is the question of whether the invention disclosed in the patent specification is deserving of the interpretation that the patentee is seeking to place on it in order to prove the claimed infringement.

In the biotechnology field this is complicated primarily because many gene-based inventions are often claimed, in part, by way of the identified gene itself. However, the gene is not the 'effector' molecule of the invention, this usually being the gene product. It follows that the characterisation of a gene is necessarily a major step away from identifying the function of the gene product subsequently identified. Further, to satisfy the industrial applicability requirement under patent law, the function of the gene product, not the underlying gene, must be disclosed within the patent specfication.

The broad principles that can be extracted from past biotechnology-specific UK case law are as follows:

- One method of achieving a result does not necessarily then give a patentee the right to claim all ways of achieving this result.[26]
- Patentees need to ensure that if their inventions are defined by reference to a biological reference (eg, a sample of naturally derived product), then this 'reference product' is sufficiently well defined to enable third parties to determine whether or not they fall within the claims of the patent.[27]
- Claims to 'isolated' DNA are limited to products containing purified fragments of DNA corresponding to the claimed sequence, unless the term is otherwise defined within the patent.[28]
- Defining an invention by means of a class of molecules can be risky if the patent does not contain a definite and reliable manner of determining

---

25   *Biogen Inc v Medeva Plc* [1997] RPC 1; *Kirin-Amgen Inc v Hoechst Marion Roussel Ltd* [2004] UKHL 46; *Monsanto Technology LLC v Cargill International SA* [2008] FSR 8; *Eli Lilly & Co v Human Genome Sciences Inc* [2010] EWCA Civ 33.

26   *Biogen* above.

27   *Amgen* as above.

28   *Monsanto* as above; *Eli Lilly* as above.

29   *Monsanto* as above.

membership of the claimed class of molecules.[29]

- Inventions based on gene sequences discovered via data mining or bioinformatics tools are patentable provided sufficient information is contained within the patent so as to satisfy the need to provide a plausible industrial application for the discovered nucleotide sequence.[30]

3.1   EU law trumps UK law

To place a more complex gloss on the UK-specific position, one has also to recognise that biotechnology is an area in which patent law has been harmonised by the European Union.[31] The Biotechnology Directive was passed on July 30 1998 and was due to be implemented by member states by July 30 2000. The implementation was not consistent across the European Union. Some of the inconsistencies have now been settled through the Court of Justice of the European Union (CJEU) ruling in the *Monsanto v Cefetra case*,[32] which held that:

- The Biotechnology Directive ushered in a new regime restricting the scope of protection of DNA-based claims to only those products where the DNA is performing its disclosed function.
- The Biotechnology Directive applied retroactively to all patents filed or granted before the Directive came into force.
- Member states are not permitted to 'gold-plate' the Directive and allow normal, product-based protection for claims to DNA molecules.
- All of the above restrictions were consistent with TRIPS and did not unfairly prejudice the rights of holders of biotechnology-related European patents.

In this way, the CJEU has restricted the scope of protection (and so the enforceability) of a large swathe of claims contained within biotechnology-related patents. The ruling does not contain a great deal by way of explanation or reasoning.

The Biotechnology Directive (which has been implemented by the EPO into the Implementing Rules of the EPC) contains a number of provisions that could be prejudicial to the protection through the patent system of cell-based inventions, particularly stem-cell-based inventions. Article 5(1) of the Biotechnology Directive prevents the patenting of any cell that can develop into any tissue (totipotent cells). Article 6(2) of the Directive also operates to prevent the patenting of any invention that in some way uses human embryos in an industrial or commercial manner.

However, historically, some stem-cell-based inventions have been protected, and both the EPO and UK IPO have developed guidelines to assist applicants in formulating claims in a manner which permits patents to be granted. In the EBA decision in the WARF case,[33] the EPO decided that, provided that an invention could be put into practice as at the filing date by a method that did not entail the destruction of an embryo, this avoided the Article 6(2) exclusion. As a result of that

---

30   *Eli Lilly* as above.
31   The Biotechnology Directive, Directive 98/44/EC, *OJ* 1998 L 213, p 13.
32   Case C-428/08 *Monsanto Technology LLC v Cefetra BV & Others.*
33   Decision G 2/06, *Wisconsin Alumni Research Foundation (WARF).*
34   See www.ipo.gov.uk/pro-types/pro-patent/p-law/p-pn/p-pn-stemcells-20090203.htm.

case, the UK IPO amended its practice such that its procedure mirrors the EPO's approach.[34]

The opinion of Advocate General Bot in the *Brüstle* case[35] now threatens this position. This case has been referred to the CJEU as a result of a referral from the Bundespatentgericht in a revocation action brought by Greenpeace. The patent is directed at the use of pluripotent stem cells and the referral seeks to clarify the meaning both of 'embryo' and also 'industrial uses' of such embryos in the Biotechnology Directive. Unfortunately for the biotechnology industry, the Advocate General provided an opinion that could extend the Article 6(2) exclusion to patentability well beyond what the EPO currently imposes. Should the CJEU follow the Advocate General's opinion, it seems likely that any cell-based invention will not be patentable if it can be traced at some stage in its development to derivation from a cell extracted from an embryo by a process that, the Advocate General considers, would "necessarily entail the destruction of the human embryo".[36] Of course, the EPO is not bound by decisions of the CJEU, because it is not an organ of the European Union. However, it would seem strange if the EPO continued to grant patents in the knowledge that all EU member state courts would find them invalid as concerning excluded subject matter.

4.   Acts of patent infringement

4.1   Infringement

Where the patented invention is a product, a patent is infringed by a person who makes, disposes of, offers to dispose of, uses, imports or keeps the product. Where the patented invention is a process, the patent is infringed by a person who uses or offers the process for use in the United Kingdom when he knows, or it would be obvious to a reasonable person, that the use of the process would amount to an infringement of the patent. A process patent is also infringed by a person who disposes of, offers to dispose of, uses, imports or keeps any product obtained directly by means of the patented process.

The claims of the patent define the patentee's monopoly and therefore determine whether a particular product or process infringes the patent. However, when construing the claims of a patent, a court must have regard to the description (specification) of the patent and the drawings in order to interpret the claims.[37] This approach is known in the United Kingdom as 'purposive' construction of the patent. The approach developed initially in the *Catnic* case.[38] In this case, the House of Lords rejected the old approach of literally construing the claims of a patent.

This was developed in the Improver case,39 where Hoffmann J (as he then was) formulated a number of questions that he suggested would assist the courts in purposively construing patent claims where the alleged infringement was in relation

---

35   Case C-34/10 *Oliver Brüstle*.
36   See paragraphs 109 and 110 of the Advocate General's opinion.
37   Article 69 of the European Patent Convention.
38   *Catnic Components Ltd v Hill & Smith Lt*d [1982] RPC 183.
39   *Improver v Remington* [1990] FSR 181.

to a variant of the patented invention. These questions were used by the UK courts for more than a decade but the approach was restated by Lord Hoffmann in the *Kirin-Amgen* case in the House of Lords.[40] In this latest decision Lord Hoffmann distanced himself somewhat from the 'Improver Questions' that he previously formulated. As it now stands, the main issue that the court should seek to determine when purposively construing the claims of a patent is "what the person skilled in the art would have understood the patentee to mean by the language of the claims".[41]

This approach, said Lord Hoffman, was consistent with that required under the EPC.[42] The *Improver* Questions were no more than guidelines to assist in this exercise. Further, the underlying principle that patent claims should be construed purposively was expressly endorsed as being consistent with the approach required by the EPC.[43]

4.2   Contributory infringement

A patent is also infringed by a person who supplies or offers to supply the means to make the patented product or use the patented process, if that person knows, or it is obvious to a reasonable person, that those means are suitable for putting, and are intended to put, the invention into effect in the United Kingdom. The means supplied must relate to an "essential element of the invention".[44]

The issue of contributory infringement has recently been reviewed by the English Court of Appeal.[45] In this judgment an agricultural machine was the subject of the claim. The issue was whether or not the sale of a machine with a non-infringing component, but which could be switched for an infringing type, could amount to an infringement under Section 60(2). The alleged infringer's appeal was based on the requirement that Section 60(2) should require proof that the supplier knew, or the circumstances made it obvious, that the user actually intended to make the switch. It was also contended that there could not be infringement under Section 60(2) unless it could be shown that an end user had, in fact switched the components thereby creating an infringing device.

The Court held that what was required under Section 60(2) was a 'reasonable man' test (ie, it was not necessary that the supplier actually knew that the product would be put to an infringing use). Further, the intention to put the product to an infringing use only has to be that of the end user and not that of the supplier. This can be made out by proving on a balance of probabilities that some end user would use the product for an infringing use. Mere possibility of infringing use, it seems, would not be sufficient as the court would tend to exclude "maverick or unlikely uses of a thing".[46]

From the above it can be seen that the Court rejected the submission that actual infringement by an end user was not required in order to make out infringement

---

40   *Amgen* as above.
41   See paragraph 71 of *Kirin-Amgen* above.
42   Article 69 and the Protocol on its interpretation.
43   Protocol to Article 69 of the EPC.
44   Section 60(2) of the Act.
45   *Grimme Maschinenfabrik GmbH & Co KG v Derek Scott (t/a Scotts Potato Machinery)* [2010] EWCA Civ 1110.
46   *Grimme* as above, para 116.

under Section 60(2). In fact, an infringement action based on Section 60(2) could be brought by a patentee upon the sale of the 'means essential', which could be a long time prior to any infringing use actually occurring.

There is a concept in English law of joint tortfeasance. Under this joint tort, a non-infringing party can be held jointly liable with the actual infringer for a patent infringement. The circumstances that could give rise to such liability are where, for example, a director or controlling shareholder of a company goes beyond the simple exercise of control over the actions of a company through the constitutional organs of that company. The principles for such infringement were reviewed in the Court of Appeal in the *MCA* case,[47] where the Court approved the principles set out in the leading cases of *CBS Songs*[48] and *Unilever*.[49] Although there is no defined tort of 'procuring the infringement of a patent', it is possible that through such procurement a party may incur liability as a joint tortfeasor. In order to prove a party is a joint tortfeasor it will be necessary to demonstrate that a party "intends and procures and shares a common design that the infringement takes place" and/or that the alleged joint tortfeasor has "made the act his own".

For example, in a recent case joint tortfeasance was proved where a director had been instrumental in negotiating the supply of patent-infringing goods and, almost single-handed, made the decision to supply two categories of infringing product.[50] Therefore, the director's procurement of the infringement by the company in this case supported a finding of joint tortfeasance.

4.3 National treatment of 'Bolar'-type provisions

The provision known as 'Bolar' is a complete defence to patent infringement if the acts complained of fall within its scope. The exemption is set out in Article 10(6) of Directive 2001/83/EC (as amended), which states:

Conducting the necessary studies and trials with a view to the application of paragraph 1, 2, 3 and 4 [of Article 10] and the practical requirements shall not be regarded as contrary to patent rights or supplementary protection certificates for medicinal products.

This was implemented in the United Kingdom through Section 60(5)(h) of the Act, which states:

An act which, apart from this subsection, would constitute an infringement of a patent for an invention shall not do so if:

....

(i) it consists of –

(i) an act done in conducting a study, test or trial which is necessary for and is conducted with a view to the application of paragraphs 1 to 5 of article 13 of Directive 2001/82/EC or paragraphs 1 to 4 of article 10 of Directive 2001/83/EC, or

(ii) any other act for the purpose of the application of those paragraphs.

The application of Bolar in the United Kingdom is clearly set out in a practice

---

47   *MCA Records Inc v Charly Records Ltd* [2001] EWCA Civ 1441.
48   *CBS Songs Ltd v Amstrad Consumer Electronics Plc* [1988] AC 1013.
49   *Unilever Plc v Gillette (UK) Ltd* [1989] RPC 583.
50   *Boegli-Gravures SA v Darsail-Asp Ltd and Andrei Ivanovich Pyzhov* [2009] EWHC 2690 (Pat).

note issued by the Intellectual Property Office and UK medicines agency (the MHRA).[51] To this end, the scope of Bolar is relatively limited, being applicable only to trials or studies performed for the purposes of obtaining a generic marketing authorisation within the United Kingdom including:

- manufacture or import of active substances and validation of manufacturing processes;
- manufacture or import of finished product and validation of manufacturing process;
- development/testing/use of analytical techniques associated with manufacture of the active and the finished product;
- conducting pre-clinical tests, clinical and bioavailability trials and stability studies on the medicinal product; and
- compilation and submission of a marketing authorisation and samples of products to regulatory authorities.

Other territories in the European Union give a broader scope to the Bolar provision (eg, the UK Bolar provision would not apply to clinical trials performed by an innovator company rather than a generic company seeking to demonstrate bioequivalence prior to making an abridged application under Article 10 of Directive 2001/53/EC). Given the inconsistent implementation of this provision through Europe, if an English court were to follow the letter of the IPO practice note, it is possible that a referral to the CJEU could be made to determine whether or not the United Kingdom has properly implemented Article 10(6).

Currently, the IPO is conducting a consultation in order to understand whether the manner in which the Bolar defence has been implemented in the United Kingdom is harming the attractiveness of the UK as a base for clinical trials. Hence IPO guidance relating to the scope of this defence to patent infringement in the United Kingdom could change in the near future.

4.4 Experimental-use exemptions

In addition to the Bolar defence to patent infringement, Section 60(5)(b) of the Act also provides that a third party using a patented invention for experimental reasons that are related to the subject matter of the invention does not infringe that patent. To be an act carried out for 'experimental purposes' it should be carried out in order to discover something unknown or to test a hypothesis. Experiments with commercial exploitation in view are not excluded. The requirement is that the experiment should be carried out in order to obtain information, irrespective of intent.

In *Monsanto v Stauffer*[52] the English Court of Appeal compared Section 60(5)(a), which has a specific limitation directed at 'private and non-commercial' purposes, and Section 60(5)(b), which does not. This implies the limitation does not apply and so Section 60(5)(b) extends to exempt experiments with both commercial and non-

---

51   See www.ipo.gov.uk/policy-issues-patents-pharmaceutical-activities.htm.
52   *Monsanto Co v Stauffer Chemical Co* (1985) RPC 515.

commercial ends. Hence 'experimental purposes' would include trials carried out "in order to discover something unknown, or to test a hypothesis, or even in order to find out whether something which is known to work in specific conditions ... will work in different conditions".

Hence this exemption concerns acts which generate new knowledge. It is therefore most likely that experiments or trials conducted only in order to demonstrate to a third party that a product works in the manner claimed would not be considered to be done for 'experimental purposes'.

The *Monsanto* case is consistent with the Court of Appeal's decision in *Auchincloss v Agricultural & Veterinary Supplies Ltd.*[53] It held that making and experimenting with a patented invention merely for the purposes of gaining official approval would not fall within the experimental-use exception. It seems, though, that if the results of experiments conducted as 'genuine experiments' were sent to a regulatory authority, they could well fall within this exemption.

In *Smith, Kline and French v Evans Medical*,[54] the English Patents Court was asked to determine where the limits of the experimental-use exemption were to be found. In this case it was ruled that whilst the exemption could be relied upon as regards attempts to improve a patented process, it did not also extend to cover the use of a product in those experiments that was also the subject of a patent.

4.5   Submitting authorisations and offers to supply

In the United Kingdom, making an offer to supply a patent-protected product after the date of expiry of the patent is not regarded as a patent infringement.[55] Therefore obtaining a marketing authorisation, absent proof of an intention to supply prior to the expiry of the patent, does not give rise to an automatic cause of action. It is therefore likely, although it has never been tested, that a patentee would require more than simply a marketing authorisation and silence on behalf of the would-be infringer in order to found an infringement action.

4.6   Summaries of product characteristics (SmPCs)

An SmPC contains a summary of the key characteristics of a medicinal product and includes the indications for which the product is authorised. It is most likely that issues of patent infringement in relation to this document will arise dependent upon the inclusion of patented uses or dosing regimens in the SmPC. There is no UK case law relating to patent infringement solely on the basis of the content of the SmPC. However, practice within regulatory authorities in the European Union permits generic companies to create unharmonised SmPCs across EU member states (through the deletion of patented uses of the product) should certain indications or dosing regimens be protected by third-party patents in member states of the European Union.

However, this guidance has no binding legal effect on courts trying issues of

---

53   *Auchinloss and Another v Agricultural and Veterinary Supplies and Others* [1999] RPC 397.
54   *Evans Medical's Patent* [1998] RPC 517.
55   *Gerba Garment v Lectra Systems* [1995] RPC 383.

patent infringement. It therefore remains unknown how an English court would approach such matters. Presumably, the more evidence that existed to establish intent in the company marketing the generic drug 'off-label' and in relation to a patented indication or dosing regimen, the more risk that particular company runs of being sued for patent infringement.

5. Patent enforcement

5.1 Obtaining information on the infringer and the infringement

A patentee may bring an action against a non-infringing third party in order to identify infringers or infringing goods. The patentee could request a *Norwich Pharmacal* order.[56] Such orders are useful to patentees aware of infringing goods entering the United Kingdom but unable to ascertain the identity of the infringer. In these cases it is likely that certain governmental agencies or other third parties could possess information that would assist the patentee. The UK courts would be sympathetic to the granting of an order placing the third parties under a duty to assist the patentee by giving full information and disclosing the identity of the infringer.

A patentee may also seek pre-action disclosure, in accordance with the provisions of the Civil Procedure Rules Part 31.16, of specified documents or classes of documents which are relevant to the issue of infringement. However, such an application to the court must be supported by evidence and will only be awarded if the court's view is that such disclosure is desirable to dispose fairly of the anticipated proceedings; to assist the dispute to be resolved without proceedings, or to save costs. An order for pre-action disclosure will not be made if the application is purely speculative.[57]

A further way to obtain information about the identity of an infringer would be to write to, for example, a supplier requesting information as to the identity of the manufacturer of the infringing goods. Such letters should be drafted carefully with regard to the threats provisions (see section 5.4 below).

5.2 Interim relief

The current view of English courts is that a party wishing to launch a generic pharmaceutical product, in the face of an innovator secondary patent, should 'clear the way'. What this means is that there is a burden on the party wishing to launch a generic product to seek either a declaration of non-infringement from the patentee or if necessary the court, or to seek to revoke the patent prior to launch. If the third party instead chooses to wait until there is insufficient time to carry out either of these two actions before notifying the patentee of its intentions, this will be taken into account by the court when deciding whether to grant an interim injunction.[58]

The principles which the court should consider when determining whether to grant an interim injunction are set out in the *American Cyanamid*[59] case. If damages would be an adequate remedy and the defendant would be in a position to pay them,

---

56   *Norwich Pharmacal Co v Commissioners of Customs & Excise* [1974] RPC 101.
57   *BSW Ltd v Balltec Ltd* [2007] FSR 1.
58   *SmithKline Beecham v Apotex Europe Ltd* [2003] EWCA Civ 137.
59   *American Cyanamid Co v Ethicon Ltd* [1975] RPC 513.

then no injunction should be granted. If damages would not be an adequate remedy, the court must then consider whether, if the defendant were to succeed at trial, he would be adequately compensated by the claimant's undertaking as to damages. If so, then there would be no grounds to refuse the interim injunction. If there is doubt as to the adequacy of damages in both scenarios, then the court must consider the 'balance of convenience'; and if other factors appear to be evenly balanced, then the court should take measures to preserve the status quo.

In many cases involving generic pharmaceuticals the High Court would tend to grant an interim injunction, to maintain the status quo and prevent a hypothesised price crash for the product in question.[60] However, where the court has evidence that the price crash is not an issue, it is possible that no injunction would be granted.[61]

If an interim injunction is wrongly granted such that a cheaper generic product should have been on the market for the period of the injunction, then a further losing party is the National Health Service (NHS). It may therefore be in the interests of the NHS or the Department of Health to be joined as a party to the action to ensure that it can benefit under the cross-undertaking for damages provided by the patentee. In the *Abbott v APS*[62] case, the Secretary of State for Health applied to be named in the cross-undertaking for damages. In the event Pumfrey J held the relevant claim to be invalid on an application for summary judgment, although Jacob LJ later suggested that it should be standard practice where there is an application for an interim injunction in respect of a pharmaceutical that the patentee should give notice of the application to the Department of Health, in case it wishes to seek an undertaking as to damages.[63]

5.3   'Springboard' relief

The courts in England and Wales are able and willing to grant post-patent-expiry injunctions against those who pre-empt the expiry of the patent, thus acquiring a commercial advantage upon patent expiry.[64] In the pharmaceuticals sector, actions resulting in such an award could be the importation of a generic pharmaceutical product prior to patent expiry to ensure that product can be sold in the United Kingdom as soon as the patent expires.

It should be remembered that an offer before patent expiry to supply a product after patent expiry is not an infringement.[65]

5.4   Unjustified threats

Threats provisions are contained in Section 70 of the Act. A person aggrieved by threats made against him by a patent proprietor may seek remedy from the court in the form of a declaration that the threats were unjustified, an injunction against the

---

60   *SmithKline Beecham v Generics [UK]* unreported, October 23 2001, Patents Court *SmithKline Beecham v Apotex* as above.

61   *Cephalon, Inc & Others v Orchid Europe Ltd & Generics (UK)* [2010] EWHC 2945 (Pat).

62   *Abbott Laboratories and Abbott Laboratories Ltd v Approved Prescription Services, Ranbaxy Europe Ltd and Generics (UK) Ltd* [2004] EWHC 2723 (Pat).

63   *SmithKline Beecham v Apotex Europe Ltd* [2006] EWCA 658 (Civ) at para 77.

64   *Dyson Appliances v Hoover (No 2)* [2001] RPC 27.

65   *Gerber* as above.

continuance of the threats and damages for any loss suffered.[66] The person threatened shall not be entitled to the remedy sought if the patentee can demonstrate that the acts complained of constituted an infringement of the patent. However, if the claimant can demonstrate that the patent is in fact invalid he will be entitled to the relief sought, unless the patentee is able to show that at the time of making the threats he did not know and had no reason to believe that his patent was invalid.

In assessing whether the patentee has 'reason to suspect' that its patent is invalid, Arnold J indicated that an objective assessment should be carried out based on what the patentee knows at the time the threat is made.[67] In the judge's view there should be a specific reason for suspecting invalidity, rather than a general awareness that the patent might be held to be invalid.

Various actions are excluded from these provisions: a patentee threatening infringement proceedings against any third party manufacturing or importing an infringing product;[68] and a patentee threatening the importer or manufacturer of infringing products with regard to any other infringing acts undertaken in relation to that product (eg, offering for sale[69] would not fall foul of the provisions).

Finally, if the patentee merely provides factual information about his patent, makes enquiries for the sole purpose of discovering whether or not his patent has been infringed and makes an assertion about the patent for the purposes of such enquiries, he shall not be deemed to have made a threat.[70]

5.5   Remedies

Section 61 of the Act sets out the remedies available to a successful patentee, being:

- an injunction restraining further infringement;
- an order for delivery-up or destruction of the infringing products;
- damages or an account of profits; and
- a declaration that the patent is valid and has been infringed.

When assessing the quantum of any damages award, the court will seek to restore the patentee to the position it would have been in but for the wrongful acts of the infringer. Factors to be taken into account by the judge include the lost profits of the patentee, moral prejudice caused to the patentee, any price reduction necessitated by the infringer's competing product being on the market, and losses in respect of sales of goods normally sold together with the patented goods. Alternatively, damages can be assessed on the basis of a reasonable royalty rate if the defendant had sought a licence to the patent.

An account of profits is a restitutionary remedy with the purpose of depriving an infringer of the profits generated by committing wrongful acts.[71] The defendant is thus able to deduct from the gross margin all allowable costs, such as manufacturing

---

66   Section 70(3) of the Act.
67   *FNM Corporation Ltd v Drammock International Ltd* [2009] EWHC 1294 (Pat).
68   Section 70(4)(a) of the Act.
69   Section 70(4)(b) of the Act.
70   Section 70(5) of the Act.
71   *Spring Form Inc v Toy Brokers Ltd* [2002] FSR 17.

and distribution costs, to arrive at the net profit.

Furthermore, the infringer must only account for the profits on those products which were made by infringement of the patent. Profits can therefore be apportioned appropriately. Clearly, in cases where the invention is essential in creating the defendant's product whatever that may be, apportionment of profits would not be appropriate.[72]

6. Compulsory licensing

Compulsory licences are not frequently granted in the United Kingdom. A compulsory licence can only be granted by the Comptroller General of Patents.[73] An application for a compulsory licence cannot be made to the Comptroller until three years after the date of grant of the patent in question. The grounds upon which a compulsory licence can be granted depend on whether the patent owner is a WTO proprietor or a non-WTO proprietor. A proprietor is a WTO proprietor if he is a national of, or is domiciled in, a country which is a member of the World Trade Organization, or he has a real and effective industrial establishment in such a country. All other proprietors are non-WTO proprietors.

There are three grounds for relief by the grant of a compulsory licence for WTO proprietors as follows:

- A demand in the United Kingdom for a patented product is not being met on reasonable terms.
- By reason of the refusal of the proprietor to grant a licence on reasonable terms:
  - the exploitation in the United Kingdom of any other patented invention which involves an important technical advance of considerable economic significance in relation to the invention for which the patent was granted is hindered; or
  - the establishment or development of commercial or industrial activities in the United Kingdom is unfairly prejudiced.
- By reason of conditions imposed by the proprietor on the grant of licences or on the use of the patented product or process, the use of materials not protected by the patent or the establishment or development of commercial or industrial activities in the United Kingdom is unfairly prejudiced.

There are five grounds of relief for non-WTO proprietors. The grounds are similar to those for WTO proprietors, but allow a compulsory licence to be granted in certain situations where the invention is capable of being commercially worked in the United Kingdom but is not being so worked. Such a ground cannot be used against a WTO proprietor because under Article 27(1) of TRIPS a patent right cannot be restricted on the basis of whether the proprietor chooses not to work his invention in the United Kingdom but instead to import the relevant products.

---

72   *Hoechst Celanese International v BP Chemicals* [1999] RPC 203.
73   Sections 48, 48A and 48B of the Act.

7. Ownership, inventors and compensation

A patent can be granted to: (i) the inventor or joint inventors; (ii) other persons entitled to property in the invention; or (iii) successors in title. The inventor is the 'deviser of the invention'; in other words, the inventor is the person who 'came up with the inventive concept'.

In the United Kingdom, an invention will belong to the inventor's employer where either (i) it was made in the course of the normal duties of the employee or in the course of duties falling outside his normal duties, but specifically assigned to him; or (ii) the invention was made in the course of the duties of the employee and he had a special obligation to further the interest of the employer. In determining whether an invention belongs to an employee inventor or his employer, the court will consider what the employee's duties actually were at the time the invention was made, rather than assume that they are accurately described in any employment contract. The court will then consider whether the circumstances under which the particular invention in question was made were such that the invention might reasonably be expected to result from carrying out those duties.

Where the patent belongs to the employer, an employee inventor can claim compensation from his employer where the invention or the patent for it is of outstanding benefit to the employer, taking into account the size and nature of the employer's business. In practice, it has been difficult for employee inventors to establish an outstanding benefit and there has only been one reported case where such a claim has been successful.[74]

8. Branding and designs

Holders of marketing authorisations are not free to choose any name they desire in order to market their products. The names that holders of marketing authorisations choose to market their products under are known as 'invented names'. Directive 2001/83/EC requires that 'invented names' should not create confusion with the common name of the drug (or the name of any other medicinal product). In addition, the chosen 'invented name' should not be misleading as regards the therapeutic effects of the product, its composition or its safety. A guideline produced by the MHRA provides further details as to the type of considerations that the MHRA will take into account when assessing the acceptability of an 'invented name'.

'Invented names' can also be registered as trade marks (and usually are when accepted by the regulatory agencies). However, one should not assume that medicinal products can be promoted in the same way as other products protected through trade marks.

Given that it is possible to register 'invented names' as trade marks, it is important that such names should not contain INN stems in the stem location as published by WHO.[75] Details of the INN system can be obtained from the World Health Organization website.[76]

---

74   *Kelly and Another v GE Healthcare Ltd* [2009] EWHC 181 (Pat).
75   World Health Assembly Resolution 46.19.
76   See www.who.int/medicines/services/inn/innquidance/en/index.html.

Generic manufacturers can incorporate the generic name of the drug into their product names along with the name of the company manufacturing the product, provided these companies comply with the necessary guidelines published by the MHRA.

Whilst brand owners can therefore protect their products though the use of trade marks, such protection is of less value than a patent as it does not prevent generic producers manufacturing and selling equivalent products. Further, parallel importers are able to 'over-sticker' a product sold under a trade mark in one EU member state and import this into another if the price differential in those member states make such exercises financially viable.[77]

As products can be protected by trade marks, so designs may also come into consideration. However, if the design in question is not protecting a proprietary drug delivery device that provides a significant advantage over the basic product, it is also unlikely to present a significant barrier to generic entry.

The promotion of medicines is illegal under the provisions of the Medicines Act 1968 (as amended). The provision is strictly enforced and promotion by third parties (even those with no commercial links to the marketing authorisation holder) can result in prosecution for those individuals concerned.[78] This is a major difference between medicines legislation in the United States and Europe, and companies need to take care that promotional efforts on one side of the Atlantic do not 'spill over' and become seen as illegal marketing of medicines in Europe. Facilitating online social networking and chat room activity in relation to particular products needs to be closely monitored if these provisions are to be respected.

9.   Protecting valuable information

The preservation of confidentiality is of paramount importance in knowledge-based industries for a number of reasons. First, it is important that patent filings are not jeopardised through a premature disclosure to the public. This can occur where disclosures are made to third parties where there is no obligation of confidence between the parties or through publication by the company itself. Often, companies will choose to protect certain parts of their business through trade secrets. (These do not need to be technical in nature – for example, most companies' lists of customers, potential business partners and details of collaboration and partnering negotiations will be kept secret.) Where elements of the business are protected in this manner, it is critical that companies have in place robust security procedures to ensure that the documents and/or knowledge that constitute the trade secrets in question do not pass outside the company's control.

Companies should establish a clear standard operating procedure to preserve confidence in their research and development and other trade secrets through a variety of measures. These would ordinarily include:

- including express confidentiality obligations in contracts of employment with all employees;

---

77   For a discussion of the obligations placed on a parallel importer, see Commission COM(2003) 839 final.
78   Case C-421/07 *Criminal Proceedings against Frede Damgaard*.

- having a clear policy of invention disclosure from employees to the company and early participation of patent attorneys in the invention protection procedure to decide how the particular invention is to be taken forward;
- never engaging in research and development or other work of a business-critical nature with third-party contractors without clear intellectual property assignment provisions and confidentiality obligations contained within the contracts with those commissioned to carry out the work;
- adhering to strict policies that stipulate that confidential information is only accessible by the majority of employees and contractors on a 'need to know' basis; and
- ensuring that the documents in which the confidential information is embodied are in secure document servers where copying is either not possible or as difficult as possible.

Pre-priority filing disclosures can destroy the novelty in the invention for which the patent is filed. Even after this date, care should be taken during the year following the initial filing as this initial filing is often used to establish the priority date for the invention and nothing more. In such circumstances it can be disastrous for a patentee to have published during the priority year where the priority date is later proven not to be made out (due to additions to the patent document over the first year that effectively make the invention a 'different invention' from that which is disclosed in the priority document).

Issues often arise where employees leave one company and move to a competitor, or themselves set up a competing company. In such cases, it is possible that the earlier employer could use the common law of confidentiality against the ex-employees should they attempt to use business confidential information or trade secrets.[79] Alternatively, in recent years patent entitlement actions have been used as a way of claiming rights in inventions filed by ex-employees following a move from an employer, if the earlier employer considers that inventions made while the ex-employees were in its previous employment have been taken and exploited elsewhere.[80]

10. Counterfeiting

10.1 Counterfeiting and IP

Counterfeit medicines are a problem in the global economy. Although it is a more serious problem in developing countries, 1% of all medicines in developed markets such as the United Kingdom are estimated to be counterfeit. Such products will not infringe patents – typically the products are dangerous mainly because they do not contain the patented active ingredients that are required to treat the medical condition for which the product has been prescribed. Therefore, counterfeit products will, in the main, infringe the trade marks and copyright in the packaging and get-

---

79   See *Faccenda Chicken v Fowler* [1985] FSR 105 for a full explanation of the necessary quality the information must possess to be protected under the common law.

80   *Markem Corporation and Another v Zipher Ltd and Markem Technologies Ltd and Others v Buckby and Others* [2005] EWCA Civ 267.

up of the original product. UK law contains criminal provisions for trade-mark and copyright infringement. These laws are regularly used against counterfeiters.

10.2   MHRA and counterfeiting

The UK medicines regulator, the MHRA, has a specialist anti-counterfeiting unit, known as the Enforcement Group, which intercepts products at various levels in the supply chain as well as coordinating product recalls where counterfeit products are discovered. An example of the cooperation between the industry and MHRA is illustrated by the recent example of three men arrested and more than £1 million worth of counterfeit material seized following a tip-off from investigators retained by Eli Lilly.[81]

10.3   EU Anti-Counterfeiting Regulation

The EU Anti-Counterfeiting Regulation[82] provides customs authorities across the European Community with the power to detain goods suspected of infringing intellectual property rights. Customs can detain these goods either of its own volition or (more usually) pursuant to a notification by the rights holder to 'watch' for potentially counterfeit goods.

Following the detention of suspected infringing goods, the rights holder has an opportunity to inspect the goods. In certain circumstances the rights holder is also provided with samples for analysis, if this is required to demonstrate infringement. The rights holder then has a 10-day period (extensible to 20 days) in which to commence national infringement proceedings. If no action is brought, the seized goods are then released.

In a recent case referred to the CJEU by the English High Court,[83] the Advocate General has delivered an opinion that gives further hope to brand owners and other IP rights holders in the battle against counterfeit products. The case concerned whether the EU Anti-Counterfeiting Regulation could be used to detain goods in transit through EU member states (rather than goods only destined for onward sale in a particular jurisdiction). This opinion recommended that the CJEU rules that goods bearing a Community trade mark and which are subject to customs supervision in a member state and are in transit from one non-member country to another non-member country may be seized by member state customs authorities provided that there are sufficient grounds for suspecting (i) that they are counterfeit goods and, in particular, (ii) that they are to be put on the market in the European Union, either in conformity with a customs procedure or by means of an illicit diversion. Should the CJEU follow this opinion, brand owners could use transit through the European Union as a way of stemming the flow of counterfeits bound elsewhere too. Patentees should therefore be alive to the possibility that this legislation could be advantageous in identifying and disrupting the transport of infringing products.

---

81   MHRA press release of March 11 2011.
82   EC Council Regulation 1383/2003.
83   C495/09 *Nokia Corporation v Her Majesty's Commissioners of Revenue and Customs.*

11. Collaborative models – 'open innovation'

11.1 Open innovation

There is no doubt that 'open innovation' has worked successfully in the computer software industry. However, this industry has relatively low start-up costs and high turnover of products, which has resulted in rapid advances in innovation. As such, open innovation has proved a highly beneficial model for this industry. For example, leaving aside the issues surrounding the patenting of software under the EPC, most software would be obsolete and/or have been subject to several new version releases in the time it would take a patent to be prosecuted and granted. Patents are therefore seen (even in those jurisdictions that grant patents on such inventions without any statutory exclusion to circumvent) as a crude and mostly inappropriate way in which to protect software. In fact the use of patents in enforcement proceedings in the software industry is mostly the domain of patent trolls[84] and many software companies own patent portfolios primarily for defensive purposes.

Although the pharmaceutical and biotechnology industry is the antithesis of those features of the software industry listed above (ie, relatively high start-up costs, low turnover of products that directly compete and hence a slower rate of innovative product release), the industry is beginning to explore the open-innovation model as a way of progressing projects that might not otherwise be able to proceed.

For example, in 2010 Pfizer agreed[85] to provide data on more than 500 drug candidates to researchers at Washington University in St Louis. This initiative is focusing on identifying new uses for the compounds. It has been reported that the agreement entitles Washington University to $22.5 million over five years and access to proprietary data not normally released to university groups. The project seems a 'win–win' one in that the university gets access to products that are well characterised already, therefore providing ready-made early-stage data and information relating to a mechanism of action that could permit the university to identify a potential new function; Pfizer on the other hand, could generate more revenue from established compounds that may have fallen out of favour, or on which research is not actively ongoing. The agreement then provides for the parties to negotiate a commercialisation agreement as and when new discoveries are made.

A more radical project is one spun out of *Eli Lilly* a decade ago – Innocentive (www.nature.com/openinnovation/index.html). This project operates on a number of levels, but at its simplest it permits companies to challenge its online population of experts and researchers to solve problems in return for cash incentives. One press release details the experience Roche had with this organisation. It seems positive; a problem that had plagued the company and its research partners for more than 15 years was solved in 60 days. However, the importance of intellectual property provisions is stressed within the press release and so the 'open innovation' in Roche's view had to lead to a product that was protectable and in which it had property rights.[86]

---

84   Also known as 'non-practising entities', such companies acquire patents usually for the purpose of enforcement and not through any desire to exploit the technology disclosed in the patent.
85   See news.wustl.edu/news/Pages/20770.aspx.
86   See www.innocentive.com/files/node/casestudy/roche-experience-open-innovation.pdf.

Perhaps the ultimate collaborative project is the European Union's Innovative Medicines Initiative (IMI). This constitutes a large research fund given over entirely to consortia that meet a certain profile. The project is a joint programme funded by the European Union and the European Federation of Pharmaceutical Industries and Associations (EFPIA). This project fits the description of an 'open innovation' model better than either of the above two examples. This is because the way in which the collaborators' intellectual property rights are managed within the standard-form agreements means that the final product is largely unencumbered by the need to obtain licences from the consortia members. Clearly, this makes the project popular with the companies aiming to market such end products. Although the IMI terms whereby consortia members appear to lose control of their intellectual property 'input' to the project in return only for the funded research have been criticised by the biotechnology industry,[87] the IMI is now on its fourth call for funding and is reporting that there is no shortage of applicants.

11.2 Patent pools

Other forms of open innovation have a more philanthropic aim. UNITAID is an organisation set up to assist in the scaling-up of disease treatments, mainly in the developing world, and primarily for HIV/AIDS, malaria and tuberculosis. Part of that thrust has included setting up the HIV patent pool.

The HIV patent pool (initially established by UNITAID and now operating out of the Medicines Patent Pool, www.medicinespatentpool.org) is collecting licences to patents to various products and intends to become a 'one-stop shop' for licensees seeking to manufacture HIV medicaments in the future. The National Institute of Health became the first organisation to contribute patents into this pool.[88] The patents concerned the product darunavir. The pool is now looking to accumulate rights to use patents directed at numerous other HIV treatments.

11.3 Conclusions

'Open innovation' seems to have a role to play in the pharmaceutical industry. However, it is clear that, outside the more philanthropic ventures where no profit is sought to be made, the structure of open innovation in this sector is different from the software industry. Gone is the recognition that innovation happens too fast for the registration of intellectual property rights. Instead, a strong focus is placed on the freedom to operate and protection of the final product. However, some comparisons stand, such as the recognition that one company is often not in a position to conquer every issue with which it is confronted. In such circumstances the industry is proving highly innovative in adapting innovation structures to further its commercial ends.

12. Hot topics

Perhaps the most pressing issues within the United Kingdom life sciences sector

---

87   See, for example, *SCRIP* December 3 2010, p 18.
88   www.whitehouse.gov/blog/2010/09/30/us-government-first-share-patents-with-medicines-patent-pool.

concern dispute resolution, both current and future, and matters of 'access to justice' for patentees who believe that their patents are being infringed.

12.1 The new Patents County Court

A recent development in England and Wales is the new procedure that has been introduced in the Patents County Court. The procedure that applied to this jurisdiction used to be to all intents and purposes the same as that for the Patents Court in the High Court. This unsatisfactory situation meant that the Patents County Court was never able fully to discharge its role as the venue of choice for smaller companies with patents disputes.

The new-look court limits both costs recovery (to £50,000, calculated on a scale basis) and damages (to £500,000). The procedure aims to manage the case actively from the case management conference (CMC) stage, and the usual trappings of patent litigation such as disclosure, experiments, expert and fact evidence, and cross-examination at trial are expected to have to be justified at the CMC before the court will make an order for their inclusion in any particular case. In this manner, whilst still permitting these elements of the litigation in appropriate cases, the court intends to dispose of patent cases as quickly and proportionately as possible. Trial lengths are expected to be curtailed and one to two days will be the maximum for a patent case in this court.

Although a transfer procedure is in place whereby cases can move between the Patents Court in the High Court and the Patents County Court, on its face this new court is much more attractive to a small company wishing to enforce a patent as the financial 'downside' of enforcement is much easier to predict.

12.2 The single EU patent jurisdiction

(a) The EU patent

One development that would change the nature of patent practice in the United Kingdom would be the establishment of either the single EU patent or the single EU patents court (or both). Currently, patents can be prosecuted centrally through the European Patent Office in Munich, but upon grant these patents splinter into national designations of those European patents and must be enforced member state by member state. In recent years this has led to overly complex multi-jurisdictional litigation strategies being required in Europe.

It has long been a dream of politicians in the European Union to have a single, unitary patent that could be prosecuted and enforced as one right. The single EU patent is now close to being a reality. Currently, an enhanced procedure is advancing comprising 25 out of the 27 EU member states – Italy and Spain have refused to join the initiative because of disputes over the language regime that would apply and these two member states are challenging the legality of the procedure at the CJEU.

(b) The single EU patents court

In parallel, EU member states are trying to find a formula for a single court that could provide the forum for a single jurisdiction within which to litigate patents in Europe.

The CJEU recently ruled unlawful a previous single-court proposal that included all European Patent Convention contracting states (many of whom are not members of the European Union). The current EU presidency has, therefore, redesigned the proposal so as only to include EU member states and also to enshrine the idea of the primacy of EU law and the CJEU within a new proposal.

(c)　*No trolls' charter*

The European Commission and most EU member states' governments see these projects as being inextricably intertwined, so that the European Union must achieve both goals, or neither. If a single EU patent comes into being before the single court, that right could be litigated anywhere in the European Union in a competent patents court and decisions – for example on validity – would dispose of the right for the whole of the European Union. Likewise, by analogy with Community trade mark law, it should then be possible to obtain a pan-EU injunction should an infringement action prove successful.

Given the enormous variation in the manner in which patent cases are tried in the civil courts of the member states, this would lead to forum shopping and the importance of obtaining the 'first strike' in a dispute being accentuated. Many practitioners fear that such an arrangement would constitute a trolls' charter, giving these non-practising entities the freedom to pick and choose victims and forums simply to place pressure on the alleged infringer to settle the action no matter how unmeritorious the action might seem (either from an infringement or an invalidity perspective).

(d)　*Conclusion*

Europe is inching towards a single patents jurisdiction. The exclusion of key markets such as Spain and Italy from the EU patent process is highly regrettable and it is hoped that the new arrangements will not consist of an 'EU minus' jurisdiction when it finally comes into being. Such a development would be troubling within a trading bloc that should be acting in unison on all matters of commercial importance.

United States

Katherine A Helm
Noah M Leibowitz
Simpson Thacher & Bartlett LLP

Intellectual Property (IP) is often a company's single most valuable asset. The United States is a strong player in the global economy, in part because of its staunch protection of IP rights. Protecting these rights, as well as defending against infringement allegations of others, is critical to any company's success in an increasingly competitive global marketplace. It is imperative for all businesses to have an effective IP rights-management strategy. Pharmaceutical and biotechnology companies are adept at recognising the value of IP portfolios and devoting substantial resources to protecting and exploiting their own IP, as well as gaining access to essential IP owned by others. As a result, the law that provides for and protects IP rights in the life sciences is both highly complex and ever evolving. It is also one of the most challenging and rewarding areas of US legal practice.

This chapter provides an outline of the major applications of IP rights in life sciences in the United States. The focus is on patents because of their heightened influence in this industry. A patent gives the owner the right to exclude others from making, using, offering for sale, selling, or importing the patented inventions.[1] The exclusive patent right allows patentees to gain economic benefit from their inventions and to fund future innovation. A strong patent system is a supremely important mechanism for encouraging and fostering pharmaceutical and biotechnological research, drug discovery, therapeutic product development, investments and ultimately future innovation. Successful companies are willing to invest significant amounts of their revenues derived from sales of patent-protected products and patent licensing royalties for future research and development efforts. In this way, the future of American innovation in the life sciences depends on the nature and effectiveness of the US patent system and, more generally, US IP laws.

1. Small molecules

Small molecules are the foundational building block of the drug discovery process. Pharmaceutical and biotechnology companies involved in drug research place significant focus on patenting small molecules. Small molecules are generally defined as low-molecular-weight organic compounds that are made by chemical synthesis and that are typically assayed to test for some desired activity against a particular target or disease. They may be administered orally or formulated for intravenous, transdermal or other means of administration. Although they are not

---

1   35 USC § 271(a).

the only patentable aspect of a drug, protection of the compound itself as a new chemical entity often represents the most basic and valuable form of patent protection available to cover a commercial drug product.

1.1 Product and process claims

Product or composition-of-matter patents that cover active compounds, in the context of small molecules, generally specify the compound's chemical structure. See, for example, *In re Papesch* (the physical properties of a chemical compound are inseparable from its structure, and "the thing that is patented is not the formula but the compound identified by it");[2] *Daiichi Sankyo Co Ltd v Matrix Labs Ltd* (affirming patentability of a chemical compound where "the structural similarities and differences between the compounds claimed and those in the prior art" did not render the compound obvious);[3] and *Takeda Chemical Indus Ltd v Alphapharm Pty Ltd* ("in cases involving new chemical compounds, it remains necessary to identify some reason that would have led a chemist to modify a known compound in a particular manner to establish obviousness of a new claimed compound").[4]

In addition to patenting a compound itself, a patent may be granted on methods or processes for manufacturing both old and new compounds.[5] Patentable processes can include steps for synthesising a compound – see, for example, *In re Ochiai* (process for making chemical compound having antibiotic properties based on non-obvious starting material).[6] But this inclusion principle only applies if the process produces a chemical compound shown to be useful: *Brenner v Manson* (holding unpatentable a process for making steroidal compounds with no disclosed use);[7] *In re Brana* (disclosure for use of end compound requires "some desirable pharmaceutical property in a standard experimental animal");[8] and see also 35 USC § 103(b)(1) and (2) (providing statutory requirements for patentability of biotechnological processes).

1.2 Scope of protection of claims and Markush formulae

Both product and process patent claims can provide meaningful protection in a commercial setting, because such claims may cover (ie, may be infringed by) the sale and use of a drug product, the way it is made and used, and the dosage form in which it is sold. A Markush group is a listing of specified alternatives of a group in a patent claim, typically expressed in the form 'a member selected from the group consisting of A, B, and C' (see *Abbott Labs v Baxter Pharm Prods Inc*).[9] Typically, one and only one member of a Markush group is required in order to infringe the claim. In some instances, a Markush-style claim will not cover multiple members of the group together.[10] The risk of a Markush claim form is that if any one member of the group is found to be anticipated or obvious, the entire claim is invalid.[11]

---

2   *In re Papesch* 315 F.2d 381 (CCPA 1971).
3   *Daiichi Sankyo Co, Ltd v Matrix Labs., Ltd* 619 F.3d 1346, 1352 (Fed. Cir. 2010).
4   *Takeda Chemical Indus, Ltd v Alphapharm Pty, Ltd* 492 F.3d 1350, 1357 (Fed. Cir. 2007).
5   See 35 USC § 101 – patentable subject matter includes "any new and useful process".
6   *In re Ochiai* 71 F.3d 1565, 1570 (Fed. Cir. 1995).
7   *Brenner v Manson* 383 US 519 (1966).
8   *In re Brana* 51 F.3d 1560, 1567 (Fed. Cir. 1995).
9   *Abbott Labs v Baxter Pharm Prods, Inc* 334 F.3d 1274, 1280 (Fed. Cir. 2003).

1.3 Metabolites

A metabolite is a compound formed in the patient's body upon ingestion of a drug. A new metabolite compound is formed when an ingested drug undergoes a chemical conversion in the digestion process.[12] A company may infringe a claim to a metabolite if it markets a product that, when ingested, metabolises to form the claimed metabolite.[13] A patent claiming either the active ingredient of a drug or a method of using that ingredient does not necessarily also cover its metabolites.[14] A metabolite may not be patentable if the metabolite is inherently formed as a natural result of a prior-art drug's administration.[15]

2. Second-generation inventions

Second-generation patents can provide an additional period of patent protection, and accordingly an increased term of market exclusivity and enhanced commercial value, for a pharmaceutical product.

2.1 Combinations

A combination of two or more drug products may be eligible for its own patent protection, where administration of the combination to treat a desired condition provides a synergistic improvement over the effect that would have been expected by administering either product alone or the two products together. Administration of a formulation comprising two or more active compounds, or the co-administration of the multiple active compounds, can be patented even if each of the underlying products were known, so long as the combination was not obvious.[16] By contrast, a combination of two active ingredients to treat co-occurring conditions may not be patentable if the prior art teaches the combination of interchangeable ingredients for at least one of the conditions.[17]

2.2 Enantiomers

Enantiomers are compounds that are mirror images of each other. They are a type of stereoisomer, which refers to one of a set of two or more compounds that are composed of the same constituent atoms, connected in the same sequence, but

---

10   *Ibid* at 1280 to 1281, holding that where a claim recites a Markush group preceded by the indefinite article 'a' and does not include qualifying language, the claim covers a single recited member of the Markush group.

11   *Ecolochem, Inc v Southern California Edison Co* 91 F.3d 169, 1996 WL 297601, *2 (Fed. Cir. 1996), citing *In re Skoll* 523 F.2d 1392, 1397 (CCPA 1975).

12   *Schering Corp v Geneva Pharms* 339 F.3d 1373, 1375 (Fed. Cir. 2003).

13   See *Hoechst-Roussel Pharms, Inc v Lehman* 109 F.3d 756, 759 (Fed. Cir. 1997); see also *Zenith Labs, Inc v Bristol-Myers Squibb Co* 19 F.3d 1418, 1421 to 1422 (Fed. Cir. 1994) (noting a compound claim could cover a metabolite formed upon ingestion).

14   *Hoechst-Roussel Pharms., Inc* 109 F.3d at 759.

15   *Schering Corp* 339 F.3d at 1382.

16   See, for example, *Knoll Pharm Co v Teva Pharms USA, Inc* 367 F.3d 1381, 1384–85 (Fed. Cir. 2004), where a combination of the opioid hydrocodone and the analgesic ibuprofen was found non-obvious over the prior art based on the unexpected result of achieving a 'surprising' benefit from the combination in treating pain relief and muscle repair after exercise.

17   *McNeil-PPC, Inc v L Perrigo Co* 337 F.3d 1362, 1369–70 (Fed. Cir. 2003), where the combination of the anti-diarrhoeal agent loperamide and anti-gas agent simethicone was judged to be obvious where conditions were known to occur together and anti-diarrhoeal agents had been prescribed with simethicone.

differing in spatial arrangement. Enantiomers can exist in mixtures of various isomers that differ in form, or they can be isolated into a pure form in which only a single enantiomer exists. An equal mixture of two enantiomers is called a 'racemate'.[18] Enantiomers, while having the same atoms as other isomeric forms, can impart particular properties and provide grounds for new patent protection.[19]

2.3 Selection inventions

A selection invention is based on selecting one or more species within a broader genus already disclosed in the prior art. The patentability of a species as a selection invention depends on it having new and non-obvious benefits over the prior-art generic disclosure, such as specific dose ranges of a pharmaceutical formulation, optimal units of enzyme, or particular functions of a subset of compounds. The reasoning behind selection inventions is that art disclosing a genus does not necessarily disclose every species that is a member of that genus, particularly if the genus is large, so the identification of a species with advantageous properties may constitute a separate invention.[20]

Recently, the Supreme Court's 'obviousness' analysis in *KSR Int'l Co v Teleflex Inc*[21] has made it more difficult to patent selection inventions even in situations where the prior art does not explicitly teach the advantageous properties of the claimed selection.

2.4 Methods of use and secondary indications

A novel, non-obvious pharmaceutical compound can be patent-protected both for its composition and for its use. Additional protection can be conferred on pharmaceutical products by patenting new methods of use, or new or secondary indications for a drug. A secondary method of using an existing drug can be patented if using the drug in that way or for that indication would be an unexpected property (see, for example, *In re Schoenwald*[22]) but not if the new use is merely the recognition of a drug's inherent properties (*Bristol-Myers Squibb Co v Ben Venue Labs* – newly discovered results of a known process directed to the same purpose are not patentable because such results are inherent).[23]

---

18   See *Pfizer, Inc v Ranbaxy Labs, Ltd* 457 F.3d 1284, 1286 (Fed. Cir. 2006).
19   See, for example, *Sanofi-Synthelabo v Apotex, Inc* 470 F.3d 1368, 1380 (Fed. Cir. 2006) (rejecting the argument that enantiomers are unpatentable over disclosures of their racemates); *Forest Labs, Inc v Ivax Pharms, Inc* 501 F.3d 1263, 1269 (Fed. Cir. 2007) (substantially pure (+)-enantiomer of citalopram not anticipated by or obvious over racemic citalopram, based on difficulty of separating the constituent enantiomers and the unexpected properties of (+)-citalopram); *Ortho-McNeil Pharm v Lupin Pharms* 603 F.3d 1377, 1381 (Fed. Cir. 2010) (the enantiomer levofloxacin is a 'different drug product' that is separately patentable from its racemate ofloxacin).
20   See *In re Jones* 958 F.2d 347, 383 (Fed. Cir. 1992) ("... [a] disclosure of millions of compounds does not render obvious a claim to three compounds, particularly when that disclosure indicates a preference leading away from the claimed compounds"); see also *In re Bell* 991 F.2d 781, 784 (Fed. Cir. 1993) (claimed DNA sequence coding for human insulin-life growth factors not obvious over known protein because of the 'nearly infinite' number of sequences that could code for the protein; but cf, *In re Petering* 301 F.2d 676 (CCPA 1962) (claimed compound was anticipated by prior-art disclosure of generic class of about 20 compounds that sufficiently described all 20).
21   *KSR Int'l Co v Teleflex Inc.* 550 US 398 (2007).
22   *In re Schoenwald* 964 F.2d 1122 (Fed. Cir. 1992).
23   *Bristol-Myers Squibb Co v Ben Venue Labs* 246 F.3d 1368 (Fed. Cir. 2001).

Method-of-use claims are also vulnerable to obviousness-type double-patenting challenges over composition-of-matter patents that describe methods for using the composition. See, for example, *Sun Pharm Indus Ltd v Eli Lilly and Co* (holding invalid for obviousness-type double patenting a method-of-use claim for a pharmaceutical product where the composition was previously claimed and the use was disclosed but not claimed in a prior patent).[24]

2.5   Methods of treatment

Method-of-treatment claims are a form of process claim. They have been permitted even where the compound and a method of using the compound are both known, so long as the claimed method of treatment is limited to a specific purpose not taught in the prior art. Importantly, someone of ordinary skill in the art for a method-of-treatment claim is not limited to one skilled as a treating physician and can include a skilled worker who develops new drugs or treatments.[25] While the Supreme Court's decision in *Bilski v Kappos* provides guidance on the patent-eligibility of process claims if they comply with the 'machine-or-transformation' test,[26] it leaves open many issues with respect to method-of-treatment patents in the life sciences arena.[27]

Post-*Bilski*, the Federal Circuit has held that method-of-treatment claims involving the administration of drugs are patent-eligible as being transformative to the human body.[28]

2.6   Formulations and physical forms

Formulation patents are one of the most popular forms of second-generation patents, typically covering both the active and inactive ingredients in tablet, capsule or other final dosage form. Active ingredients are typically mixed with inactive ingredients to make a pharmaceutical formulation that can be suitably administered to a patient. New formulations are important to drug design and development and may come in numerous physical forms, such as solid tablets, liquid or gel capsules, liquids, ointments and aerosols.

Formulation patents may also encompass a new route or schedule of administration, such as controlled- or sustained-release forms.[29] Notably, formulation patents may be vulnerable to 'design around' opportunities by competitors, particularly if the patent claims narrowly capture specific ingredients, drug release rates or routes of administration.[30]

---

24   *Sun Pharm. Indus, Ltd, v Eli Lilly and Co.* 611 F.3d 1381 (Fed. Cir. 2010).

25   *Daiichi Sankyo Co v Apotex, Inc.* 501 F.3d 1254 (Fed. Cir. 2007).

26   *Bilski v Kappos* 130 SCt 3218 (2010).

27   *Ibid.* at 3228, noting "new technologies may call for new inquiries" for process claims.

28   *Prometheus Labs, Inc v Mayo Collaborative Services* 628 F.3d 1347, 1356–7 (Fed. Cir. 2010), cert granted; *Mayo Collaborative Services et al. v Prometheus Labs, Inc* (US June 20 2011) (No 10-1150).

29   See, for example, *Abbott Labs v Sandoz, Inc* 544 F.3d 1341, 1352 (Fed. Cir. 2008) (extended-release formulations of the antibiotic drug clarithromycin having the pharmacokinetic properties in the claims were not taught in prior art); *Alza Corp v Mylan Labs* 464 F.3d 1286, 1293–4 (Fed. Cir. 2006) (claims to sustained-release oxybutynin formulation found obvious, based on the reasonable expectation that oxybutynin would be colonically absorbed and thus motivation existed to produce the claimed extended-release formulation).

30   *Ibid.* at 1297, for example: in vitro dissolution rates were sufficiently dissimilar from *in vivo* extended-release properties to avoid infringement.

2.7 Reach-through claims

Reach-through claims refer to claims that cover products obtained by the use of research tools (ie, research or screening techniques that can be used to identify and evaluate drug candidates). The goal of patents covering research tools is to have the claims 'reach through' to apply to the ultimate drug product that is sold, so as to collect royalties from that sale or at least from pre-clinical drug discovery efforts.

However, US courts have cast doubt on the patentability and enforcement of claims to products identified only by reference to the material or means used to find or identify them.[31] This case law also intersects with both the common-law research exemption[32] and the statutory experimental-use exemption under 35 USC § 271(e)(1).[33] There is an ongoing debate over the extent to which life sciences companies should be allowed, as a matter of public policy, to operate within the scope of research tool patents in order to develop new drug products.

3. DNA, biologicals and personalised medicine

Biologics are large organic molecules that distinguish themselves from small molecules by virtue of their having been synthesised from living organisms. They are also typically administered by injection or intravenous infusion. This feature means that the legal framework both for approving a follow-on biologic drug and for finding infringement of a biologic drug product is different from that of a small-molecule drug.

3.1 Discoveries

Discoveries in the biotech area generally begin with the study of genes, nucleic acid sequences of DNA or RNA, amino acid sequences of proteins and resulting biological products. While the field of biology differs from chemistry, DNA is considered "a chemical compound, albeit a complex one" for the purposes of patenting.[34] The breadth of patentable subject-matter for biological inventions, as famously described in *Diamond v Chakrabarty*,[35] a case relating to genetically modified bacteria, includes "anything under the sun that is made by man".

3.2 Gene patents and industrial application

There is no explicit industrial application requirement for patents under US law. The utility requirement under 35 USC § 101 can be analogised to the industrial-use requirement that exists for patenting genes in other jurisdictions. US case law has held that an isolated and purified DNA sequence that is complete and encodes for a specific (eg, human) protein constitutes patentable subject matter,[36] whereas an incomplete gene sequence does not.[37] In 2010, however, a US district court held

---

31   See, for example, *Univ. of Rochester v GD Searle & Co* 375 F.3d 1303 (Fed. Cir. 2004); Housey Pharms, Inc v Astrazeneca UK Ltd 366 F.3d 1348, 1350 (Fed. Cir. 2004); *Bayer AG v Housey Pharms, Inc.* 340 F.3d 1367 (Fed. Cir. 2003).
32   See *Madey v Duke Univ.* 307 F.3d 1351 (Fed. Cir. 2002).
33   See *Merck KGaA v Integra LifeSciences I Ltd* 545 US 193 (2005).
34   *Amgen, Inc v Chugai Pharm Co* 927 F.2d 1200 (Fed. Cir. 1991).
35   *Diamond v Chakrabarty* 447 US 303, 309 (1980).
36   See, for example, *Amgen, Inc v Chugai Pharm Co* 927 F.2d 1200 (Fed. Cir. 1991) and *Fiers v Revel* 984 F.2d 1164 (Fed. Cir. 1993).

invalid claims to isolated DNA sequences of two breast cancer genes on the basis that isolated DNA is no different from DNA that exists in nature and thus constitutes unpatentable subject matter under Section 101.[38] The case was widely watched and a recent Federal Circuit decision reversed the district court's decision on the DNA claims, reasoning that US law permits patents on human-engineered gene sequences such as cDNA because it recognises that, like other chemical compounds, the purification of DNA transforms it into something different in character.[39]

The case to watch regarding these types of diagnostic claims will be the *Prometheus* case,[40] now pending before the US Supreme Court (and scheduled for oral argument during the October 2011 Term).

3.3 Stem cells and other organic material

There is no explicit *ordre public* or morality restriction on patentable inventions in the United States. Aside from the judicial prohibitions on patenting natural phenomena, laws of nature, abstract ideas and naturally-occurring substances, US law generally allows for a broader scope of patentable invention than many European countries. For example, the Wisconsin Alumni Research Foundation (WARF) obtained a number of US patents directed to purified and isolated stem cells – for example, US Patent Nos 5,843,780 (purified preparation of primate embryonic stem cells), 6,200,806 (purified preparation of pluripotent human embryonic stem cells) and 7,029,913 (proliferating and stably undifferentiated human embryonic stem cells cultured *in vitro*). These and other WARF patents have been contested in *inter partes* re-examination proceedings before the Board of Appeals and Interferences of the US Patent and Trademark Office, with some claims being withdrawn and others being upheld. Proceedings are ongoing and the patentability of these types of claim remains uncertain.

3.4 Bioinformatics systems

One burgeoning area of law relating to the life sciences industry is that of biotechnology inventions derived from non-wet lab techniques such as computer-assisted screening of new drugs, genes and other biological materials. The patentability of such subject matter depends on the extent to which the computer application can be claimed in terms of an apparatus rather than a type of mathematical algorithm.[41] The patentability of biological screens and related diagnostics will undoubtedly continue to be the subject of much litigation in the wake of *Bilski v Kappos*.[42]

---

37 See *In re Fisher* 421 F.3d 1365 (Fed. Cir. 2005) (partial sequence in form of expressed sequence tags lacks utility unless gene function is identified) and *Regents of the University of California v Eli Lilly & Co* 119 F.3d 1559 (1997) (claim to a plasmid containing cDNA coding for human insulin held invalid where only rat cDNA sequence was disclosed).

38 *Association for Molecular Pathology et al. v United States Patent and Trademark Office et al.* 94 USPQ 2d 1683 (SDNY March 29 2010).

39 See *The Association For Molecular Pathology v US Patent and Trademark Office et al.* No 2010–1406, 2011 WL 3211513, *20–21 (Fed. Cir July 29 2011) (holding claims to isolated DNA patent-eligible, while diagnostic/method claims involving screening DNA sequences are patent-eligible only if they involve a transformative step beyond simply 'comparing' or 'analysing' two gene sequences).

40 *Mayo Collaborative Services et al. v Prometheus Labs, Inc.* (US June 20 2011) (No 10-1150).

3.5 Copyright and sequence information

Copyright law is addressed in sections 6, 7, 9.3 and 10 below.

3.6 Sufficiency issues

Sufficiency issues are covered elsewhere in this chapter, in other sections discussing patentability under US law.

4. Acts of patent infringement

The nature and effectiveness of the exclusive patent right depends in part on the remedies available for its infringement. There are several acts of patent infringement prohibited by statute.

4.1 Direct infringement

Under 35 USC § 271(a), anyone who makes, uses, offers to sell, sells or imports into the United States a patented invention is deemed an infringer. And under 35 USC § 271(g) a patent claiming a process for making a product may also be directly infringed by the importation into the United States of a product made by the claimed process. There is no intent element to direct infringement and knowledge of the patent is not required.[43]

Patent claims that are not 'literally' infringed because the accused product or process does not include all of the limitations contained in the claim may nevertheless be infringed under the doctrine of equivalents, if the differences between the accused product or process and the claim limitations that are not literally present are insubstantial.[44] The doctrine prevents an accused infringer from avoiding infringement by changing only minor or insubstantial details of a claimed invention while retaining its essential functionality.[45] However, a patentee may not invoke the doctrine of equivalents to recapture subject matter that was surrendered before the US Patent and Trademark Office in an effort to obtain allowance of the patent claims. Such surrender may arise either by amending claims or making arguments that limit claim scope during prosecution of the patent application.[46] In other words, the doctrine of equivalents is tempered by the doctrine of prosecution history estoppel.

4.2 Indirect infringement

Liability for patent infringement is not confined to those who commit acts of direct infringement. There are also two types of indirect infringement: inducement of infringement, proscribed by 35 USC § 271(b), and contributory infringement,

---

41   See *Gottschalk v Benson* 409 US 63, 71–2 (1972) (treating a mathematical algorithm as an unpatentable 'idea'); *Diamond v Diehr* 450 US 175, 186 (1981) (use of mathematical formulae in an industrial application is patentable); and *In re Alappat* 33 F.3d 1526 (Fed. Cir. 1994) (*en banc*) (computer software claims to an apparatus are patentable).
42   *Bilski v Kappos* 130 SCt 3218 (2010).
43   *Intel Corp v US International Trade Commission* 946 F.2d 821, 832 (Fed. Cir. 1991).
44   *Warner-Jenkinson Co v Hilton Davis Chem Co* 520 US 17, 38–40 (1997).
45   *Ibid* at 28–9.
46   See *Festo Corp v Shoketsu Kinzoku Kogyo Kabushiki Co* 535 US 722 (2002) (detailing the doctrine of prosecution history estoppel).

proscribed by 35 USC § 271(c). Neither type of indirect infringement can exist in the absence of an act of direct infringement.[47]

A person induces infringement by actively and knowingly aiding and abetting another's direct infringement. There is an intent requirement for inducement, which requires more than just intent to cause the acts that produce direct infringement; the inducer must have an affirmative intent to cause direct infringement.[48] There is also a knowledge requirement; inducing infringement requires knowledge that the induced acts constitute infringement. Wilful blindness, however, can constitute knowledge.[49]

Contributory infringement arises when one "sells within the United States ... a component of a patented machine ... knowing the same to be especially made or especially adapted for use in an infringement" of the patent at issue.[50] In order to prove contributory infringement, a patentee must show that an alleged infringer "knew that the combination for which its components were especially made was both patented and infringing".[51] In addition, the patentee must show that the alleged infringer's component has no "substantial non-infringing use".[52]

4.3   National treatment of 'Bolar'-type provisions.

The 'safe harbour' research exemption to patent infringement is the US equivalent of a 'Bolar'-type provision in other jurisdictions. Congress created the statutory safe harbour in response to the Federal Circuit's refusal to create a common-law exemption to infringement for the testing of generic drugs.[53] 35 USC § 271(e)(1) provides an exemption to patent infringement for "uses reasonably related to the development and submission of information" to the Food and Drug Administration (FDA). Congress passed Section 271(e)(1) as part of the Drug Price Competition and Patent Term Restoration Act of 1984 (also known as the 'Hatch-Waxman Act') to facilitate the entry of generic drugs into the market upon expiration of the branded patent on the compound.[54] The statutory safe harbour exemption under 35 USC § 271(e)(1) permits generic pharmaceutical manufacturers to use a patented invention in order to develop a generic equivalent prior to a branded drug's patent expiration. The rationale behind the safe harbour is to facilitate the entry of generic drugs into the market upon patent expiry with as little artificial extension of the patent monopoly as possible.

Since the enactment of the safe-harbour exemption under Section 271(e)(1), US

---

47   See, for example, *Micro Chem Inc v Great Plains Chem Co* 103 F.3d 1538 (Fed. Cir. 1997).
48   See *DSU Med Corp v JMS Co* 471 F.3d 1293, 1305–6 (Fed. Cir. 2006) (inducement requires specific intent, with evidence of culpable conduct directed to encourage another's direct infringement).
49   See *Global-Tech Appliances, Inc v SEB SA* 131 S.Ct. 2060, 2070–1 (2011) (wilful blindness "surpasses recklessness and negligence" and requires that: "(1) the defendant must subjectively believe that there is a high probability that a fact exists and (2) the defendant must take deliberate actions to avoid learning of that fact.").
50   35 USC § 271(c).
51   *Golden Blount, Inc v Robert H Peterson Co* 365 F.3d 1054, 1061 (Fed. Cir. 2004).
52   35 USC § 271(c).
53   See *Roche Prods v Bolar Pharms* 733 F.2d 858, 863 (Fed. Cir. 1984) (refusing to construe experimental use exemption to encompass testing of a branded pharmaceutical by a generic drug maker to demonstrate the equivalence required for regulatory approval).
54   *See Pub.* L. No. 98-417, 98 Stat. 1585 (1984).

courts have construed the exemption broadly. See, for example: *Eli Lilly & Co v Medtronic Inc* (holding that Section 271(e)(1) is not limited to regulatory submissions relating to drugs or veterinary biological products);[55] *AbTox Inc v Exitron Corp* (as long as research is 'reasonably related' to obtaining FDA approval, the data can also be used for other purposes);[56] and *Merck KGaA v Integra LifeSciences I Ltd* (holding that Section 271(e)(1) provides a 'wide berth' for the use of patented drugs in activities that necessarily include preclinical studies for the development of a potential drug candidate).[57]

4.4 Experimental-use exemptions

The experimental-use exemption is a common-law exemption to patent infringement for experimental activities that have no commercial purpose. The exemption is narrow and even the slightest commercial implication will render it inapplicable.[58] In *Madey v Duke University*,[59] the Federal Circuit addressed the experimental-use exemption in the context of scientific research performed in an academic setting. The court reasoned that a research university's goals are inherently commercial in nature and thus the research activities infringe if they further the institution's business objectives of educating students and/or increasing the status of the institution.[60]

4.5 Submitting authorisations and offers to supply

In the United States, an offer for sale is an act of direct infringement. 35 USC § 271(a) provides that: "Except as otherwise provided in this title, whoever without authority makes, uses, offers to sell, or sells any patented invention, within the United States or imports into the United States any patented invention during the term of the patent therefor, infringes the patent." Anyone who offers to supply a patented invention will therefore be liable for infringement under US law. However, an offer to sell infringes a patent only when the offer involves an actual sale that would occur before the patent expires – "an 'offer for sale' or an 'offer to sell' by a person other than the patentee or any assignee of the patentee, is that in which the sale will occur before the expiration of the term of the patent".[61]

Although infringement liability may result when an entity supplies components of inventions to be assembled outside the United States under certain circumstances,[62] there is no liability for the mere offer to supply such components.[63]

---

55   *Eli Lilly & Co v Medtronic, Inc.* 496 US 661, 666 (US 1990).
56   *AbTox Inc v Exitron Corp* 122 F.3d 1019, 1030 (Fed. Cir. 1997).
57   *Merck KGaA v Integra LifeSciences I Ltd* 545 US 193 (2005).
58   See, for example, *Embrex, Inc v Serv Engineering Corp* 216 F.3d 1343, 1353 (Fed. Cir. 2000) (affirming the finding of an infringement where the patented immunisation method was being used in an attempt to design around it and create a new inoculating machine).
59   *Madey v Duke University* 307 F.3d 1351 (Fed. Cir. 2002).
60   *Ibid.* at 1362–3.
61   35 USC § 271(i).
62   35 USC § 271(f)(1) and (2).
63   See *Rotec Industries, Inc v Mitsubishi Corp.* 215 F.3d 1246, 1257 (Fed. Cir. 2000) (rejecting argument that 35 USC § 271(f)(2) imposes liability on those who 'offer to supply' in or from the United States any component of a patented invention thereunder).

4.6 Summaries of product characteristics (SmPCs)

This aspect of patent law is not applicable in the United States.

5. Patent enforcement

The term of enforcement of patents is provided under 35 USC § 154, which permits a patent owner to exclude others from practising the patented invention, as defined by the claims, during the term of the patent. This right to exclude is considered the bargained-for exchange in return for disclosing to the public a new, non-obvious and useful invention that will become part of the public domain upon the patent's expiration.

A contentious area of litigation involves research-based pharmaceutical companies enforcing their patents against competitor drug companies seeking to market generic versions of approved drugs while the patents on the approved drug or its use are still in place. The Hatch-Waxman Act provides the regulatory framework for such litigation. Under Hatch-Waxman, generic companies wishing to enter the market before expiration of the patent(s) covering a commercial drug product or its use are permitted to file Abbreviated New Drug Applications (ANDAs) that piggyback on the branded drug maker's safety and efficacy data. The generic company need only show 'bioequivalence' to the approved commercial drug product. ANDA filers must include in their ANDA a certification that there are no patents covering the branded drug, that it will not market its product until all relevant patents have expired, or a 'Paragraph IV certification' that the relevant patents are invalid or will not be infringed by the manufacture, use, sale or offer for sale of the generic drug product.[64] Thereafter, a patentee may sue the generic manufacturer for infringement and impose an automatic 30-month stay on the approval of the allegedly infringing generic drug.[65] As an incentive to bring generic versions to market early by challenging branded drug patents, the generic manufacturer that first files an ANDA with a Paragraph IV certification for a particular drug product is awarded 180 days of generic market exclusivity, during which time the FDA will not approve other generic drug applications.

A different statutory scheme provides the framework by which biopharmaceutical manufacturers can enforce their patents against competition seeking to market follow-on biologics. In March 2010, Congress passed the Biologics Price Competition and Innovation Act, commonly known as the 'Biosimilars Act'. That Act sets up a regulatory scheme for biologics distinct from Hatch-Waxman, but with the same intent of balancing the need to accelerate the approval of 'biosimilar' versions of already marketed biologics, while encouraging biopharmaceutical companies to make the investments necessary to develop new biologic products. The details and complexities of this new legislation will undoubtedly be clarified in future biosimilar patent litigation and are beyond the scope of this chapter.

5.1 Obtaining information on the infringer and the infringement

Generally, there is no special discovery process under US law for patent litigation.

---

64   21 USC § 355(j)(2)(A)(vii)(IV).
65   See 35 USC §271(e)(2) (filing a Paragraph IV certification constitutes an act of constructive patent infringement) and 21 USC § 355 (regulatory provisions for 30-month stay).

The Federal Rules of Civil Procedure and applicable local rules govern the discovery process in patent litigation for seeking and obtaining information about infringement claims. Several federal district courts have enacted their own sets of patent local rules, such as the Northern District of California and the Eastern District of Texas. These rules typically require early disclosures of claims, infringement contentions, and supporting documents in patent litigation.[66]

Legal-privilege rules may also have special application in patent enforcement.[67] The privilege does not apply to all patent prosecution-related information *per se*, however, and the law is mixed on the extent to which the privilege applies after information has been disclosed to non-attorney patent agents and/or the US Patent and Trademark Office.

Likewise, the work-product privilege,[68] which provides protection from discovery of documents and things prepared in anticipation of litigation, may extend to a variety of patent-related materials but typically does not cover documents related to patent prosecution.

5.2   Interim relief

Preliminary injunctive relief against patent infringement depends on the same equitable test that is used for other areas of IP and, indeed, general law. The equitable remedy of a preliminary injunction is governed by a four-part test much like those outlined for permanent injunctions in the Supreme Court's case of *eBay Inc v MercExchange*.[69] These four factors include: (i) the likelihood of the patentee's success on the merits; (ii) irreparable harm if the injunction is not granted; (iii) the balance of hardships between the parties; and (iv) the public interest. The first two factors must each be established before preliminary relief may be granted, whereas the first factor may have less relevance for permanent relief because the injunction remedy is only addressed after the patent has been held valid and infringed – see *Automated Merchandising v Crane*,[70] where the judgment noted that, post-*eBay*, the presumption of irreparable harm is not assumed based on proof of infringement; the patentee must demonstrate its potential losses cannot be compensated by monetary damages.

In the case of pharmaceutical patent litigation governed by Hatch-Waxman, a branded drug company can obtain a 30-month stay of the approval of the generic drug manufacturer's ANDA. The stay creates a *de facto* injunction to prevent the approval, sale or use of the generic drug. The 30-month stay may be extended or reduced by a court if it finds either party is failing to cooperate in reasonably expediting the patent infringement action,[71] and will expire upon a final court order that the patent is invalid, unenforceable or not infringed.[72]

---

66   See, for example, N.D. Cal. Patent Local Rules 3-1, 3-2.
67   See, for example, *In re Spalding Sports Worldwide, Inc.* 203 F.3d 800 (Fed. Cir. 2000) (holding the attorney–client privilege extends to invention records submitted to a patent attorney for legal advice about the patentability of an invention).
68   See Federal Rule of Civil Procedure 26(b)(3).
69   *eBay Inc v MercExchange, LLC* 547 US 388 (2006).
70   *Automated Merchandising Sys, Inc v Crane Co* 357 Fed. Appx. 297, 301 (Fed. Cir. 2009).
71   21 CFR § 314.107(b)(3)(i)(A).
72   21 CFR § 314.107(b)(3)(i)(B)(ii).

5.3 'Springboard'/post-patent-expiry injunctions

The US Supreme Court has rejected the extension of a patent beyond its term.[73] Thus, US law generally does not grant post-patent-expiry injunctions. Courts will, however, award damages after patent expiration for pre-expiration infringement. In addition, some regulatory exclusivities may actually run longer than the patent term – for example in the case of FDA orphan drug exclusivity, the data exclusivity provisions under the Hatch-Waxman Act and the Biosimilar Act, and the 180-day generic exclusivity, as discussed above.

5.4 Unjustified threats

There are no statutory provisions in the United States directly analogous to the prohibitions against unjustified threats of enforcement in some other countries, such as the United Kingdom and Australia, but US antitrust laws provide similar defences. For example, an accused infringer may bring a *Walker Process* type of counterclaim against the patentee, asserting that the patentee is violating the antitrust laws by attempting to enforce an illegally obtained patent that was secured by fraud, or is engaging in 'sham' (ie, objectively and subjectively baseless) litigation.[74] An enforcing party's use of patents to maintain monopoly power may also present additional defences of 'unclean hands' and/or patent misuse.[75]

Another common claim brought by accused infringers against patentees is patent unenforceability on the basis of inequitable conduct – see 35 USC § 282, where defences in any action involving the validity or infringement of a patent include unenforceability. Inequitable conduct is a judicially constructed, equitable defence to patent infringement that, if proven, will render the entire patent unenforceable. To prevail on a claim of inequitable conduct, an accused infringer must establish that the patentee misrepresented or omitted material information related to a patent application's prosecution with the specific intent to deceive the US Patent and Trademark Office (USPTO).[76] In *Therasense Inc v Becton, Dickinson and Co*,[77] the Federal Circuit raised the standard for proving inequitable conduct by eliminating the sliding-scale approach to intent and materiality that had allowed strong evidence of materiality to compensate for weak evidence of intent to deceive, and vice versa. Now, the accused infringer must prove both elements – intent and materiality – by clear and convincing evidence. Moreover, except in cases involving affirmative egregious misconduct, 'but-for' materiality must be proven, meaning that the court must find that the USPTO would not have allowed the claim but for the deception, and there must be evidence of a 'deliberate decision' to deceive.[78]

5.5 Remedies

The traditional remedy for patent infringement is a 'reasonable' royalty,[79] which a

---

73   See *Brulotte v Thys Co* 379 US 29 (1965).
74   See *Walker Process Equipment, Inc v Food Machinery & Chemical Corp* 382 US 172 (1965); also 15 USC § 2.
75   35 USC § 271(d).
76   *Star Scientific Inc v RJ Reynolds Tobacco Co* 537 F.3d 1357, 1365 (Fed. Cir. 2008).
77   *Therasense, Inc v Becton, Dickinson and Co* 649 F.3d 1276 (Fed. Cir. 2011) (en banc).
78   *Ibid.* at *8–9.
79   35 USC § 284.

patentee typically establishes by presenting proof related to the patent's use and profitability in the market, hypothetical licence negotiation rates, and the extent of infringement – see in particular *Georgia-Pacific Corp v United States Plywood Corp*, which lays out 15 factors to determine the type of monetary payments that would compensate for a patent infringement (and which are often referred to as the 'Georgia-Pacific factors').[80]

A patentee that establishes causation between infringement and loss of profit through lost sales or price erosion may recover lost-profit damages instead. If the patentee proves by clear and convincing evidence that its opponent's infringement was wilful or in bad faith, a court may award treble damages.[81] Additionally, courts have discretionary power to award permanent injunctions against continued infringement;[82] the *eBay* case, referred to above, sets out a four-factor test for granting a permanent injunction.

6. Compulsory licensing

The United States has no compulsory patent licensing regime *per se*. As a general rule, a patent owner has the exclusive authority to decide whether to license its rights as well as the parties to, and terms of, the licence agreements it enters, as long as they are within the bounds of US antitrust laws. There are, however, a few exceptions to this general rule.

First, 28 USC § 1948 allows the US government – including federal employees, federal agencies, contractors, and even private parties that the federal government has authorised – to use a patent without any formal licensing agreement or negotiation. An owner seeking compensation for a compelled licence may bring an action against the United States for reasonable recovery but cannot seek to enjoin or bring an infringement action against the user – see, for example, *Crater Corp v Lucent Technologies*,[83] holding an alleged infringer, a private company, not liable because the company's work was performed for the government.

Second, particular federal statutes provide for compulsory licensing of patents. Under The University and Small Business Patent Procedures Act of 1980 (also referred to as the 'Bayh-Dole Act'), the government may 'march in' and compel not-for-profit organisations and small businesses to license patents that they fail to practise or which the government deems necessary for public health.[84] No federal government agency has ever exercised its 'march-in' rights, however, despite petitions to do so for patents related to stem cell therapy, HIV/AIDS and glaucoma drugs, and a treatment for Fabry disease.[85] The Clean Air Act mandates that an owner concede to licensing if the Attorney General determines that a patent is necessary for compliance with the Act's emission standards.[86] Under the Plant Variety Protection Act, the US government may require a plant breeder who refuses to license a patent at a fair price to do so, if

---

80   *Georgia-Pacific Corp v United States Plywood Corp.* 318 F. Supp. 1116 (SDNY 1970).
81   35 USC § 284.
82   35 USC § 283.
83   *Crater Corp v Lucent Technologies* 255 F.3d 1361 (Fed. Cir. 2001).
84   See 35 USC § 203, which outlines requirements for 'march-in' rights.
85   See, for example, "NIH March-In Responses", available at www.ott.nih.gov/policy/Reports.html.
86   42 USC § 7608.

necessary to ensure adequate supplies of "fibre, food, or feed".[87] Similar compulsory provisions exist for patents related to atomic energy,[88] and black lung disease.[89]

State sovereign immunity prevents a patent holder from suing an allegedly infringing state, thus producing an additional form of *de facto* compulsory licensing. Furthermore, Federal Trade Commission guidelines anticipate the possibility of court-ordered compulsory licences as a remedy for antitrust violations.[90]

Compulsory licensing regulations exist for certain other forms of intellectual property. The Copyright Act, for example, contains a compulsory licensing scheme for making and distributing 'cover' arrangements of non-dramatic musical works and for "published nondramatic musical works and published pictorial, graphic, and sculptural works in connection with noncommercial broadcasting".[91]

Finally, courts may deny the issuance of a permanent injunction against a patent infringer, thus creating *de facto* compulsory licences by allowing infringement to continue without the patentee's consent (see again the *eBay* case, which heightens the equitable test for permanent injunctions).[92] Such a situation may occur when a drug has been brought to market and its removal would frustrate the public interest, particularly when the patentee has failed to practise its invention. The *eBay* decision is important for life sciences patent holders to keep in mind, as refusal to license or practise patents may restrict access to medical therapies that the courts deem crucial to public health.

7. Ownership, inventors and compensation

An inventor is the person (or persons) who first conceived of the invention as well as a definite plan for carrying it out.[93] In contrast to many other countries, US patent law has a unique 'first-to-invent' system, which means that the inventor who first conceived of the invention and then diligently reduced it to practice (or filed a patent application to constructively reduce the invention to practice) is considered the first inventor.

Patent reform legislation, widely known as the 'America Invents Act', seeks to change the US standard to a 'first-to-file' regime, in which the first inventor is the one who first files a patent application, regardless of the actual date of invention. Both houses of the US Congress have passed similar Bills that are now being reconciled. Should the America Invents Act pass and be signed into law, it will bring the United States into harmony with the international first-to-file standard and will substantially reform a number of other aspects of US patent law as well.

---

87   7 USC § 2404.
88   42 USC § 2183.
89   30 USC § 937.
90   See the US Department of Justice and Federal Trade Commission's *Antitrust Enforcement and Intellectual Property Rights: Promoting Innovation and Competition*, pages 21–25 (2007), available at www.usdoj.gov/atr/public/hearings/ip/222655.pdf.
91   17 USC §§ 115, 118.
92   *eBay Inc v MercExchange*, LLC 547 US 388 (2006).
93   See *Hybritech Inc v Monoclonal Antibodies Inc* 802 F.2d 1367, 1376 (Fed. Cir. 1986) (conception is the "formation in the mind of the inventor, of a definite and permanent idea of the complete and operative invention, as it is hereafter to be applied in practice") and *Fiers v Revel* 984 F.2d 1164, 1168 (Fed. Cir. 1993) ("[u]nless a person contributes to the conception of the invention, he is not an inventor").

In the United States, an inventor receives initial ownership rights in a patent and is not divested of these rights unless he affirmatively assigns full or partial ownership rights to another entity. In the case of joint inventors, each joint inventor owns an undivided partial interest in the entirety of the patent. An inventor who received government funding for his research may be required to share rights with the US government.[94] Ownership of a patent grants a patent owner the so-called 'negative' right to exclude others from making, using, offering for sale, selling or importing into the United States the invention claimed in the patent.[95] It does not, however, grant the positive right to practise the invention.

Particularly in the life sciences industry where exclusivity is an essential part of business models, companies typically require employee researchers to assign the entirety of their patent rights to the company as a condition of employment – and often before the invention process begins. If the assignment does not cover the full scope of rights, remaining joint owners retain the ability individually to license full patent rights to external entities "without the consent of and without accounting to the other owners".[96] In addition to the scope of assignment, companies acquiring life sciences technology must exercise caution when transacting with patent holders under the Bayh-Dole Act's ambit. The Act forbids assignment of patents developed through federally funded research by small businesses, not-for-profit institutions, and universities, although it does permit exclusive licensing agreements.[97] The Supreme Court recently held that "the Bayh-Dole Act does not confer title to federally-funded inventions" on US government contractors or authorise the US government "to unilaterally take title to those inventions" and the Act cannot supersede a third party's assigned rights to a patent invention that was developed using federal funding.[98] Compensation for patent ownership or inventorship typically occurs through direct profits or through licensing. Under the terms of a licence agreement, royalty payments may be structured in a variety of ways and can include payments upfront, periodically, upon occurrence of milestone events, and/or through royalty schemes or stock rights.

Under copyright law, an author or authors own the copyright in a protected work, but may transfer ownership by any means of conveyance.[99] An employer is by default the owner of any work made for hire, which includes work that an employee creates in the scope of his employment as well as commissioned work.[100] An employee who wishes to retain copyright in his work must contract out of this default rule or else demonstrate that he is an independent contractor rather than an employee.[101]

---

94   35 USC § 261.
95   35 USC § 154(a)(1).
96   35 USC § 262.
97   18 USC §§ 200 et seq.
98   See *Board of Trustees of Leland Stanford Jr University v Roche Molecular Sys et al.* 131 S.Ct. 2188, 2197–8 (2011) (the holding of university ownership rights in federally funded inventions does not divest inventors of their rights to assign inventions contractually to third parties).
99   17 USC § 201.
100   *Ibid.*
101   *Ibid.*

8. Branding and designs

8.1 Trade marks in life sciences

The US Trademark Statute (also known as the Lanham Act) defines a trade mark as a word, phrase, symbol or design, or a combination thereof, that identifies and distinguishes the source of the goods of one party from those of others.[102] 'Trade dress' is a type of trade mark that protects a product's overall image and distinctive features such as size, colour, shape and texture (involving product packaging and 'look and feel' that is often not registered). Trade mark protection generally safeguards a business's valuable goodwill. For life sciences businesses, trade mark law is complicated by the role of the FDA and the Federal Trade Commission (FTC) in regulating pharmaceutical and medical-device naming, labelling and advertising. For example, a drug manufacturer's choice of name for a new drug must satisfy requirements under trade mark law to receive trade mark protection, but also must receive approval from the FDA before the drug may enter the market. The FDA may reject a proposed proprietary drug name if it believes consumers will confuse the name with other drugs.

8.2 Extending protection using trade marks and designs

In the United States, a trade mark's owner need only use the mark in commerce to establish rights in the mark. Enforcement of trade mark rights is through state laws or the federal Lanham Act. Public notice of ownership of federal trade-mark rights is given via ™ or ® insignia. Trade mark owners may bring suit against sellers who employ marks, descriptions or designations in commercial advertising that are likely to cause confusion, to mislead or to deceive in a manner that injures the trade mark owner.[103] Other persons who might be damaged by these unfair practices may also bring claims, subject to prudential standing limitations.[104]

A trade mark owner may register its mark with the USPTO. Registration confers the advantages of being able to bring suit for trade mark infringement in a federal court, being able to use the federal registration symbol ® on the mark and being listed on the USPTO's online databases, which grants a presumption of ownership and the nationwide exclusive rights to use the mark in federal commerce. To register a trade mark, the owner must specify that the mark is in use in federal commerce or establish *bona fide* intent to use the mark in commerce. A USPTO examiner will deny registration only if the mark is generic (ie, is in common parlance, such as 'aspirin'), is primarily functional, or for other reasons forbidden by statute.[105] An applicant may appeal against a denial of registration to the Trademark Trial and Appeal Board (TTAB), which also hears opposition proceedings.

Attacks on a trade mark's validity typically arise in opposition proceedings before the TTAB or in the context of litigation, as a defence against a claim of trade mark infringement. A party challenging the validity of a trade mark may prevail by

---

102   15 USC § 1127.
103   15 USC § 1125 (false advertising).
104   *Phoenix of Broward, Inc v McDonalds Corp.* 489 F.3d 1156 (11th Cir. 2007).
105   15 USC § 1052.

showing that a trade mark's owner has abandoned the mark or that the mark is generic or descriptive and without a secondary meaning that customers associate with a specific product.

Absent a showing that a mark has become generic or has been abandoned, trade mark protection is unlimited in duration. Accordingly, the threat of 'genericisation' is of great concern to life sciences industry manufacturers, particularly those that rely on a trade mark's goodwill to preserve market power once patent exclusivity has expired. Pharmaceutical companies often engage in active education and trade-mark protection programmes to maintain their band's value and protected status.

Trade dress is also important in life sciences, as it extends to packaging, sales techniques and advertising themes. An example of trade dress in the life sciences is AstraZeneca's 'Purple Pill'. While the colour alone is not inherently distinctive, AstraZeneca's marketing efforts have led consumers to associate purple pills with the specific product Prilosec®, which confers a secondary meaning on the colour purple. Pfizer's Viagra, the 'Little Blue Pill', is another illustration of trade dress. While a number of courts have recognised shape and colour of a drug as protected trade dress,[106] others have been reluctant to recognise trade dress protection in this realm. Particularly where branded manufacturers have brought infringement actions against generic drug makers for imitating the shape and colour of the plaintiff's branded drug, courts have considered colour and shape to be functional attributes that assist with accurate dosing and administration to protect patients and therefore not protectable elements of trade dress.[107] Public-policy concerns may drive this rationale more so than strict application of trade mark law.

8.3 Protection using unfair competition

Federal trade-mark law is just one variety of a greater body of unfair competition law that encompasses federal, state and common-law protections against, *inter alia*, deceptive trade practices, dilution, internet domain name misuse, misrepresentation, misappropriation, and copying or imitation of distinctive advertising and merchandising efforts. For example, an implied falsity claim may be brought in instances where a drug advertisement is literally true but is nevertheless likely to mislead or confuse consumers.[108]

More commonly, drug manufacturers will seek relief from unfair competition by bringing false-advertising claims against competitors under the Lanham Act.[109] Under this provision, a plaintiff must prove that the competitor's advertisement, which

---

106   See, for example, *SK & F Co v Premo Pharmaceutical Laboratories* 625 F.2d 1055 (3d Cir. 1980) and *Ciba-Geigy Corp v Balor Pharmaceutical Co* 747 F.2d 844 (3d Cir. 1984).
107   See, for example, *Ives Laboratories, Inc v Darby Drug Co* 488 F. Supp. 394 (EDNY 1980) and *Shire US, Inc v Barr Labs, Inc* 329 F.3d (3rd Cir. 2003).
108   See *Johnson & Johnson v Merck Consumer Pharms Co v Smithkline Beecham Corp*, 960 F.2d 294, 297 (2d Cir. 1992) (claim that television commercials communicated a literally true but misleading message with respect to the safety of aluminium-based antacids, based on exploiting public misperception that ingestion of aluminium causes Alzheimer's disease).
109   15 USC § 1125.
110   See, for example, *Schering Corp v Pfizer, Inc.* No 98-cv-2000, 2000 WL 718449, *9 (SDNY June 5 2000) (granting Schering a preliminary injunction to prevent Pfizer detailers from telling physicians that ZYRTEC® is non-sedating or is as non-sedating as Schering's CLARATIN®).

may include print and media presentations, as well as oral statements of drug sale representatives ('detailers') to physicians, is false or will mislead consumers.[110] A plaintiff will not prevail by showing a lack of clinical support for a competitor's claim of a drug's efficacy, since ensuring the sufficiency of clinical data and enforcing the labelling and promotional practices of drugs is the province of the FDA and FTC. Plaintiffs may offer violations of the Food, Drug and Cosmetic Act (FDCA), which the FDA and FTC jointly enforce, as evidence of an advertisement's misleading nature or falsity,[111] but may not offer such as *per se* proof of a Lanham Act violation.[112]

8.4 Comparative advertising and advertising restrictions

Under the Lanham Act, the fair use of a famous mark by another person in comparative commercial advertising is not actionable.[113] Thus, drug manufacturer A may stop competitor B from advertising that "drug B is more effective than drug A" only if manufacturer A proves that drug B is not more effective than A or that competitor B's advertisement misleads consumers.[114]

Federal agency regulation of drug and medical device advertising supplements this minimal protection. The FDA regulates prescription drug advertising and labelling, while the FTC oversees non-prescription (over-the-counter) drug labelling.[115] The FTC and FDA share responsibilities for regulating medical-device advertisements. As under the Lanham Act, the FDCA prohibits false or misleading advertising, or 'misbranding', of drugs and medical devices.[116] Off-label promotion of drugs is prohibited by the FDCA, whereas advertisements that promote use within FDA-approved indications are permissible. These advertisements may target healthcare providers and also consumers ('direct-to-consumer advertising'), the latter of which is not permitted in many other countries. In most forms of advertising, the FDA requires manufacturers to include an oral or written statement of all risks included on the drug's label.

8.5 Protection of online content

In the United States, online advertisements of drugs and devices must conform with the same regulatory restrictions as print and broadcast advertisements. The FDCA contains no explicit internet provision. Rather, the FDA applies its general provisions to online content on a case-by-case enforcement basis. The Lanham Act, however, contains a cyberpiracy prevention provision (the AntiCybersquatting Consumer Protection Act) that prohibits registering, with the bad faith intent to profit, a domain name that is confusingly similar to a registered or unregistered mark, or dilutive of a famous mark.[117]

---

111   See *Zeneca, Inc v Eli Lilly & Co* No 99-cv-1452, 1999 WL 509471 (SDNY July 19 1999).
112   See, for example, *Sandoz Pharmaceuticals v Richardson-Vicks, Inc.* 902 F.2d 222 (3rd Cir. 1990).
113   15 USC § 1125.
114   *Ibid.*
115   21 USC §§ 301 *et seq.*
116   21 USC § 352.
117   15 USC § 1125.

9. Protecting valuable information

9.1 Preserving confidentiality

Trade secrets are protected by the individual laws of the several states. Most states have adopted the Uniform Trade Secrets Act (UTSA), which affords protection to any information that derives economic value from its secrecy and its inability to be ascertained in the market. Manufacturers might seek this form of IP protection for inventions that they do not wish to disclose publicly; for inventions, identifiers or works that do not meet the requirements for trade-mark, patent, or copyright protection; or when the manufacturer desires unlimited duration of protection. Holders of trade secrets lose protection if they do not make reasonable efforts to maintain secrecy.

Manufacturers in the life sciences industry rarely utilise trade secret protection with respect to pharmaceutical products themselves, because such products are generally subject to clinical testing, ongoing disclosure requirements and extensive scrutiny by the FDA, which requires a level of disclosure that is often inconsistent with trade secret protection. However, some features of the manufacturing process or specific components of a drug product may warrant trade secret protection.

Under the UTSA or equivalent state statute's misappropriation provision, a manufacturer that properly maintains a trade secret may bring a legal action against a party (typically a competitor, employee or former employee) that has wrongfully misappropriated, disclosed or threatened to disclose a trade secret.[118] To prevail on this theory, a trade-secret owner must establish that it took affirmative steps to maintain secrecy and prevent disclosure. In anticipation of this burden, life sciences companies should do some or all of the following, as appropriate: give notice to and enter into confidentiality agreements with employees, licensees and other parties with which they contract; utilise non-compete agreements; mark sensitive information as 'confidential' and restrict employee access to such information; institute policies to curtail copying; and set up physical and electronic security measures. In addition to remedies for misappropriation, an owner of an exposed trade secret may seek remedies under state unfair-competition statutes or common-law breach of contract.

9.2 Know-how and ex-employees

Employees in the life sciences industry, particularly those involved with research and development, may have gained technical skills and knowledge over the course of employment at an institution. An employer may employ narrowly tailored non-compete agreements to curtail the loss of general know-how and goodwill that results when ex-employees seek employment with institutional competitors. The enforceability of these agreements varies from state to state.

Of greater concern to life science businesses is ex-employee disclosure of trade secrets or the loss and competitive use of know-how that a former employee derived specifically through his access to trade secrets. Misappropriation offers a remedy only

---

118   See, for example, *PepsiCo, Inc v Redmond* 54 F.3d 1262, 1269 (7th Cir. 1995).

for disclosure that has already occurred; the 'inevitable disclosure' doctrine, on the other hand, offers institutions a potential means of enjoining a former employee from working for a competitor, on the theory that the ex-employee is incapable of separating the know-how he learned while working with trade secrets from his general knowledge and training.[119]

The doctrine is rarely successful, however, absent a prior contractual relationship between the ex-employee and first employer regarding protection of trade secrets – see, for example, *Schlage Lock Company v Whyte* and *Bayer Corp v Roche Molecular Systems, Inc.*[120]

9.3   Copyright, patient information leaflets and packaging

US copyright law protects "original works of authorship fixed in any tangible medium of expression".[121] This includes, *inter alia*, literary, musical, dramatic, graphic, sculptural, audiovisual and architectural works.[122] The owner of copyrightable expression enjoys exclusive reproduction, adaptation, distribution, public performance and display, and digital performance and sound recording rights. Copyright law does not protect purely factual information.[123]

A fixed expression that conveys disease, product or promotional information about a drug or medical device – for example in patient information leaflets or drug product packaging – may be protected by copyright. As long as a hint of originality exists in the expression, it is protectable. Therefore, a competitor's exact imitation or copy of patient leaflets or product packaging could constitute copyright infringement.[124] However, the strength of a copyright infringement claim against a competitor is weakened substantially if the wording or presentation of the competitor's leaflets and packaging is altered even slightly.

If only the non-protectable ideas, concepts or raw facts are copied, the competitor has not copied any protectable expression and thus has not infringed. Furthermore, under the merger doctrine, expression that can only be communicated in a very limited number of ways, such as a blank patient medical history form that a healthcare provider uses for record keeping, also may not receive copyright protection.[125] The merger doctrine is not likely to preclude a drug manufacturer from prevailing against an accused infringer, however, because patient information leaflets and product packaging tend to include original stylistic elements and may be communicated in many different ways.

---

119   See, for example, *PepsiCo, Inc v Redmond* 54 F.3d 1262 (7th Cir. 1995) (affirming an injunction based on the 'inevitable disclosure' doctrine).

120   *Schlage Lock Company v Whyte* 101 Cal. App. 4th 1443, 1447 (2002) (rejecting the inevitable disclosure doctrine as an 'after-the-fact covenant not to compete' that is incompatible with state law prohibitions on 'non-competes'); *Bayer Corp v Roche Molecular Systems, Inc.* 72 F. Supp. 2d 1111, 1120 (N.D. Cal. 1999) (noting that "California law does not recognise the theory of inevitable disclosure").

121   17 USC § 102.

122   17 USC § 102(a).

123   See *Harper & Row Publishers, Inc v Nation Enterprises* 471 US 539 (1985).

124   See, for example, *FMC Corp v Control Solutions, Inc.* 369 F. Supp. 2d 539 (EDPA 2005) (granting a branded pesticide manufacturer's motion for a preliminary injunction against a generic pesticide manufacturer's use of an identical product label).

125   See *Baker v Selden* 101 US 99 (1879).

10. Counterfeiting

No criminal liability exists for patent infringement in the United States. However, varieties of copyright infringement, theft and misappropriation of trade secrets, and the trafficking of counterfeit trade-marked goods, carry criminal penalties. The No Electronic Theft Act of 1997 criminalised wilful copyright infringement that occurs for the purpose of commercial advantage, involving reproduction or distribution to the public of a copyrighted work via a computer network.[126] The Economic Espionage Act of 1996 criminalised theft or misappropriation of trade secrets related to products in commerce generally and also created a special provision for theft or misappropriation of trade secrets conducted for the purpose of benefiting a foreign government.[127]

The Trademark Counterfeiting Act of 1984 criminalised trafficking in counterfeit goods.[128] Under this Act and the subsequent Anticounterfeiting Protection Act of 1996, the Stop Counterfeiting in Manufactured Goods Act of 2006, and the PRO-IP Act of 2008, US Immigration and Customs Enforcement (the principal investigative arm of the US Department of Homeland Security) collaborates with US Customs and Border Protection to seize imported goods that infringe copyright and trade mark rights and to investigate and prosecute infringing parties.[129] Joint enforcement with the FDA, the Federal Bureau of Investigations and the Department of Justice occurs frequently for investigation and prosecution of parties who intentionally traffick or attempt to traffick in counterfeit pharmaceuticals. Offenders dealing in counterfeit pharmaceuticals face enhanced criminal penalties, as authorities typically consider counterfeit drug sales to constitute known, reckless or attempted infliction of bodily harm or death under the PRO-IP Act of 2008.

11. Collaborative models – 'open innovation'

11.1 Using the internet to innovate and the issues raised

There are no US laws that specifically regulate allocation of IP rights in new collaborative models, yet 'open innovation' often creates opportunities for collaboration between large, sophisticated organisations and individuals that may not have experience with IP rights transactions.

One example of a successful 'open innovation' model that relies on creative IP policies is InnoCentive's online marketplace.[130] First developed by Eli Lilly but now an independent corporation, InnoCentive hosts an online platform for organisations seeking innovation ('seekers'), typically in the form of solutions to specific problems or ideas for addressing unmet needs, to post 'challenges' and solicit solutions from the general public. The 'solver' that best addresses the challenge receives an award. The nature of the innovation sought determines the terms of IP ownership. For example, InnoCentive's guidelines require that a seeker soliciting solutions to a

---

126   17 USC § 506.
127   18 USC §§ 1831, 1832.
128   18 USC § 2320.
129   See *US Customs and Border Protection & US Immigration and Customs Enforcement, Intellectual Property Rights: Fiscal Year 2010 Seizure Statistics* 2 (2011), available at www.ice.gov/doclib/news/releases/2011/110316washington.pdf.
130   See www.innocentive.com.

problem in the form of general ideas receives a non-exclusive, perpetual licence to use any of the solutions submitted. Solvers retain IP rights and the ability to negotiate with the seeker as the collaborative effort further defines the solution. When seekers solicit detailed solutions, data or physical innovations, however, InnoCentive's terms of participation typically require solvers to assign full IP rights to the seeker or else to accept a lesser reward in exchange for a licensing agreement. Solvers – often university or industry researchers and engineers – are responsible for determining whether their employment poses a barrier to a transferral of rights and, if necessary, for submitting an employer-signed waiver.[131]

11.2 Patent pooling and 'pre-competitive' data pooling

'Patent pooling' is a collaborative model that incentivises research by making a greater array of IP available to a greater number of users. The method is especially advantageous for the development of therapeutics that hold limited market value but great social value, such as orphan drugs. An example of this strategy is The Pool for Open Innovation against Neglected Tropical Disease.[132] The project, initiated by GlaxoSmithKline in 2009, asks IP holders to grant pool 'users' – scientists who utilise the contributions – royalty-free, non-exclusive licences to research, develop, manufacture and export therapies for 16 neglected tropical diseases for use in the world's least-developed countries. IP owners that contribute to the pool retain the right to negotiate royalty terms for use of therapies in more developed countries, to continue to use the IP they contributed in their own research, and to execute confidentiality agreements with users regarding any know-how they contribute. A non-profit organisation oversees administration of the pool and several pharmaceutical companies, universities and government agencies currently contribute to, and use, the contents of the pool.

By no means is IP pooling limited to philanthropic, royalty-free models. Pooling may refer to any sort of cross-licensing of patents between holders. Different competitive strategies typically dictate the terms of pooling. For example, patent pooling may offer a solution for companies holding mutually blocking patents or a means of meeting international technology standards for companies holding complementary patents. Agreements to pool complementary and/or competitive IP for purposes of price fixing and exclusion of competition are susceptible to intense antitrust scrutiny. The Department of Justice has issued Antitrust Guidelines for the Licensing of Intellectual Property for determining whether patent pools comply with the antitrust laws. The guidelines effectively seek to limit patent pools to situations where blocking patents, transaction costs and the threat of costly patent litigation pose substantial obstacles to the commercial development of a technology. However, interest in patent pools has increased recently among pharmaceutical companies as well as among medical device and diagnostic companies, especially in the area of diagnostic genetics. Healthcare reform in the United States is likely to incite further interest in life sciences IP pools.

---

131   See "Frequently Asked Questions" at www.innocentive.com/faq/seeker.
132   See http://ntdpool.org/.

12. Hot Topics

Any so-called 'hot topics' that apply at present within the United States in relation to intellectual property have been highlighted in earlier sections of this chapter.

The authors gratefully acknowledge the contributions of Rachel Farnsworth, a Simpson Thacher summer associate, who assisted in the preparation of this chapter.

About the authors

Pravin Anand

Managing partner, Anand and Anand

Pravin@anandandanand.com

Pravin is ranked as the number-one IP attorney in India. He has been a counsel in several landmark IP cases involving the first Anton Piller order, the first Mareva injunction order, the first Norwich Pharmacal order, rights of privacy, trade mark dilution, recognition of market survey evidence, domain names, punitive and exemplary damages, phishing, patent compulsory licensing and numerous other cases which have transformed the IP landscape in India.

Pravin serves on the editorial board of several international IP journals. He has authored several papers on intellectual property and has co-authored the two IP volumes in *Halsbury's Laws of India*. He is a regular speaker at various forums including WIPO, AIPPI, INTA, LES, IBA, LAW ASIA, UN Conferences etc.

Pravin serves on several international and national expert groups/committees and parliamentary committees have invited him to give evidence on amendments to IP laws. He is chairman of the IPR promotion advisory committee (IPAC) constituted by the government of India and the IT committee set up by FICCI. He has been president of the Asian Patent Attorneys Association (APAA India) and director of the INTA Board. He is current president of the AIPPI (Indian group) and vice-president of the Society of Indian Law Firms.

Sarah Bailey

Solicitor, Simmons & Simmons LLP

sarah.bailey@simmons-simmons.com

Sarah specialises in intellectual property, dealing with all the aspects involved in the creation, exploitation and protection of intellectual property rights, and is responsible for the trade mark filing service in Paris. Sarah also deals with data protection and advertising matters.

Sarah is a French-qualified lawyer as well as a solicitor of the Supreme Court of England and Wales. She holds a Certificate in Intellectual Property (trade marks, designs and models) from the Centre Robert Schumann of the University of Strasbourg.

Nick Bassil

Partner, Kilburn & Strode LLP

nbassil@kstrode.co.uk

Nick qualified as a Chartered Patent Attorney in 1998 and as a European Patent Attorney in 1999 and became a partner in 2001. Described as "brilliant at his job" by the *Legal 500* (2006 edition) and a 'Leader in their Field' by *Chambers UK* (2011 edition), Nick's practice concentrates on pharmacology and biochemistry, particularly cell and molecular biology. Nick has been significantly involved in patent applications relating to embryonic stem cells and aspects of cell cloning (nuclear transfer) technology.

His practice ranges from drafting original patent applications and international filing programmes to advising on due-diligence matters and opinion work. His European portfolio includes post-grant opposition proceedings at the European Patent Office, contentious proceedings in other foreign jurisdictions, as well as obtaining supplementary protection certificates in the pharmaceutical area.

Greg Beach

Associate, Belmore Neidrauer LLP
greg.beach@belmorelaw.com

Greg was called to the Ontario Bar in 2007 and practises intellectual property litigation, with a focus on pharmaceutical patent litigation. Prior to joining Belmore Neidrauer, Greg practised with a boutique law firm specialising in intellectual property.

Before commencing his legal training, Greg completed a degree in metallurgical engineering at McGill University and worked as a management consultant in Montreal and Boston.

Greg is a member of the Canadian Bar Association and the Intellectual Property Institute of Canada, and he has co-authored an article on pre-trial discovery in intellectual property litigation which was published in the *Canadian Intellectual Property Review* in 2005.

Neil Belmore

Co-founding partner, Belmore Neidrauer LLP
Neil.Belmore@belmorelaw.com

Neil was admitted to the Law Society of Upper Canada in 1985 after receiving his LLB from Osgoode Hall Law School.

As lead counsel in patent, pharmaceutical, copyright, trade mark, trade libel, trade secret and misleading advertising cases, Neil's practice focuses on trial and appellate advocacy. Noteworthy clients where he has acted as lead counsel in pharmaceutical patent cases include: Bayer Shering Pharma AG, Janssen Inc (Johnson & Johnson), Nycomed Canada Inc, Alcon Research Ltd, Kyowa Hakko Kirin Co Ltd, Novo Nordisk Canada Inc and Boehringer Ingelheim GmbH.

Neil is internationally recognised as a leading intellectual property lawyer in peer-selected reviews. In 2009, he co-founded the Toronto-based intellectual property advocacy firm Belmore Neidrauer LLP. The firm is dedicated to creative and strategic advocacy and works on prominent files in the field of intellectual property law.

Bas Berghuis van Woortman

Partner, Simmons & Simmons LLP
bas.berghuis@simmons-simmons.com

Educated at the Radboud University of Nijmegen and the Univerisity of Brussels, Bas has more than 15 years' experience in patent proceedings before national and international courts, including the Netherlands Supreme Court and the European Court of Justice, as well as the Dutch and European Patent Offices. Bas has a particular focus on the life sciences sector, handling patent and technology matters regarding medical devices, biotech and pharmaceuticals. He has been involved in all aspects of litigation, as well as other IP-related issues such as patent prosecution strategies, R&D agreements and licensing issues.

Alexander Christophoroff

Partner, Gowlings International Inc
alexander.christophoroff@gowlings.com

Alexander is a partner in Gowlings' Moscow office, specialising in intellectual property litigation and administrative procedures.

His practice includes general and arbitration courts, anti-monopoly authorities, and Board of Patent Disputes of the Russian Patent & Trademark Office, and he has been involved in many precedent-setting cases.

Since 1997, Alexander has been an attorney for private practice, admitted as attorney at law (advocate), Russian patent attorney and Eurasian patent attorney. Prior to this, he worked for the Russian Patent & Trademark Office as General Counsel and member of the Board.

Wayne Condon

Partner, Griffith Hack

wayne.condon@griffithhack.com.au

Wayne is the practice head at Griffith Hack Lawyers. He has particular experience and expertise in the pharmaceuticals, biotechnology and agricultural chemicals sector. He led the team in the Australian patent proceedings that clarified the law relating to the patentability of methods of medical treatment and he has represented clients in many of the leading pharmaceutical patent cases conducted in Australia in the last 20 years.

Wayne is recognised as a leader in his field by industry publications such as *The Best Lawyers in Australia* and *Chambers Global*. He is especially noted for being a 'top tier patent talent' as seen by his recent endorsement as a Tier 1 leading individual for patent litigation by *PLC Which Lawyer?*

Mattie de Koning

Lawyer, Simmons & Simmons LLP

mattie.dekoning@simmons-simmons.com

Mattie is a graduate from the Erasmus University where he obtained his LLM, and the Technical University Eindhoven where he obtained his MSc in chemical engineering. He specialises in intellectual property law, with a focus on patents and technology. He advises and litigates on behalf of clients in a variety of technical areas, in particular the field of life sciences. In addition, Mattie regularly advises investors and start-up companies in various technical fields on a wide range of technical, corporate and commercial issues.

Gualtiero Dragotti

Partner, DLA Piper

gualtiero.dragotti@dlapiper.com

Gualtiero co-heads the IP practice at DLA Piper and is based in the firm's Milan office.

Gualtiero has extensive experience in IP matters, with a focus on patents. In this connection, he has advised leading Italian and international clients on various contentious IP cases, including some of the key patent litigation in Italy. In recent years he has dealt with many pharmaceutical patent cases. In addition, he has experience in a variety of corporate matters relating to protection and management of IP rights, for both domestic and international clients. *Chambers & Partners Europe* and a number of other client surveys have identified Gualtiero as a key individual in the IP field.

Catherine Drew

Associate, Winston & Strawn

cmdrew@winston.com

Catherine is an intellectual property associate at Winston & Strawn, London. She has experience in large-scale, multi-jurisdictional patent litigation and has been involved at all stages of litigation in the English High Court, Court of Appeal, and Supreme Court.

Catherine has also worked on non-contentious aspects of intellectual property law, including assisting with the IP aspects of commercial agreements, advising on regulatory matters such as product liability and data exclusivity, and advising on legal issues around the marketing of medicinal and cosmetic products.

Paul England

IP professional support lawyer

Simmons & Simmons LLP

Paul.England@simmons-simmons.com

Paul's principal expertise is in life sciences patent litigation and associated regulatory issues. He has a particular interest in multijurisdictional patent litigation strategy, the preparation of case theory and the handling of expert evidence. Paul is the author of *Expert Privilege in Civil Evidence* (Hart Publishing, Oxford) and publishes regularly in practitioner publications and peer-reviewed journals.

Prior to joining Simmons & Simmons LLP in

2010, Paul spent more than 10 years at another patent practice in the City of London. Before turning to law, he gained an honours degree in chemistry at Edinburgh University, followed by a doctorate in biochemistry and molecular biology from Linacre College, Oxford University. Paul also has a Diploma in Intellectual Property Law from Bristol University.

Paul has directly experienced the life sciences sector, having worked for the Human Genome Organisation and the former Wellcome Foundation (now merged with GSK).

Joachim Feldges

Managing partner, Field Fisher Waterhouse
Joachim.Feldges@ffw.com

Dr Joachim Feldges is Managing Partner for Germany for his firm, based in FFW's Munich office. He specialises in the field of intellectual property.

Joachim has practised complex patent litigation across all technical fields for more than 20 years, with a particular focus on the fields of pharmaceuticals and biotechnology. He advises clients in developing and implementing international patent litigation strategies in the various European jurisdictions, including the requirements for parallel litigation in the United States.

He represents clients before all major German courts on infringement and validity cases, including those in Düsseldorf, Mannheim and Munich, the Federal Supreme Court and the Federal Patent Court. He is regularly involved in opposition proceedings at the European Patent Office. Joachim's trial experience also includes arbitration and mediation. He has negotiated, drafted and revised and has been involved in dispute resolutions relating to numerous types of technology transfer agreements in many regions, including Europe, the United States and the CIS countries.

Kasper Frahm

Attorney-at-Law, Plesner
kfr@plesner.com

Kasper is an attorney-at-law in the IP department at one of the leading Danish law firms. His primary areas of practice are trade marks, domain names, marketing law and patents. He has litigated various IP cases in court and advises several large international clients on the above-mentioned areas of law. He has written a number of articles on IP.

Kasper is a Master of Law from the University of Copenhagen and, in the academic year 2011 to 2012, will study for a Master of Law (LLM) at King's College London. He is a member of the Danish Association for the Protection of Industrial Rights and AIPPI.

Afif Hamid

Associate, Belmore Neidrauer LLP
Afif.Hamid@belmorelaw.com

Prior to joining Belmore Neidrauer, Afif articled at a full-service national law firm where his work focused primarily on intellectual property litigation, particularly in the pharmaceutical patent context.

Afif obtained an LLB degree from Osgoode Hall Law School in 2008. During law school, he was awarded first place in the first-year moot competition. He was also a participant in the 14th annual VIS International Commercial Arbitration Moot in Vienna, Austria in 2007. At the end of his second year, Afif summered with the in-house legal department of one of Canada's largest banks. Prior to attending law school, Afif obtained a Bachelor of Arts degree in Economics from the University of Toronto in 1999.

Afif is a member of the Canadian Bar Association and the Intellectual Property Institute of Canada.

Katherine A Helm
Associate, Simpson Thacher & Bartlett LLP
khelm@stblaw.com

Katherine is an associate in the intellectual property group at Simpson Thacher & Bartlett LLP in New York. Her practice focuses on biotechnology and pharmaceutical patent litigation. She is a former law clerk to the Honorable Alvin A Schall, US Court of Appeals for the Federal Circuit and to the Honorable Marilyn Hall Patel, US District Court for the Northern District of California.

Katherine previously worked as a technical advisor at another major law firm in New York, where she focused on contested proceedings in biotechnology and pharmaceutical patent law, including US interferences, European oppositions and US litigation. She has published numerous articles and commentary on legal issues and is a monthly columnist for Law.com, where she writes on a variety of legal and ethics topics.

Katherine currently represents Human Genome Sciences in patent litigation against Genentech and City of Hope over Benlysta®, the first FDA-approved treatment for systemic lupus erythematosus in more than 50 years.

Mami Hino
Partner, Abe, Ikubo & Katayama
hino@aiklaw.co.jp

Mami is a partner at Abe, Ikubo & Katayama, Tokyo, Japan. She has practised as a patent attorney both in Japan and in the United States and represents clients in intellectual property matters including patent litigation, patent references, strategic counselling, and due diligence. She has taken a leading role in prosecuting and enforcing patents for pharmaceutical, biotechnology, and medical device companies located in various jurisdictions.

Mami registered as a Japanese patent attorney in 1992 and was admitted to practise law in New York in 2000. She worked as an associate at Pennie & Edmonds LLP from 1999 to 2002, at Baker & McKenzie, Tokyo, from 2002 to 2004, and joined Abe, Ikubo & Katayama as an of-counsel in 2004.

Lewis Ho
Partner, Simmons & Simmons LLP
Lewis.Ho@simmons-simmons.com

Lewis is recognised as one of the leading IP and life sciences lawyers in China. He is acknowledged by *Chambers Asia* and *Legal 500* and named by IAM as one of the leading 250 patent litigators, life sciences litigators and licensing lawyers in the world.

Lewis was the vice chairman of the IP working group (Shanghai) of the EU Chamber of Commerce in China (2009 to 2011). He is also the Honorary Legal Advisor to HKAPI, the trade association representing research-based pharmaceutical companies in Hong Kong.

Lewis has extensive experience in patent litigation, technology transfer and commercial intellectual property across various industries, including the life sciences and TMT sectors.

He has represented various international pharmaceutical companies setting up R&D centres and has negotiated more than 35 collaboration, outsourcing and joint-venture projects with Chinese academic institutes, hospitals and contract research organisations in the last two years.

Andrew Hutchinson
Associate, Simmons & Simmons LLP
andrew.hutchinson@simmons-simmons.com

Andrew Hutchinson is a supervising associate at Simmons & Simmons LLP and part of its international dispute resolution group.

Andrew is a member of the firm's core life sciences team and the main focus of his practice is on patents and regulatory work, covering both contentious and non-contentious matters. Andrew has acted in many patent litigation matters before the UK Patents High Court in the life sciences and other fields. Much of his patent litigation work involves advising clients involved in multi-jurisdictional litigation, including the United States, Europe and elsewhere worldwide. Andrew has advised clients in parallel with Simmons' patent litigation team spanning its London, Dusseldorf, Amsterdam and Milan offices.

Andrew is experienced in all aspects of intellectual property law and has advised clients on matters relating to trade marks, passing off and data protection. He regularly writes articles and updates on intellectual property law, including in particular the recent developments in Europe relating to supplementary protection certificates.

Sarah Innes
Associate, Winston & Strawn
sinnes@winston.com

Sarah is an associate in the intellectual property group at Winston & Strawn in London.

Sarah has assisted in large-scale, multi-jurisdictional patent litigation in the High Court (Patents Court) and the Court of Appeal. She provides advice on patent and trade-mark issues to companies in a variety of industries, including pharmaceuticals, medical devices, specialist chemicals and engineering.

Birgit Kramer
Associate, Field Fisher Waterhouse
Birgit.Kramer@ffw.com

Dr Birgit Kramer is an associate in Field Fisher Waterhouse's Munich office. She specialises in the field of intellectual property, with a particular focus on patent law.

Birgit has gained experience advising and representing clients in complex international patent litigation, including infringement and validity cases across various technical fields with particular focus on pharmaceuticals and biotechnology. In addition to representing clients before German courts, Birgit focuses on the negotiation and drafting of various types of

agreements such as licence agreements and research agreements.

Birgit has worked for law firms in Munich, Zurich and Cape Town and in the legal departments of major international firms. She is the author of many publications in the field of patent law.

András Kupecz
Managing associate, Simmons & Simmons LLP
andras.kupecz@simmons-simmons.com

András read Law at the University of Amsterdam (*cum laude*), and has a Masters degree in molecular biology/biochemistry (Utrecht University, BSc *cum laude* University College Utrecht).

András acts both as a lawyer and as a European and Dutch patent attorney for clients in the Dutch courts and the European Patent Office in a variety of technical areas, in particular in the field of life sciences and energy resources. He has represented clients in patent cases regarding biotech, medical devices and pharmaceuticals, and has extensive experience in conducting pan-European patent litigation.

Noah M Leibowitz
Partner, Simpson Thacher & Bartlett LLP
nleibowitz@stblaw.com

Noah is a partner in Simpson Thacher & Bartlett's intellectual property group and litigation department. He represents clients in patent litigation, licensing disputes and in matters involving scientific, medical and technical expertise. Noah has experience litigating all aspects of patent cases in courts across the United States, including discovery, *Markman* hearings, dispositive motion practice, pre-trial, trial and appeal. He has litigated cases involving a variety of technologies, including pharmaceuticals, medical devices, global positioning systems, semiconductor fabrication, computer hardware and software, videoconferencing and data conferencing systems, and consumer products.

Noah has particular experience in the pharmaceutical, healthcare and biotechnology industries and in Hatch-Waxman patent litigation. He represented Daiichi Sankyo in litigation to enforce Daiichi's patent on its multibillion-dollar-per-year blockbuster anti-infective Levaquin®. He successfully defended the Levaquin® patent at trial and on appeal. He also successfully represented Daiichi Sankyo at trial in Hatch-Waxman challenges to its successful Benicar® and Benicar® HCT products. Noah also represents Human Genome Sciences in patent litigation against Genentech and City of Hope over Benlysta®, the first FDA-approved treatment for systemic lupus erythematosus in more than 50 years.

Andrew McIntosh

Partner, Bereskin and Parr
amcintosh@bereskinparr.com

Andrew graduated from the University of Western Ontario in 1990 with a BSc Honours in chemistry. He then concurrently attended the University of Windsor and the University of Detroit, where he obtained law degrees from each (LLB and JD respectively) in 1993. Andrew was called to the Ontario Bar in 1995, following which he practised patent prosecution and intellectual property litigation with a major intellectual property law boutique for 14 years.

Andrew's practice focuses on intellectual property litigation relating to patents, trade marks and trade secrets. He has extensive experience in complex patent litigation, including multi-jurisdictional proceedings and pharmaceutical patent litigation.

Andrew is also a registered Canadian and US patent agent, and a registered Canadian trade-mark agent.

Miquel Montañá

Partner, Clifford Chance
miquel.montana@cliffordchance.com

Miquel has been a partner at Clifford Chance since 2000, specialising in intellectual property and competition law. Since joining Clifford Chance, he has worked on multi-jurisdictional intellectual property and commercial disputes involving patents, trade marks, unfair competition, distribution and agency contracts, copyright and breach of confidential information. He joined the firm after graduating from Harvard University (LLM, Laylin Prize).

He has published a book and more than 40 articles on commercial and intellectual property matters worldwide in law journals such as *Patent World*, *Journal of World Intellectual Property*, *Trademark World*, *American Journal of International Law*, *Journal of World Trade*, *Columbia Journal of Transnational Law*, *Sport and the Law Journal* and the *German Yearbook of International Law*.

Gareth E Morgan

Partner, Winston & Strawn
gmorgan@winston.com

Gareth is a partner in Winston & Strawn's London IP practice with experience in all areas of contentious and non-contentious IP law, and a particular focus on the life sciences and healthcare sectors. He advises on both exploitation and enforcement of intellectual property, having assisted clients to negotiate and draft complex licences, distribution and other technology-related agreements in addition to advising clients on the interpretation of contracts and (where necessary) litigation on the same.

Gareth has acted in many landmark IP (contract (licence), patents, copyright and SPC) matters before the English courts, particularly in the biotechnology/pharmaceutical field. He advises clients in relation to EU medicinal regulatory law and has assisted clients in judicial review proceedings in the medicines regulatory field in the English and European courts.

Gareth is listed as a leading individual in *2011 Chambers* for life sciences regulatory work and the *IAM Global 250* directory of patent litigators.

Lindsay Neidrauer

Co-founding partner, Belmore Neidrauer LLP
lindsay.neidrauer@belmorelaw.com

Lindsay practises in all areas of intellectual property litigation, including patents, trade marks and copyright, and has a particular interest and experience in pharmaceutical patent litigation.

In 2002, Lindsay received both a JD and a Masters degree in criminology from the University of Toronto. She received a Bachelor of Science (Honours) degree in biology from Queen's University in 1998 and was called to the Ontario Bar in 2003.

Lindsay co-founded the Toronto-based intellectual property advocacy firm Belmore Neidrauer LLP in 2009. The firm is dedicated to creative and strategic advocacy and works on a number of prominent files in the field of intellectual property law.

Marjan Noor

Partner, Simmons & Simmons LLP
marjan.noor@simmons-simmons.com

Marjan specialises in patent and regulatory law in the life sciences sector. On the regulatory side she has advised pharmaceutical and biotech companies on a range of issues, including SPCs, data and market exclusivity, orphan drugs, paediatric extensions and product life cycle management generally. On the contentious side, she has represented clients in judicial review proceedings before the UK court and CJEU.

Marjan graduated in 1991 with a degree in pharmacology (having completed the preclinical years of a degree in medicine) before going on to study law.

She writes and speaks extensively on a range of patent and regulatory topics, including giving the "Regulatory Law for IP practitioners" lecture at Oxford University's IP Diploma course.

Scott Parker

Managing associate, Simmons & Simmons LLP
scott.parker@simmons-simmons.com

Scott is a managing associate in the international intellectual property practice of Simmons & Simmons, based in the firm's London office.

Scott specialises in intellectual property law generally, with particular emphasis on litigating in the pharmaceutical, biotechnology and medical-device industries. He has considerable experience of successfully representing clients in complex pharma and biotech patent litigation – in particular, defending the patents of research-based pharmaceutical companies against attack by their generic competitors. Typically, this involves working closely with the client's legal teams in other jurisdictions.

Scott has a strong technical background in the biological sciences and experience of working in-house at a major pharmaceutical company. As well as litigation, he is regularly instructed to give patent validity and infringement opinions and advises clients on licence agreements and other commercial agreements. He writes articles and speaks regularly on patent issues.

Alexandre Regniault

Partner, Simmons & Simmons LLP
alexandre.regniault@simmons-simmons.com

Alexandre specialises in arbitration, litigation and dispute resolution, principally on behalf of life sciences (pharmaceutical, medical-device and biotech) companies and significant players in highly regulated industries. His experience as a litigator mainly covers product liability, breach of contract, break-off of negotiations, comparative advertising and claim recovery. He is also a member of the firm's crime, fraud and investigations group.

Alexandre has significant expertise in

transactional and licensing negotiations, compliance audits and advice on regulatory matters, advising notably sponsors of clinical trials (including on data protection issues), and major companies in the sector on pricing, reimbursement, advertising and commercial practices. He is experienced in crisis management and frequently liaises with the French regulatory authorities on behalf of industry players.

David Rosenberg

Vice president IP policy, GlaxoSmithKline plc

david.h.rosenberg@gsk.com

After several years in the IP group of Clifford Chance, David joined GSK to undertake an IP policy function. He has worked on many aspects of IP policy both within and outside the European Union, including issues relating to access to medicines.

David was a member of the group comprising representatives of industry and academic stakeholders that, in conjunction with the European Commission, drew up the IP policy of the Innovative Medicines Initiative. Furthermore, he represents GSK on the IP committees of several pharmaceutical and cross-industry organisations. He also chairs the IP committee of the European Federation on Pharmaceutical Industries and Associations (EFPIA) and was heavily involved in the industry-led inquiry into DG competition.

David is currently deeply involved in the discussions to create a unitary European patent and unified patent court.

David is a regular speaker (and very occasional author) on all of these issues.

Sture Rygaard

Partner, Plesner

sry@plesner.com

Sture is partner in the IP department at one of the leading Danish law firms. Sture's primary areas of practice are patents, life sciences law, trade marks, designs, and marketing law, as well as

distribution and technology transfer agreements. He is an experienced litigator, and litigation and preliminary injunction proceedings in these fields of law constitute a large part of his work.

Sture is chairman of the Board of the Danish Anti-Counterfeiting Group. He is Master of Laws (LLM) from Cambridge University, England. He is a member of the Danish Association for the Protection of Industrial Rights and AIPPI.

Denis Schertenleib

Partner, Cabinet Schertenleib

ds@schertenleib-avocats.com

Denis is dual-qualified to practise as a lawyer at the Paris Bar and as a solicitor in England and Wales. He specialises in patent litigation, arbitration, technology licensing, European competition law applied to intellectual property, and regulatory law in the pharmaceutical field.

His practice involves multi-jurisdictional litigation in the life sciences and in other areas of technology such as chemistry, mechanics, telecommunications and physics.

Denis has previously worked in an international law firm in the City of London and was, for several years, a partner in a leading IP law firm in Paris before founding Cabinet Schertenleib.

Robert Shapiro

Lawyer

rshapiro79@hotmail.com

Robert practises in the area of intellectual property litigation with an emphasis on pharmaceutical patents. He focuses on proceedings under the Patented Medicines (Notice of Compliance) Regulations and patent and trade-mark infringement matters. Robert is a registered trade-mark agent.

Robert has written an article on patent rights during times of emergencies, focusing on the anthrax scare, which was published in the *Windsor Review of Legal and Social Issues*, and a Master's thesis on competition and antitrust issues of authorised generics. He has appeared before the Federal Court of Appeal, Federal Court and the Ontario Divisional Court.

Adrian Smith

Partner, Simmons & Simmons LLP
adrian.smith@simmons-simmons.com

Adrian is a partner in the international intellectual property practice of Simmons & Simmons, based in London. He advises life sciences companies and 'household name' clients in other sectors on creation, clearance and protection of new trade marks and on resolving international brand conflicts. He also advises on complex commercial transactions involving brands and other IP.

Adrian publishes articles in international journals and regularly speaks on trade mark issues and related IP topics. He was a speaker at the 2011 INTA Annual Meeting in San Francisco.

Adrian has for many years been active in MARQUES – the Association of European Trade Mark Owners. He is a member of INTA and PTMG (the Pharmaceutical Trade Marks Group). Adrian is described as "practical, pragmatic and commercial" by *Legal 500 UK 2011*, which also refers to the firm's brands practice led by Adrian, together with David Stone, as "winning praise for its remarkable efficiency, appropriate advice and utmost courtesy".

Julianna Tabastajewa

Partner, Gowlings International Inc
julianna.tabastajewa@gowlings.com

Julianna is a partner in Gowlings' Moscow office. Her practice focuses on intellectual property protection and litigation, pharmaceutical law, competition law, media law, and franchising. She also has extensive experience in domestic litigation and arbitration in Russia.

Julianna has represented major Russian and international companies in sophisticated copyright, trade mark, patent, and domain disputes, including several successful precedent cases for Quelle AG, Volkswagen AG and Audi AG.

She has contributed to new domain regulations in the Russian Federation and has written extensively on the subject. She is also involved in intellectual property legislation development initiatives in Russia as an authority of the Anti-piracy Committee of the State Duma (legislative assembly) of the Russian Federation. She is currently working in the Russian Franchising Association on changes to the franchising laws. Julianna also advises media, pharmaceutical and medical-device companies on related issues in Russia.

Vladislav Ugryumov

Partner, Gowlings International Inc
vladislav.ugryumov@gowlings.com

Vladislav is a partner in Gowlings' Moscow office, specialising in intellectual property. He has extensive experience in patent drafting and prosecution in the pharma/biotech areas before Russian and Eurasian Patent Offices and in the CIS. He has appeared as Counsel in numerous trials before the Chamber for Patent Disputes and the Russian courts. Vladislav has regularly been involved in preparing and prosecuting patentability and freedom-to-operate opinions, and conducting due-diligence investigations. He has also drafted and registered licence agreements, and advised on regulatory pharma laws and regulations.

Roberto Valenti

Partner, DLA Piper
roberto.valenti@dlapiper.com

Roberto is a partner responsible for DLA Piper's IP practice group in Italy, based in Milan. He has extensive experience in IP litigation, including patents, trade marks, copyrights, designs, unfair competition, and misleading and comparative advertising.

Roberto also has key experience in dealing with IP non-contentious matters, such as drafting licence and transfer agreements relating to patents, trade marks, copyrights and designs. He has experience of analysing IP portfolios and evaluating intangible assets in merger-and-acquisition scenarios.

He frequently organises conferences and seminars on intellectual property matters. He also collaborates with the University of Pavia and the University of Urbino and is a member of the editorial board of the legal magazine *AIDA*. Roberto has been shortlisted as one of the top five Italian IP lawyers in the Top Legal Awards for 2009 and 2010.

Liad Whatstein

Partner, Dr Shlomo Cohen & Co
liadoffice@shlomocohen.co.il

Liad is head of litigation at Dr Shlomo Cohen & Co and has participated in some of Israel's most complex and widely publicised patent cases. He is regularly acclaimed by international directories for his litigation skills (for example, *Chambers*: "superb litigator – one of the best in Israel"; *IAM250 Life Science Patent Litigation*: "Liad Whatstein dominates life science patent litigation on behalf of innovators"; *IAM250*: "has a prodigious reputation as a litigator"; *WTR 1000*: "a truly great litigator").

Liad has successfully represented major pharmaceutical companies in some of the most important patent litigation in Israel. He has acted, among others, for Sanofi-Aventis, Merck, Eli Lilly, Procter & Gamble, Wyeth, Genzyme, Novartis, Gilead, Abbott, Amgen, Johnson & Johnson, Lundbeck and Schering-Plough. He has conducted numerous complex experiments for patent litigation in a wide array of techniques, and cooperated with some of the world's leading experts in numerous scientific disciplines.

Liad is also current Chairman of the Intellectual Property Committee of the Israeli Bar Association.

Neeti Wilson

Managing associate, Anand and Anand
neeti@anandandanand.com

Neeti is the managing associate of Anand and Anand Advocates in the firm's patents and designs department. She specialises in molecular biology and her main areas of practice are patents, plant variety protection and biological resources. She advises clients on prosecution strategies, regulatory approvals and contentious matters for representing them at various forums including the patent office, Intellectual Property Appellate Board, Plant Variety Registry and the National Biodiversity Board of India.

Neeti is enrolled in the Bar Council of India and is a member of the Delhi High Court Bar Association, the International Association for the Protection of Intellectual Property (AIPPI), the Indian Science Congress Association, the Asian Patent Attorney Association (APAA) and the Governing Board of Zonal Technology Management and Business Planning and Development Unit, IARI.

She is a regular speaker at conferences/forums and is also involved in IP courses for university students.

Marian Wolanski

Associate, Belmore Neidrauer LLP
marian.wolanski@belmorelaw.com

Marian practises in the area of intellectual property litigation, with a focus on pharmaceutical patents. She has assisted in a number of Patented Medicines (Notice of Compliance) Regulations proceedings and patent infringement actions before the Federal Court. Marian's practice also includes providing opinions on patent validity and freedom to operate.

While attending law school, Marian completed a clerkship at the Provincial Court and was the managing editor of the *Windsor Review of Legal and Social Issues*.

Marian also has a BSc Honours in genetics

and an MSc in molecular biology. After completing her graduate degree, Marian worked as a research associate at a developmental genetics laboratory at the University of Windsor, where she contributed to projects related to brain, heart and eye development.

Yulia Yarnykh

Associate, Gowlings International Inc
yulia.yarnykh@gowings.com

Yulia is an associate in Gowlings' Moscow office. Her practice focuses on intellectual property and general corporate matters. She advises on intellectual property rights protection, including copyright, right to company names, trade marks, know-how, media law, e-commerce, data protection and anti-monopoly law.

Yulia is also involved in projects on structuring franchise schemes for major foreign investors in Russia. She drafts license agreements and advises international trade-mark holders on trade mark protection in Russia and on domain name registration in the RuZone. She also advises Russian and international companies on information and personal data protection compliance and works on regulatory issues with regard to non-commercial organisations in Russia.